MOBILE AND
WIRELESS INTERNET
Protocols, Algorithms and Systems

MOBILE AND
WIRELESS INTERNET
Protocols, Algorithms and Systems

Kia Makki, Ph.D.
Telecommunications & Information Technology Institute
Florida International University

Niki Pissinou, Ph.D.
Telecommunications & Information Technology Institute
Florida International University

Kami (Sam) Makki, Ph.D.
Department of Information Technology
Queensland University of Technology

E.K. Park, Ph.D.
School of Interdisciplinary Computing & Engineering
University of Missouri - Kansas City

Editors

KLUWER ACADEMIC PUBLISHERS
Boston / Dordrecht / London

Distributors for North, Central and South America:
Kluwer Academic Publishers
101 Philip Drive
Assinippi Park
Norwell, Massachusetts 02061 USA
Telephone (781) 871-6600
Fax (781) 871-6528
E-Mail: < kluwer@wkap.com>

Distributors for all other countries:
Kluwer Academic Publishers Group
Post Office Box 322
3300 AH Dordrecht, THE NETHERLANDS
Telephone 31 78 6576 000
Fax 31 78 6576 254
E-Mail: < services@wkap.nl>

 Electronic Services < http://www.wkap.nl>

Library of Congress Cataloging-in-Publication Data

Mobile and Wireless Internet
Protocols, Algorithms and Systems
Kia Makki, Niki Pissinou, Kami (Sam) Makki, E.K. Park (Eds.)
ISBN 0-7923-7208-5

DEDICATION

To Alex

Kia Makki and Niki Pissinou

To Michelle, James, and Jonathan

E.K. Park

To our Parents
To our Brothers and Sisters

Our parents not only gave us life, but also the opportunity of a superb education. Without their support and love, this book would not have been written.

Kami (Sam) Makki
Kia Makki

CONTENTS

viii

PREFACE

Recent advances in mobile and wireless communication and personal computer technology have created a new paradigm for information processing. Today, mobile and wireless communications exit in many forms, providing different types of services. Existing forms of mobile and wireless communications continue to experience rapid growth and new applications and approaches are being spawned at an increasing rate. Recently, the mobile and wireless Internet has become one of the most important issues in the telecommunications arena. The development of the mobile and wireless Internet is the evolution of several different technologies coming together to make the Internet more accessible. Technologies such as the Internet, wireless networks, and mobile computing have merged to form the mobile and wireless Internet. The mobile and wireless Internet extends traditional Internet and World Wide Web services to wireless devices such as cellular phones, Personal Digital Assistants (PDAs) and notebooks. Mobile and wireless Internet can give users access to personalized information anytime and anywhere they need it, and thus empower them to make decisions more quickly, and bring them closer to friends, family, and work colleagues.

Wireless data communication methods have been around for sometime. However, lack of standards, robust infrastructure and enabling technologies, and prohibitive costs have limited the use of wireless networks for information exchange. However, with the advent of several new technologies and standards over the past few years, things have changed dramatically. The Internet Protocol architecture was originally conceived with the primary objective of accessing the network by means of wired links. It is now becoming clear that Internet may not be suitable for addressing the emerging needs of today's mobile and wireless Internet users, who desire access to multimedia services regardless of location. Projections have been made by Industry analyst that by the year 2003, there will be more than a billion wireless subscribers, thus over a billion wireless communication devices in use.

In recent years, there has been tremendous growth in two technological sectors: the Internet and mobile and wireless communications. The combination of both developments, the growth of the Internet and the success of mobile and wireless networks, suggests that the next trend will be an increasing demand for mobile and wireless access to Internet applications.

Even with millions of users worldwide, the mobile and wireless Internet market is still in its infancy. The use of wireless networks to access the Internet represents a fundamental paradigm shift in the way we view and access information. The task set before us is to eliminate the shortcomings of mobile computers and wireless media so that the inherent convenience of mobility will no longer suffer the burden of inadequate or inappropriate system design.

This book brings together a number of papers, which represent seminal contributions underlying mobile and wireless Internet. It is our hope that the diverse algorithms and protocols described in this book will give the reader a good idea of the current state of the art in mobile and wireless Internet access. The authors of each chapter are among the foremost researchers or practitioners in the field.

Kia Makki
Niki Pissinou
Kami (Sam) Makki
E.K. Park

ACKNOWLEDGEMENTS

This book would not have been possible without the wisdom and cooperation of the contributing authors. Thanks go to Eduardo Serrano and Allan Harvey at the Telecommunications and Information Technology Institute. Special thanks to President Modesto Maidique, Provost Mark Rosenberg, and Dean Vish Prasad at Florida International University for providing us with such a stimulating environment for writing this book.

We would also like to thank Alex Greene, Senior Publisher and his staff, specifically Melissa Sullivan at Kluwer Academic Publishers for their strong support and encouragements. It was a pleasure working with Alex and Melissa, who were incredibly patient, very responsible, and enthusiastic about this book. We also would like to express our sincere appreciation to the reviewers of this book, whose suggestions were invaluable.

This book would not have been possible without the indulgence and infinite patience of our families during what often appeared to be an overwhelming task. They graciously accommodated the lost time during evenings, weekends, and vacations. As a small measure of our appreciation, we dedicate this book to them.

Chapter 1

DYNAMIC CONFIGURATION OF MOBILE DEVICES FOR WIRELESS INTERNET ACCESS

Sandra Thuel, Luca Salgarelli, Ramachandran Ramjee and Thomas La Porta
Bell Laboratories, Lucent Technologies

1.1   INTRODUCTION

The dynamic configuration of devices plays a pivotal role in supporting plug-and-play operation in IP networks. For devices that are not mobile, dynamic configuration is typically supported by the widespread Dynamic Host Configuration Protocol (DHCP) [5], which replaces burdensome manual configuration procedures and enables the address space to be managed efficiently. The configuration state allocated to devices by DHCP includes the device's IP address and a series of DHCP options [7] such as routing parameters (e.g., a subnet mask and default gateway), identity information (e.g., the device's domain name) or service related parameters (e.g., server addresses for DNS, IMAP, HTTP proxy, NTP, etc.). However, since DHCP was designed to serve the configuration needs of trusted clients in a local LAN, it does not support the configuration of devices connected to a remote network, a situation that arises when the devices are mobile.

Mobile devices gain and maintain network connectivity through Mobile IP, the internet mobility management protocol standard [4]. Thus, it is desirable that dynamic configuration solutions for mobile devices interwork with Mobile IP while leveraging the use of existing DHCP services. Although recent extensions [3] to Mobile IP provide it with preliminary support for dynamically configuring the IP address of mobile devices, several issues remain open. Specifically, it is not clear how Mobile IP home agents and DHCP servers would interoperate, nor the mechanisms through which DHCP options can be delivered to mobile devices.

This chapter provides a taxonomy of dynamic configuration solutions for mobile devices and focuses on the study of two proposals which use the home agent to mediate a mobile device's access to DHCP, namely, Transient Tunneling and DHCP proxy [8,12,13]. In Transient Tunneling, a home agent is used to temporarily establish an IP tunnel to carry messages between a home DHCP server and a remote client executing on the mobile device. The home agent does not intervene in the DHCP transactions between the server and the client but simply provides a conduit for those transactions to take place. In contrast, the DHCP proxy mechanism uses home agents to serve as proxies or surrogate clients that assume the identity of remote mobile devices and locally transact with DHCP servers in the home network to acquire addresses on their behalf. In this case, home agents assume responsibility for client transactions while relegating the remote mobile devices to the passive role of installing configuration state acquired by their proxies. Additionally, since DHCP configuration state is leased, it is referred to as soft-state. This state needs to be periodically defended or renewed by the proxies to prevent expiration. Regarding the location of DHCP client functions, Transient Tunneling is a *device-centric* solution whereas DHCP proxy is an *agent-centric* solution. Both mechanisms call for implementation enhancements to home agents while requiring no changes to the DHCP and Mobile IP standards. However, their different design viewpoints raise interesting tradeoffs on performance, implementation complexity, and access to configuration state that constitute the subject of this chapter.

1.2   TAXONOMY OF SOLUTIONS

In general, the dynamic configuration of mobile devices requires a mechanism for devices to request and obtain network access and a mechanism for them to get state from a configuration allocator like DHCP. Typically, the network access mechanism is only needed when the mobile device is located away from its home network. We distinguish three models for providing these access and configuration services, illustrated in *Figure 1.1*. Model (a) uses the network access protocol to also provide configuration services. In model (b), configuration allocation is performed as a back-end process that is triggered by and logically subsumed within the network access protocol. A third model, shown as (c), is for the mobile device to independently invoke the services of a network access protocol and a configuration allocator. It is also possible to envision hybrid solutions where, for instance, some configuration state is provided by the network access protocol and other configuration state is provided by an external configuration allocator. The key point is that from the mobile device's perspective, solutions of types (a) and (b) opt to couple configuration allocation with the access protocol whereas solutions of type (c) strive to keep these services separate.

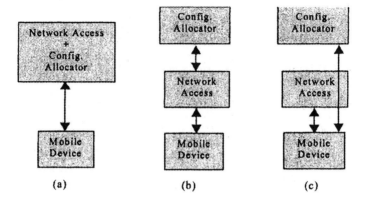

Figure 1.1. Dynamic Configuration Models, where configuration allocator is: (a) within the network access protocol; (b) a back-end service to network access; and (c) separated from network access

Mobile IP is the network access protocol standard defined for mobile devices. IP security (IPSec, [18]) tunnels with VPN access is an effective network access protocol for nomadic devices but the lack of mobility support in VPNs makes it unsuitable for mobile devices. So the open question is how to provide configuration support for devices that use Mobile IP for network access.

Model (a) suffers from the serious drawback that it requires some element involved in providing network access services to be modified to duplicate part or all of the functionality already provided by DHCP. One example is to equip Mobile IP home agents with dynamic addressing capabilities. This introduces address space management inefficiencies, as the network address pool has to be partitioned between DHCP servers and home agents and sized appropriately in each case to avoid depletion. It also adds undesirable complexity to the home agent, which is worsened if it must also undertake responsibility for allocating other configuration options. In addition, providing soft-state support for any configuration state allocated by a home agent would entail changes to the Mobile IP standard to require mobile devices to register even while they are at home in order to defend their allocated state. A similar example is to introduce dynamic configuration support into the IP security infrastructure (i.e., AAA [15]). Since security services need to be invoked for mobile devices as part of their network access procedures, the security service provider upon their successful authentication could allocate configuration state to devices. A hybrid solution combining addressing support on home agents with configuration support on AAA servers is another possible option. However, AAA does not provide any support for dynamic configuration to date and if it did it would suffer from the same pitfalls of the home agent solution. In conclusion, dynamic configuration solutions that exclude DHCP appear costly and are too immature to warrant serious consideration at this time. It is unclear whether there are any benefits to justify their inherent duplication of DHCP functionality and encourage their eventual deployment.

As a result, solutions that leverage the use of DHCP in models (b) and (c) are preferred. The *DHCP Proxy* mechanism proposed in [8] (similar to the Proxy Agent approach suggested in [12]) is a good example of a solution of type (b) whereas the *Transient Tunneling* mechanism proposed in [12,13] exemplifies a solution of type (c).

1.2.1   Mobile IP Basics

Mobile IP is designed to make a mobile device appear to be connected to a LAN, referred to as the mobile's home network, even though the device may be attached to a remote or foreign network. This location transparency is enabled by two mobility agents: a *home agent* and a *foreign agent* [4]. A home agent gives mobility support to hosts that belong to the same home network, while a foreign agent serves hosts that are visiting from a remote home network. Each mobile host that is roaming on a foreign network must have two adresses: a *home address* and a *care-of address (COA)*. Typically, the COA is assigned by the foreign agent and is the address of one of its interfaces. Alternatively, if the foreign agent functionality is integrated with the mobile device, the COA may be acquired from the local subnet using techniques such as DHCP; this is referred to as the *co-located COA* option.

Packets sent to the mobile device are always addressed to its home address. While the device is attached to its home network, packets reach it following conventional Internet routing. When the mobile device is resident in a foreign network, the home agent intercepts packets addressed to the mobile device, and encapsulates them in packets addressed to the care-of address of the mobile device. The encapsulated packets are then routed as usual until they reach the device's foreign agent. These *tunneled* packets are decapsulated by the foreign agent and the original packet is forwarded to the mobile device. Each time the device moves between points of attachment crossing a network or subnet boundary, it acquires a new care-of address and re-registers it with its home agent. Home agents associate a lifetime to the state they install for a host, requiring periodic registrations to avoid state expiration and removal. When a mobile device returns to its home network, it de-registers with its home agent since it no longer requires its traffic forwarding services.

1.2.2   DHCP Basics

In DHCP, clients and servers are either on the same subnet or connected through a *relay*. Relays remove the need for a DHCP server on every subnet by passing messages between clients and servers on different subnets of the same LAN. An address *lease* is obtained by a client through the exchange of two client-server message pairs, namely DISCOVER-OFFER and REQUEST-ACK. The first pair allows the client to discover a DHCP server and receive an address offer from a server while the second pair is the client's acceptance of a particular offer and the server's acknowledgement of a successful allocation. Address leases have an associated expiration time, requiring periodic lease *renewals* to prevent expiration and state removal. Packets destined to the server are always sent as IP broadcasts when the client does not know the address of a server. Otherwise, the client may unicast its requests to the server's IP address (e.g., during a lease renewal). Until the client has a configured IP address, responses from the server to the client are usually sent as IP broadcasts. In the case when a relay is used, server responses are unicast to the relay's IP address and the relay forwards the response to the client.

1.3   DYNAMIC CONFIGURATION SOLUTIONS BASED ON DHCP

1.3.1 Goals and Assumptions

The Transient Tunneling and DHCP proxy mechanisms described in this chapter suggest different ways to dynamically configure a mobile device lacking a home address, when the device powers up in a foreign network. Both mechanisms have the common goal of leveraging the use of the DHCP and Mobile IP protocol standards without violating compliance. Transient Tunneling has the additional distinctive goal of enabling mobile devices to have unlimited access to all configuration parameters managed by DHCP while minimizing the functional enhancements required at home agents and mobile devices. Conversely, DHCP proxy minimizes configuration latency. Both mechanisms assume that on power-up the mobile device: a) has some way to determine whether or not it is in its home network; and b) knows its Network Access Identifier or NAI (similar in format to an email address [6]) or be equipped with some way of inferring it. There are known mechanisms for making such inferences but their discussion is out of the scope of this chapter.   Moreover, both mechanisms are built upon the recent NAI extension to the Mobile IP protocol [3], which allows a home agent to authenticate a mobile device using its NAI even if it does not have a home address. In addition, if the home address field of this message is null, the home agent infers that the mobile device needs a home address and may return one to it in its registration reply.

Before delving into the details of each mechanism, we distinguish two states for a mobile device: *remote configuration* state during which the device gets configured with parameters received from a home DHCP server and *remote lease renewal* state during which the mobile device is defending its lease. Additionally, if and when a mobile device returns to its home network while connected, the procedures for *home-de-registration* and *home-based lease renewals* must be executed to relinquish Mobile IP services and continue defending the DHCP lease, respectively. Procedural details for each mechanism are presented as follows.

1.3.2   Transient Tunneling

Transient Tunneling uses Mobile IP to set up a temporary or transient IP tunnel between a mobile device and its home agent, to carry DHCP messages between the device's client and a server in its home network. The tunnel is transient because it only exists for the time required for the device to successfully acquire a permanent home address and configuration options from a server. This tunnel is established by using a temporary home address given to the mobile device by the home agent in its initial registration reply. This address is allocated from a small address pool on the home agent managed by a very lightweight process referred to as a *bootstrapping agent*.

Figure 1.2 illustrates the remote configuration procedures for Transient Tunneling. It works as follows: the mobile device sends a registration request to its home agent with the NAI extension and a null home address field. The home agent authenticates the mobile device using the NAI and determines it needs a home address. It then assigns a temporary home address to the mobile device from its address pool (managed by the bootstrapping agent) and sets up a tunneling entry for the device with this temporary address and the care-of address specified in the registration message. The home agent then returns the temporary home address to the mobile device in its registration reply.

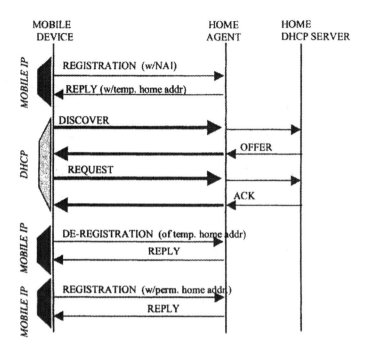

Figure 1.2. Transient Tunneling Procedure for Remote Configuration of Mobile Devices

When the foreign agent serving the mobile device receives the reply, it establishes its end of the tunnel and forwards the temporary address to the mobile device. If the foreign agent is integrated with the mobile device, as assumed in the figure, the forwarding is trivial; otherwise it must follow standard procedures specified in [3,4]. The mobile device then installs the temporary home address on one of its interfaces. The transient tunnel is now in place and the mobile device is connected to its home network.

Using this tunnel, the DHCP client on the mobile device may interact with a DHCP server in its home network as if the device were locally connected. These messages are tunneled between the foreign and home agents, as depicted by the thick arrows. As per the DHCP standard, the DISCOVER-OFFER and REQUEST-ACK client-server message pairs are exchanged. After the mobile device successfully acquires an address (along with any DHCP options), it sends a de-registration message to its home agent to tear down the transient tunnel and release its temporary home address and uses the newly acquired address as its home address for the rest of the session[1]. At this time, the mobile device registers with its home agent via standard Mobile IP procedures using its permanent home address.

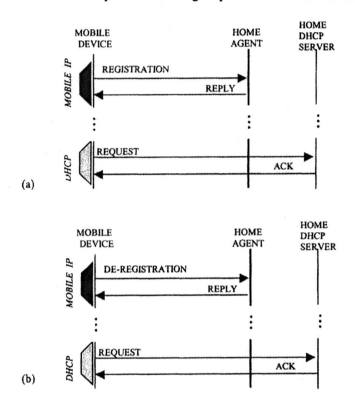

Figure 1.3. Transient Tunneling Procedures for: (a) Re-registrations and Remote Lease Renewals; and (b) Home De-registration and Home-Based Lease Renewals

In Transient Tunneling, Mobile IP re-registrations and remote DHCP lease renewals follow standard procedures. As shown in *Figure 1.3* (a), the mobile device periodically sends registrations to its home agent and unicasts DHCP requests to the home server that allocated its permanent home address.   These re-registrations and lease renewals are conducted independently from each other.   As shown in *Figure 1.3* (b), when the mobile device enters its home network, it follows standard de-registration and lease renewal procedures.

Several observations are in order. First, the requirement that the home agent maintains an address pool does not conflict with the goal of using DHCP for address allocation because permanent home addresses are still

---

[1] Alternatively, it is possible to replace the need for this explicit de-registration with a time-out, provided the home agent gives the registration a short lifetime.

allocated from DHCP's public address pool. Addresses allocated by the home agent are used by mobile devices for a very brief period of time and are only used to send and receive DHCP messages. The home agent address pool can consist of either private addresses (e.g., of type 10.* or 192.168.* [9]) or public addresses. The type of address that is used makes no difference to packet routing from home agents to foreign agents, because transit routers only see the COA in the outer header of tunneled packets. However, since by definition private addresses are not globally unique, their use in this scheme can cause routing problems along the path between the foreign to the home agent (uplink), and address collisions in the downlink when an external foreign agent is used. Uplink routing problems are resolved by using a *reverse tunnel* [16] between the foreign and home agents, therefore preventing the exposure of the mobile device's private address in the external IP header of the uplink packets. Downlink address collisions can be resolved at the foreign agent by using the home agent address and the mobile device's link-layer address to deliver packets to the mobile device. Second, note that Transient Tunneling maximizes configuration flexibility because the mobile device transacts directly with the home DHCP server without the direct intervention of the home agent. This allows the mobile device to easily get any desired configuration state besides an IP address, such as DHCP options. Third, the configuration latency is proportional to 5 round-trip times (RTTs) between the mobile device and its home network plus the processing overheads pertaining to Mobile IP and DHCP server and client processing[2]. However, this latency is only present during the initial mobile device power-up; for lease renewals, the performance is proportional to 1 RTT. Performance measurements will be provided in Section 1.4.

Finally, we observe that DHCP imposes an interesting restriction on client-server interactions: a server receiving a broadcast message from a client typically responds with a broadcast. In order for a broadcast reply to reach the mobile device, the home agent has to be enabled to forward broadcast traffic, causing all broadcast packets on the home network to be forwarded, not just the desired DHCP replies. This would introduce a costly traffic burden, especially over the wireless link. A solution to this problem is to slightly modify the DHCP client on the mobile device to behave like a *co-located relay*, that is, a client that mimics the operation of a joint client and relay. By sending messages to the server as if they were passing through a relay, the server is forced into responding with IP unicast messages, thus eliminating the need for home agent forwarding of broadcast packets.

---

[2] Recall that 1 RTT can be eliminated if the explicit de-registration is allowed to time-out.

1.3.3   DHCP Proxy

The DHCP proxy mechanism is based on the notion of proxy or surrogate DHCP clients that reside on the home agent and perform all client transactions on behalf of any mobile device lacking a home address when it attempts to register. A proxy assumes the identity of the mobile device it serves, generates client messages on its behalf, and processes server replies, thereby transacting with a local DHCP server to acquire a home address. The home agent places the acquired address in the registration reply it sends back to the mobile device for its subsequent use as a permanent home address. A proxy client is also responsible for defending the address by periodically sending renewal messages to its DHCP server to prevent lease expiration. Proxy services are relinquished when a mobile device enters its home network. The mobile device assumes control for defending its address as long as it remains home.

Remote configuration procedures for DHCP proxy are illustrated in *Figure 1.4* (a) and are described as follows: the mobile device sends a registration message to its home agent with the NAI extension and a null home address field. After the home agent authenticates the mobile device and finds it needs a home address, it allocates a proxy client to it and gives it the device's NAI. The proxy client then uses the NAI to assume the identity of the mobile device and constructs client messages that appear to have been generated by the mobile device itself. Since the home agent, by definition, has an interface on the mobile device's home link, the proxy client can send messages on the home subnet. These messages reach a DHCP server in the same way they would if the device were directly connected to the home network. A standard exchange of DHCP client-server messages takes place with a local server until the proxy client successfully acquires an address. The acquired address is then given to the home agent, which places it in a registration reply message and sends it back to the mobile device.

After installing the address on one of its interfaces, the mobile device is connected to its home network and can start receiving and sending packets addressed to its allocated home address. Periodically, the mobile device must send a re-registration message to its home agent as shown in *Figure 1.4* (b). This triggers the proxy client to send a DHCP lease renewal message (i.e., a REQUEST). After the proxy gets an ACK from the server indicating that it successfully renewed the lease, the home agent sends a registration reply to the mobile device.

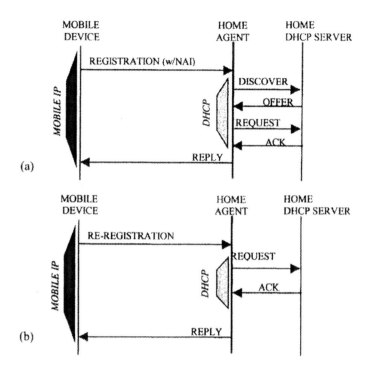

Figure 1.4. DHCP Proxy Procedures for: (a) Remote Configurations; and (b) Remote Lease Renewals

Examining *Figure 1.4* (a) it is easy to see that the configuration latency of DHCP proxy is proportional to 1 RTT between the mobile device and its home network plus the processing overheads of Mobile IP and DHCP. It is also important to observe that the only way for the mobile device and the proxy client to communicate is through the Mobile IP registration and reply messages. These messages only allow the mobile device to request and obtain an address from its proxy client, not other configuration parameters offered by DHCP. Getting additional configuration state from the proxy client would require extensions to the registration and reply messages to allow the piggybacking of such state, introducing an undesirable evolutionary coupling between the Mobile IP and DHCP standards. To circumvent this problem, after acquiring a home address through a proxy, a mobile device may launch its own DHCP client in order to send a DHCPINFORM message to request additional configuration parameters. This message was designed to enable a client with an externally configured address to only ask for local configuration parameters. Unfortunately, it lacks widespread support to date.

Figure 1.4 (b) shows that re-registrations and DHCP lease renewals are coupled. The proxy client is triggered to send a lease renewal to its server when a re-registration arrives at the home agent. Since DHCP leases usually have lifetimes on the order of hours or days while registration lifetimes are typically on the order of minutes, it would be highly inefficient to trigger a lease renewal on the arrival of every registration. Moreover, it is especially undesirable for a lease renewal to be triggered during a handoff-related registration update, as this would increase time-sensitive handoff latencies. To avoid such inefficiencies, intelligence needs to be placed on the home agent to decide on which registrations to trigger lease renewals so as to minimize unnecessary renewals while avoiding renewals during handoffs.

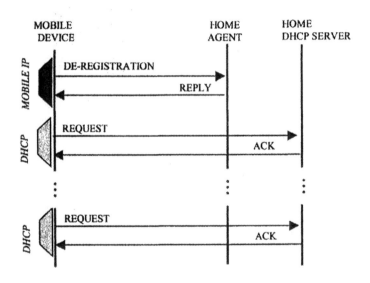

Figure 1.5. DHCP Proxy Procedures for Home De-Registrations and Home-Based Lease Renewals

The procedures that follow a mobile device's return to its home network are illustrated in *Figure 1.5*. First, the mobile device de-registers with its home agent since it no longer requires its traffic forwarding services while at home. This de-registration causes the home agent to relinquish the DHCP proxy client that was serving the device while it was away from home. As a result, the mobile device is thereafter responsible for defending its home address. It does so by immediately launching a DHCP client to renew its address lease since it does not know when the lease expires. Neither does it know the address of the DHCP server that allocated the lease in order to send to it a unicast lease renewal message (i.e., a DHCP REQUEST), as is typically done. Instead, it must broadcast its lease renewal request in the hopes of discovering the server (possibly from a set of multiple servers) that granted the lease to its (former) DHCP proxy client. This should work, provided some sort of inter-process communication between the Mobile IP and DHCP clients is added to properly launch the DHCP client for a lease renewal (e.g., giving it the address it needs to renew)[3]. After the DHCP control hand-over, the client on the mobile device periodically unicasts lease renewals to its server as per standard operation.

1.4 EVALUATION AND COMPARISON

DHCP proxy and Transient Tunneling differ in their implementation complexity and configuration latency. Though no quantitative data on implementation complexity is available for comparison, measurements on expected configuration latencies are available from experiments that were conducted on a wireless testbed.

Referring to *Figure 1.2*, it is easy to observe that four terms contribute to the configuration latency yielded by Transient Tunneling ($T_{TT\text{-}CONFIG}$), namely, the time to process the initial Mobile IP registration ($T_{MIP\text{-}REG}$), the time to complete a DHCP transaction with a home server (T_{DHCP}), the time to de-register the temporary home address ($T_{MIP\text{-}DEREG}$), and the time to register the permanent home address ($T_{MIP\text{-}REG}$). This is represented as:

$$T_{TT\text{-}CONFIG} = 2 \times T_{MIP\text{-}REG} + T_{DHCP} + T_{MIP\text{-}DEREG} \cong 5\,RTT + t_{DHCP} + t_{MIP},\quad (1)$$

where t_{DHCP} and t_{MIP} represent the combined server and client processing time of DHCP and Mobile IP, respectively. Typical values measured for these terms on a wireless LAN environment are shown in

Table 1.1 They yield a configuration latency, $T_{TT\text{-}CONFIG}$, of 184 ms., where $t_{DHCP} + t_{MIP} \cong 159$ ms. In a Wide Area Network (WAN) scenario, where RTT's in the range of 100 to 300 ms. are typical, one could expect configuration latencies to be about 659 ms. to 1.66 sec. by Equation (1). This configuration latency should be tolerable since it only affects mobile devices during power-ups, where device startup latencies are measured in tens of seconds, including boot time, radio modem configuration, and wireless radio access times.

---

[3] This corrects an error in the requirements specified in [9], which call for a DHCP DISCOVER message to be sent; sending a DISCOVER message is doomed to fail because servers, by design, cannot allocate an address that is already in use (in this case, by the mobile device itself).

Table 1.1 Configuration Latency Measurements on a LAN

TERM	VALUE
$T_{MIP\text{-}REG}$	2.9 ms.
$T_{MIP\text{-}DEREG}$	3.27 ms.
T_{DHCP}	150 ms.
RTT	5 ms.

On the contrary, the DHCP proxy mechanism minimizes configuration latency to about 1 RTT. Specifically, its configuration latency is given by:

$$T_{DP\text{-}CONFIG} = T_{MIP\text{-}REG} + T_{DHCP} \cong 1\ RTT + t_{DHCP} + t_{MIP}. \qquad (2)$$

where t_{DHCP} and t_{MIP}. represent the aggregate server and client processing time of DHCP and Mobile IP, as before ($t_{MIP} < t_{MIP}$. because only one Mobile IP transaction is performed). Using the performance measurements in

Table 1 and Equation (2), configuration latencies range from about 253 ms. to 453 ms. for WAN RTT's of 100 to 300 ms. This is clearly a significant performance improvement over the configuration latency of Transient Tunneling, as shown in *Figure 1.6* (a). However, as depicted in *Figure 1.6* (b), for an overall power-up latency of about 10 seconds for current typical mobile devices to boot and connect to their access networks, the relative performance disadvantage of Transient Tunneling is not that significant.

DHCP proxy's performance gain is balanced by the need for implementation enhancements to home agents in order to provide proxy services. The addition of proxy functionality to home agents is a non-trivial effort for several reasons. First, placing a DHCP proxy between a mobile device and a DHCP server introduces exception conditions that need to be managed. For instance, error conditions occurring in DHCP transactions between proxy and server have to be mapped or translated to an appropriate registration reply error message. Second, the home agent needs to allocate and maintain a DHCP client state machine (i.e., the proxy client) for every mobile device that powers up away from its home network. Therefore, care must be taken to ensure that the scalability of home agents is not compromised as they undertake the responsibility of providing proxy services to a potentially large number of mobile devices.

For example, consider B_i to be the processing bandwidth required for the home agent to provide proxy client services for the ith mobile device. B_i has two components: one related to the client processing time needed to acquire an address, and the other related to the periodic task of renewing the lease.

Ignoring the relatively small one-time processing overhead for sending an initial DISCOVER message and processing an OFFER, B_i can be approximated by C_i/T_i; C_i is the time it takes a client to generate a REQUEST and process an ACK message from the server (optimistically assuming there are no retries nor failures) and T_i is the device's lease renewal period. Typical DHCP performance measurements suggest C_i to be about 36 milliseconds. If DHCP lease renewals take place on a 1-hour basis, the home agent processor can support a maximum of 10,000 devices.

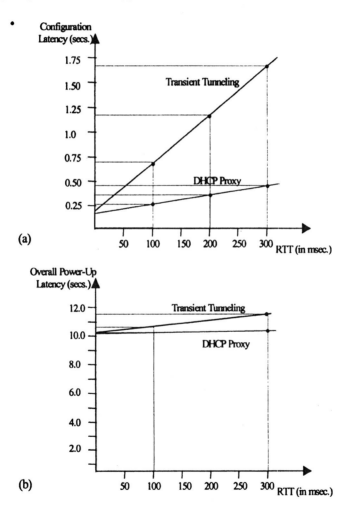

Figure 1.6. (a) Configuration latencies of Transient Tunneling vs. DHCP proxy for typical WAN RTTs and (b) the impact to overall device power-up latencies

Furthermore, if every Mobile IP re-registration and DHCP lease renewal is linked, T_i would be on the order of 10 minutes, leading to a maximum number of devices of less than 1,700. Clearly, the processing burden imposed by proxy services on the home agent cannot be neglected. Hence, scalability also points to the need for logic to avoid triggering lease renewals on every registration. In brief, home agent scalability is a real concern when registrations and lease renewals are tightly linked but it may also be a problem even after performance optimizations, depending on the average required lease renewal frequency for the mobile devices. While all these implementation requirements are feasible, they are a stumbling block to deploying the DHCP proxy mechanism.

In summary, the benefits of Transient Tunneling are: a) implementation simplicity; b) unlimited access to all configuration state managed by DHCP; c) scalability, since it does not impose any processing burden on the home agent; and d) the separation of Mobile IP and DHCP protocols. Its deficiency lies in its 5 RTT configuration latency overhead. Conversely, the benefit of DHCP proxy is its minimized configuration latency (1 RTT) while its deficiencies are: a) home agent implementation complexity and scalability concerns; b) limited access to the acquisition of a home address; and c) the introduction of a protocol coupling dependency for lease renewals, the degree of which is prone to increase if it needs to serve additional configuration state to the mobile devices.

1.5   RELATED WORK

Although previous work had studied issues related to the support for DHCP on mobile devices, except for the recent work on DHCP proxy and Transient Tunneling, no prior effort had explicitly addressed the problem of dynamically configuring mobile devices. Perkins and Luo [1] discuss problems concerning the use of DHCP on nomadic devices such as the lack of security and the need for a dynamic DNS mechanism to enable the location of dynamically addressed devices. Vatn and Maguire [2] examine the use of DHCP to dynamically acquire a co-located COA. They measure handoff latencies and suggest means to eliminate address conflict checking overheads in order to reduce the protocol's configuration latency. On a more related note, Calhoun and Perkins [3] suggest the use of a "Home Domain Allocation Function" to dynamically allocate home addresses to mobile devices through their home agents but neglect to provide a specification for this function.

Though not designed for mobile hosts, Patel et al [14] describe a method for fixed hosts to connect to a remote corporate network via an IPSec tunnel over which they conduct DHCP transactions to configure the host.   Their model relies on the existence of a VPN server in the home network that the host must know about, which has been modified to also serve as a DHCP relay.   While mobility and dynamic home addressing are outside the scope of their work, their approach also endorses the use of DHCP and suggests a tunneling mechanism for a remote host to contact its home addressing server.   Whereas their approach relies on tunneling support from IPSec, the configuration solutions presented in this chapter rely on the tunneling support provided by Mobile IP.

1.6   SUMMARY AND CONCLUSIONS

The dynamic configuration of mobile devices for wireless Internet access offers interesting research issues. This chapter described a taxonomy of possible architectural approaches to the problem, and examined the advantages and pitfalls of each of them.   Two possible solutions that integrate DHCP and Mobile IP were described and contrasted in detail. While both solutions are based on and are compliant with the Mobile IP and DHCP protocol standards, the *Transient Tunneling* mechanism adheres to a device-centric design in its location of DHCP client functions contrasting with the agent-centric design of the *DHCP proxy* mechanism. Transient Tunneling uses Mobile IP home agents to enable a DHCP client on the mobile device to communicate with a server in its remote home network. This leads to benefits in implementation simplicity and configuration flexibility at the cost of initial configuration latency.   The proxy approach places (proxy) DHCP clients on the home agent to locally transact with servers on behalf of remote mobile devices.   It minimizes configuration latency at the cost of implementation complexity and limited capability for dealing with DHCP options.

These solutions illustrate some general tradeoffs in protocol design. A device-centric solution guarantees flexibility and full control on the device, but it sometimes incurs performance overheads. On the other hand, a solution built on support from a network-based agent can optimize performance by exploiting service locality, although it requires costly enhancements to the infrastructure and entails limitations to its flexibility. In addition, the solution space shows that the provision of an existing service (e.g., dynamic configuration) to a new user base (e.g., mobile users) that has a specialized network access protocol (e.g., Mobile IP) tends to lead to polarized viewpoints. The required service is either transparently subsumed (physically or logically) within the specialized network access protocol or transparency is sacrificed to minimize infrastructure changes and to prevent functionality duplication or inter-protocol dependencies.

Although the configuration mechanisms presented do not introduce any additional security weaknesses beyond those already known to exist in the Mobile IP and DHCP protocols, future work should examine the impact of integrating emerging security enhancements to these protocols into the proposed configuration process. For example, the use of IPSec tunnels instead of standard IP-in-IP tunnels would allow the communication between the client and its home network to be secure, and would in turn secure also the DHCP transactions that happen on top of such tunnel. Similarly, the impact of the recently introduced authentication options in DHCP [17] should be evaluated, especially in the DHCP proxy model, where transactions with the DHCP server can be handled by different entities at different times, therefore making the authentication process more complex. Finally, the role of stateful dynamic configuration procedures in IPv6 networks (e.g., DHCPv6 [10]) supporting stateless address auto-configuration mechanisms [11] is another topic for future work.

REFERENCES

[1] Charles Perkins and Kevin Luo, *Using DHCP with Computers that Move*, Wireless Networks Journal, vol. 1, pp. 341-353, 1995

[2] Jon-Olov Vatn and Gerald Maguire Jr., *The effect of using co-located care-of addresses on macro handover latency*, in Proceedings of Nordic Teletraffic Seminar, August 1998

[3] Pat Calhoun and Charles Perkins, *Mobile IP Network Access Identifier Extension for IPv4*, IETF RFC 2794 Proposed Standard, March 2000

[4] Charles Perkins, Editor, *IP Mobility Support*, IETF RFC 3220 Standards Track, January 2002

[5] R. Droms, *Dynamic Host Configuration Protocol*, IETF RFC 2131 Proposed Standard, March 1997

[6] B. Aboba and M. Beadles, *The Network Access Identifier*, IETF RFC 2486, Standards Track, January 1999

[7] S. Alexander and R. Droms, *DHCP Options and BOOTP vendor Extensions*, IETF RFC 2132 Proposed Standard, March 1997

[8] S. Glass, *Mobile IP Agents as DHCP Proxies*, Work in progress, Internet Draft presented at IETF-50, Minneapolis, March 2001, http://www.watersprings.org/pub/id/draft-glass-mobileip-agent-dhcp-proxy-01.txt

[9] Y. Rehkter, et al., *Address Allocation for Private Internets*, IETF RFC 1918 Proposed Standard, February 1996

[10] J. Bound, M. Carney, and C. Perkins, *Dynamic Host Configuration Protocol for IPv6 (DHCPv6)*, Work in progress, Internet Draft presented at IETF 50, Minneapolis, March 2001, http://www.ietf.org

[11] S. Thomson and T. Narten, *IPv6 stateless address autoconfiguration*, IETF RFC 2462 Draft Standard, Dec. 1998

[12] S. Thuel, L. Salgarelli, R. Ramjee, K. Varadhan and T. LaPorta, *Dynamic Home Addressing in Mobile IP using Transient Tunnels*, Work in progress, Internet Draft presented at IETF 48, Pittsburgh, July 2000, http://www.bell-labs.com/user/thuel/draft-thuel-mobileip-tt-00.txt

[13] S. Thuel, L. Salgarelli, R. Ramjee, and T. LaPorta, *Dynamic Home Addressing in Mobile IP using Transient Tunnels*, Work in progress, Internet Draft presented at IETF 50, Minneapolis, March 2001, http://www.ietf.org, draft-thuel-mobileip-tt-01.txt, also at http://www.bell-labs.com/user/thuel/draft-thuel-mobileip-tt-01.txt

[14] B. Patel, B. Aboba, S. Kelly and V. Gupta, *DHCP Configuration of IPSEC Tunnel Mode*, http://search.ietf.org/internet-drafts/draft-ietf-ipsec-dhcp-04.txt, Work in progress, Internet Draft

[15] B. Aboba et al., *Criteria for Evaluating AAA Protocols for Network Access*, IETF RFC 2989, November 2002

[16] G. Montenegro, Editor, *Reverse Tunneling for Mobile IP, revised*, IETF RFC 3024, Standards Track, January 2001

[17] R. Droms, Editor, *Authentication for DHCP Messages*, IETF RFC 3118, Standards Track, June 2001

[18] S. Kent and R. Atkinson, *Security Architecture for the Internet Protocol*, IETF RFC 2401, Standards Track, November 1998

Chapter 2

FAST SOFT HANDOFF SUPPORT AND DIFFSERV RESOURCE ALLOCATION IN WIRELESS MOBILE INTERNET

Yu Cheng, Xin Liu, and Weihua Zhuang
Centre for Wireless Communications
University of Waterloo, Waterloo, Ontario, Canada N2L 3G1
{ycheng, xliu, wzhuang}@bbcr.uwaterloo.ca

2.1   INTRODUCTION

Provision of various real-time multimedia services to mobile users is the main objective of the next-generation wireless networks, which will implement Internet Protocol (IP) in the network layer and can interwork with the Internet backbone seamlessly [9, 28]. On the radio interface, the wideband code-division multiple access (CDMA) techniques are used, aiming to provide mobile users a reliable, high-speed, wireless Internet connection. The establishment of such wireless mobile Internet is technically very challenging. Two major tasks are the support of fast soft handoff and the provision of quality-of-service (QoS) guarantee over IP-based wireless access networks.

The next-generation, 3rd generation (3G) or 4th generation (4G), wireless networks will adopt micro/picocellular architectures for various advantages including higher data throughput, greater frequency reuse, and location information with finer granularity [30]. In this environment, the handoff rate grows rapidly and fast handoff support is essential. Especially for real-time traffic, the handoff call processing should be fast enough to avoid high loss of delay sensitive packets. A unique feature of CDMA is the use of soft handoff [22], which can effectively increase the capacity, reliability, and coverage range of the wireless system. Therefore, the implementation of fast soft handoff is critical for the efficient delivery of real-time services to mobile users.

To achieve fast handoff requires both a fast location/mobility update scheme and a fast resource allocation scheme. The popular scheme for fast location update is a *registration-domain-based* architecture, where the radio cells or the related base stations (BSs) in a geographic area are organized into a registration domain, and the domain connects to the Internet through a gateway or a domain root router. In such an environment, Mobile IP [38] is used to support *macromobility*, the inter-domain mobility. When a mobile node (MN)[1] moves into a registration domain for the first time, it will register the new care-of-address to its home agent (HA). While MNs migrate within the domain, various *micromobility* protocols [14, 41, 23, 19] have been proposed to complement Mobile IP by offering fast and seamless local handoff control. When MNs move across multiple BSs or multiple subnetworks within a domain, mobility updates messages will only be sent at farthest to the gateway, without interaction with the Mobile IP enabled Internet. In this way, the registration between the MNs and the home agent (which often locates far away) is eliminated, leading to an obviously reduced signaling overhead for location update, and a considerably less delay and packet loss during handoff. However, the previously proposed schemes focus on enabling hard handoff and do not explicitly address IP soft handoff issues. So far, to our best knowledge, no protocol has been developed to support fast soft handoff in IP-based wireless networks. In this chapter, we propose a new domain-based architecture, called Mobile Cellular IP (MCIP), which not only inherits the ability of supporting fast, seamless, local hard handoff, but also integrates newly designed mechanisms to support fast soft handoff.

A handoff will be successful only when the new BS has enough resources to support the traffic. Therefore, fast handoffs require fast resource allocation. For this purpose, we establish a resource allocation scheme over a registration domain to achieve fast call admission control, QoS guarantee, and high utilization of the scarce wireless frequency spectrum.

The integrated services (IntServ) approach [10] and the differentiated services (DiffServ) approach [8] are the two main architectures for QoS provisioning in IP networks. The IntServ approach uses the Resource Reservation Protocol (RSVP) [11] to explicitly signal and dynamically allocate resources at each intermediate node along the path for each traffic flow. In this model, every change in an MN attachment point requires new RSVP signaling to reserve resources along the new path, which incurs latency in the call admission control and is not suitable for fast handoff. Also, the heavy signaling overhead reduces the utilization efficiency of the wireless bandwidth. On the other hand, the DiffServ approach uses a much coarser differentiation model to obviate the above disadvantages, where packets are classified into a small number of service classes at the network edge. The packets of each class are marked and traffic conditioned by the edge router, according to the resource commitment negotiated in the service level agreement (SLA). In each core router, QoS for different classes is differentiated by different *per-hop behaviors* [8]. Resource allocation is performed by the *bandwidth broker* [42] in a centralized manner, without dynamic resource reservation signaling and reservation status maintaining in the core routers. In this chapter, the registration domain will be modeled as a DiffServ administrative domain, with the gateway router as the edge router connecting to the Internet backbone and the BSs as the edge devices providing MNs wireless Internet access. A bandwidth broker will manage the resource allocation over the DiffServ registration domain.

The rest of this chapter is organized as follows. In Section 2, we give a brief review of the popular micromobility protocols and point out why they can not support soft handoff. Section 3 describes the system architecture for the soft handoff protocol and the DiffServ resource allocation. Details of the MCIP protocol are described in Section 4. In Sections 5, 6 and 7, we discuss fast resource allocation in the DiffServ environment. Section 5 explains how to integrate the DiffServ QoS scheme with a registration domain. Section 6 investigates the call admission control over the DiffServ registration domain. After that, we present an adaptive assured service in Section 7, where multimedia applications experience bandwidth degradation and compensation, depending on the resource availability in wireless links. Section 8 presents numerical results to demonstrate the performance of the proposed MCIP protocol and the DiffServ resource allocation techniques. Finally, we conclude this research and discuss possible future research topics on IP-based soft handoff and wireless QoS in Section 9.

2.2   PREVIOUS MICROMOBILITY PROTOCOLS

Over the past several years a number of IP micromobility protocols [13] have been proposed that complement the Mobile IP by providing fast, seamless, and local handoff control. There have been a lot of studies [19, 15, 40 12] of the protocol complexity, processing requirements, and the handoff performance of the micromobility protocols, especially of the two key protocols, Cellular IP [14] and HAWAII [41]. In this section, we review the micromobility protocols with the focus on their possibility to support soft handoff.

Soft handoff allows an MN to communicate with multiple BSs simultaneously during a handoff in CDMA systems. Therefore, the chance for successful data transfer can be increased, as the probability that all data copies from different BSs involved in the soft handoff are corrupted at the same time is significantly reduced. Soft handoff is typically very fast (on the order of 20 ms) [27]. There are three key requirements for realizing fast soft handoff:

- **Soft handoff signaling:** The serving BS (or the serving base station controller (BSC) in MCIP) needs to inform in time the new target BSs (or BSCs) and the MN to make resource preparations for the coming soft handoff, through a properly designed signaling protocol.
- **Data distribution and selection** [47]: Separate copies of the same data need to be sent via multiple BSs to the same MN in the downlink, or from an MN to multiple BSs in the uplink.
- **Data content synchronization** [47]: In the downlink, data segments arriving from multiple BSs to an MN at the same time should be copies of the same data content in order for the MN to correctly combine these copies into a single copy. In the uplink, only one copy of the data sent by the MN to multiple BSs should be selected for delivery to the destination.

All the present micromobility protocols have been designed without explicitly considering the support of soft handoff. They can not fulfill all or some of the above requirements for fast soft handoff. Despite the apparent differences between various micromobility protocols, the operational principles that govern them are largely similar [12]. Therefore, we take Cellular IP and HAWAII as examples, and give detailed analyses of their deficiencies to support soft handoff. Other micromobility protocols suffer from the similar problems.

2.2.1 Cellular IP

2.2.1.1 Protocol Overview

As the name suggests, Cellular IP inherits cellular principles for mobility management such as passive connectivity, paging, and fast handoff control, but implements them in the IP paradigm. The universal component of a Cellular IP access network is the *base station* which serves as a wireless access point and router of IP packets while performing all mobility related functions. The BSs are built on a regular IP forwarding engine with the exception that IP routing is replaced by Cellular IP routing and location management. Cellular IP networks are connected to the Internet via *gateway* routers. MNs attached to an access network use the gateway IP address as their Mobile IP care-of-address. Assuming Mobile IPv4 [38] and no route optimization [39], packets are first routed to the host's home agent and then tunneled to the gateway. The gateway intercepts packets and forwards them toward the destination BS. Inside a Cellular IP network, MNs are identified by their home addresses, and data packets are routed to the host's actual location via Cellular IP routing protocol. Packets transmitted by MNs are first routed toward the gateway and from there onto the Internet.

An important concept in Cellular IP design is simplicity and minimal use of explicit signaling, which enable low-cost implementation of the protocol. In Cellular IP, location management and handoff support are integrated with routing. To minimize control messaging, regular data packets transmitted by MNs are used to refresh host location information. For example, *uplink* packets are routed from an MN with source IP address X to the gateway on a hop-by-hop basis. The path taken by these packets is cached by all intermediate BSs as soft-state mappings (X , BS$_i$), where BS$_i$ is the neighbor from which packets enter the node. To route *downlink* packets addressed to MN X, the path used by recently transmitted packets from the MN is reversed. When the mobile node has no data to transmit, it sends small, special *route-update* packets toward the gateway to maintain its downlink routing state. Otherwise, the route information will be deleted without refreshing in a system-specific time, called *route-timeout*. The same routing approach is used for *paging*. As an idle MN moves into a new *paging area*, it sends *paging-update* packets regularly to BSs which have better signal quality. As in the case of data and route-update packets, paging-update packets are routed toward the gateway on a hop-by-hop basis. The intermediate BSs maintain a *paging cache* in the same format and operation as the routing cache, with a longer timeout period called *paging-timeout.*

Cellular IP supports two types of handoff. Cellular IP *hard handoff* is based on a simple approach that trades off some packet loss for minimizing handoff signaling rather than trying to guarantee zero packet loss. Cellular IP *semisoft handoff* prepares handoff by proactively notifying the new access point before actual handoff. Semisoft handoff minimizes packet loss, providing improved TCP and UDP performance over hard handoff. However, semisoft handoff is in fact a kind of hard handoff, where the MN receives packets only from one BS at any time. In Cellular IP, both kinds of handoffs are mobile-controlled forward handoffs (MCFHOs) [5][2]. MNs listen to beacons transmitted by BSs and initiate handoff based on signal strength measurements. To perform a handoff, a mobile tunes its radio to a new BS and sends a route-update packet. The route-update message creates routing cache mapping en route to the gateway, configuring the downlink route cache to point toward the new BS.

2.2.1.2 Soft Handoff Support

Cellular IP is targeted at providing high-speed packet radio access to the Internet with the design principle of lightweight nature. The development of the protocol is independent of the radio interface. Its mobility management functionalities, including the routing and handoff management, are implemented in the IP layer. However, without the participation of Layer 2, the Medium Access Control (MAC) and the Radio Link Control (RLC), at the radio interface, the space diversity combining (which is the fundamental of any soft handoff scheme) is impossible [47, 27, 25]. Specifically speaking, Cellular IP lacks the ability to support soft handoff because of the following reasons:

- There is no scheme for the BSs to communicate with each other in a Cellular IP access network. The BSs implement Cellular IP routing, where they only maintain mobile-specific routing entries. The packets with destination addresses other than an MN's IP address will be routed to the gateway. However, to support soft handoff, different BSs must have the ability to exchange data with each other. The old serving BS has to exchange handoff signaling with the new target BS(s), and to forward the resource profile of the MN under consideration, all link layer and physical layer parameters to the new BS(s) so that the new air interface channel(s) between the MN and the new BS(s) can be established.

- The IP connection between the BS and the MN makes data content synchronization difficult. As an example, consider a downlink soft handoff of MN X involving two BSs. During the soft handoff, there are two traffic flows. One is from the old serving BS to MN X; the other is from the new BS to MN X, if we assume that the old BS can send the packets to the new BS without any delay. At both BSs, packets need to be buffered and scheduled to guarantee QoS in the IP layer. If the scalable DiffServ scheme is used, then at each BS traffic flow to MN X will be queued together with other flows subscribing to the same service class, in the same buffer [24, 18]. In this situation there is no way to guarantee that the two copies of a single packet destinated to MN X from the old BS and the new BS will be scheduled out simultaneously. With per-flow IntServ QoS model, the synchronization can be achieved by complicated per-flow scheduling algorithms [46], but it adds a large computation load to the already heavily burdened BSs. Furthermore, the scarce wireless bandwidth is not used in an efficient way when each IP packet is transmitted over the wireless link as it is. A detected error will lead to the retransmission of the whole packet, which will lead to a long delay and a waste of bandwidth when the packet is long.

- The MCFHO is not appropriate for seamless fast handoff in a non-random access system due to the associated large processing delay. In a CDMA network, the most common case is that the MN connects to the network via a dedicated channel to receive real-time services [25]. When the MN moves to the new BS's coverage area, it has to send the mobility update message via a Random Access Channel (RACH) [25]. An RACH usually has a rather low bit rate and uses a slotted-ALOHA like mechanism for bandwidth sharing among all the users in the cell. Thus the mobility update message is likely to experience an access delay due to collision and a long transmission delay due to the low bit rate. After the new BS receives the registration message, the BS performs a call admission control operation. If radio resources are available, the MN is assigned a dedicated channel by signaling through a downlink shared channel. This also takes time. During a hard handoff, if the connection with the old BS turns off before the new BSs makes the call admission decision, and the last decision is "reject", the handoff call will be dropped. During a soft handoff, the long delay may also cause the connection to be dropped, if the handoff occurred due to bad radio link quality [37]. The work in [5] points out that the mobile assisted backward handoff should be a better choice for CDMA wireless networks.

2.2.2 HAWAII

HAWAII is similar in spirit to Cellular IP. HAWAII relies on Mobile IP to provide wide-area inter-domain mobility. A mobile entering a new foreign domain is assigned a collocated care-of address using Dynamic Host Configuration Protocol (DHCP). Packets are tunneled to the care-of address by the home agent in the home domain. The MN retains its care-of address unchanged while moving within the foreign domain, and the connectivity is maintained using dynamically established paths of HAWAII. Nodes in a HAWAII network are enhanced IP routers, which both execute the generic IP routing protocol and maintain mobility-specific routing information, where MN-based routing entries are added to the legacy routing table. Such MN location information is created, updated and modified by explicit signaling messages sent by MNs. HAWAII defines four alternative path setup schemes that control handoff between access points. The two forwarding schemes, the multiple stream forwarding (MSF) and the single stream forwarding (SSF), are for seamless hard handoff, where the old BS buffers packets and forwards them to the new BS. The multicast nonforwarding (MNF) scheme works in a way similar to the semi-soft handoff in cellular IP. The unicast nonforwarding (UNF) considers the situation that MNs can listen/transmit to multiple BSs in the CDMA network, but the ability is mainly used to continue data transmission between the MN and the old BS when the routing information and resource allocation are processed along the new path. HAWAII also uses IP multicasting to page idle MNs when incoming data packets arrive at an access network and no recent routing information is available.

HAWAII is different from Cellular IP mainly in two aspects. Firstly, the BSs and other nodes in the HAWAII network are enhanced IP routers. They have the ability to communicate with each other. It is attributed to this ability that HAWAII can build up its four path setup schemes to achieve seamless handoff. Actually, HAWAII nodes make soft handoff possible with the enhancement from the MCIP protocol. Secondly, Cellular IP relies on the gateway to act as a foreign agent that decapsulates the packets before delivering them to the user. The presence of the gateway foreign agent can complicate per-flow QoS management, since the interior nodes cannot distinguish easily the packets destinated to different MNs but tunneled through the same gateway. However, if DiffServ QoS scheme is used, this would not be a problem for per-class QoS.

In HAWAII, BSs keep IP connections with MNs, and MCFHO schemes are used. As explained earlier, due to these two points, HAWAII can not be used directly for soft handoff.

2.2.3 Other Related Work

HAWAII and Cellular IP are local mobility protocols implemented in the IP layer. Protocols supporting local mobility in data link layer or by IP tunneling have also been proposed.

In [32] off-the-shelf Ethernet switches and wireless LAN cards are used to build wireless access networks. The learning feature of Ethernet switches is used for location management, which monitor (i.e., "snoop") mobile originated packets to update or fresh the bindings between host MAC address and the ports through which packets are received. Although this approach results in simple, low-cost, and efficient access networks, it is only suitable for small networks with the coverage radius limited to a few miles. Here, we consider outdoor cellular networks with a much larger coverage, where IP layer protocols are a better solution.

In [23, 21], Mobile IP is extended by arranging a hierarchy of foreign agents to locally handle Mobile IP registration. When necessary, MNs send Mobile IP registration messages to update their respective location information. Registration messages establish tunnels between neighboring foreign agents along the path from the MN to a gateway foreign agent at the top of the hierarchy. Packets addressed to the MN travel in this network of tunnels, which can be viewed as a separate routing network overlay on top of IP. Although the IP tunneling makes the protocols ready to be deployed over legacy IP routers, it brings complex operational and security issues [19]. The Hierarchy foreign agent architecture is not appropriate for soft handoff, where the signaling path (from the MN to the new BS, and then to a certain foreign agent) will be different from the data path (between the old BS and the new BS) in such an architecture. Also one BS has to tunnel copies of the packets to the other. The larger operational delay compared to that in Cellular IP and HAWAII systems makes the data content synchronization more difficult. The TeleMIP scheme [19] extends the protocol in [23] to a more comprehensive architectural framework for supporting intra-domain mobility in a cellular wireless network. The *mobility* agent (MA) is introduced to the foreign agent hierarchy to achieve flexible address management, simple security management, and some traffic engineering functions [33]. But the mobility agent does not bring any enhancement regarding the soft handoff support.

The work in [27] also points out that the present micromobility protocols are not appropriate for soft handoff, from the perspective of the design of the CDMA radio access network (RAN). In the present CDMA RANs, soft handoff happens between BSs within the same RAN or in directly connected RANs. The extremely tight real time constraints (5ms to 80ms) on traffic in the RAN require that RAN traffic be quickly processed in the form of radio frames for efficient delivery to and from the radio medium by the BSs. In 3G wireless networks, when IP traffic is introduced, the IP layer processing should be terminated at the RAN gateway, also called BSC, which will be responsible for splitting packets into radio frames or combining radio frames into packets. However, the protocol to support soft handoff in such an "IP core, Layer 2 RAN" environment has not been designed yet. We propose such a protocol in Section 4.

2.3   SYSTEM ARCHITECTURE

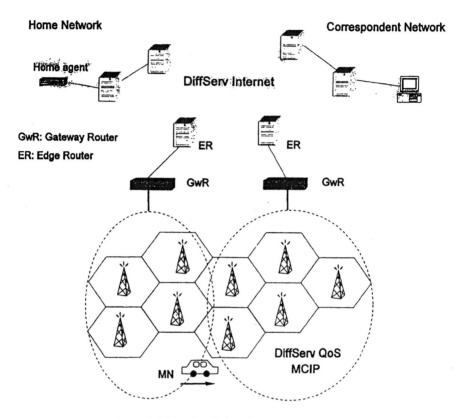

Figure 2-1. The domain-based system architecture.

We consider a domain-based architecture, as shown in Fig. 2-1. The radio cells and the related BSs in a geographic area are organized into a registration domain. Within the domain, the MCIP scheme is deployed for micromobility management and soft handoff support, and the DiffServ approach is used for QoS provision. In the following, we first focus on the MCIP protocol, and then discuss the DiffServ resource allocation techniques. The detailed structure of an MCIP domain is shown in Fig. 2-2, while how to implement DiffServ over such a domain is explained in Section 5.

As shown in Fig. 2-2, the proposed MCIP scheme adopts a 3-tier domain architecture to overcome the structural deficiency of present IP-based micromobility protocols for soft handoff. The first tier is a gateway router which connects the 3G wireless subnet [3] to the backbone Internet. All MNs within the gateway domain use the gateway's IP address as their care-of IP addresses. User traffic originating from or destined to the correspondent node (CN) is routed by Mobile IP outside the gateway domain and by MCIP routing inside the gateway domain. The gateway may collocate with a certain BSC(e.g., BSC 9 as in Fig. 2-2). The single gateway approach considerably simplifies the address management over the collocated care-of address assignment using DHCP. Although the single gateway may potentially impact reliability and complicate per-flow QoS management [41], the problems can be avoided by using backup devices and the DiffServ QoS approach, respectively.

The second tier consists of a mesh of BSCs, which are connected as a *core network*. A BSC has two roles: a router built on top of both regular IP and MCIP routing engines, and a radio network controller which is responsible for the radio resource management of all the cells, i.e., base transceiver stations (BTSs) [4], within its jurisdiction as defined in the 3G system proposals [25]. Each BSC and its associated BTSs constitute a cell cluster. Adjacent BSCs are connected by direct links to facilitate the inter-BSC (inter-cluster) soft handoff.

The third tier consists of hundreds of BTSs, which are organized into clusters; each cluster connects to a BSC. The BTS processes data only in the form of radio frames, and forwards them to its serving BSC for upper layer processing. A BTS has two roles: First, it takes part in the radio resource management within its cell under the control of the BSC, such as downlink close-loop fast power control, measurement of air interface traffic load, etc.. For this purpose, a BTS keeps an IP layer connection with its controlling BSC to obtain system parameters and to submit measurement reports; Second, a BTS works as a bridge in the radio interface between an MN and its serving BSC, delivering the basic MAC sublayer protocol data units (PDUs) which are received or to be transferred over the air. This function is implemented by the *bridge logic* at the BTS. Note that, in the system model, there is no direct IP layer connection between a BTS and an MN in the cell.

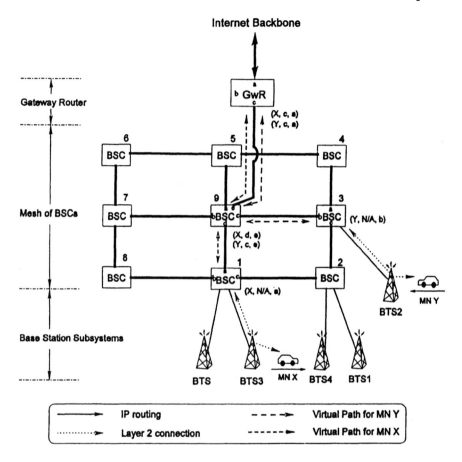

Figure 2-2. The 3-tier MCIP domain.

In summary, the 3-tier architecture is basically an "IP core, Layer 2 RAN" architecture, as suggested in [27]. Tier 1 and tier 2 have a structure similar to that of Cellular IP or HAWAII, and therefore, both the protocols can be applied directly to support the mobile controlled forward hard handoff between BSCs. The introduction of tier 3 makes soft handoff achievable by processing data in Layer 2. In the next section, we design protocols for inter-cluster soft handoff and mobile assisted backward hard handoff.

2.4   MOBILE CELLULAR IP (MCIP)

2.4.1 Protocol Architecture

The MCIP protocol architecture is shown in Fig. 2-3. Two interactive and overlapping protocol stacks are defined in MCIP: the radio interface protocol stack, and the core network protocol stack. They work in conjunction with each other to ensure fast and seamless microscopic mobility, with the capability to support the macroscopic diversity combining required by soft handoffs.

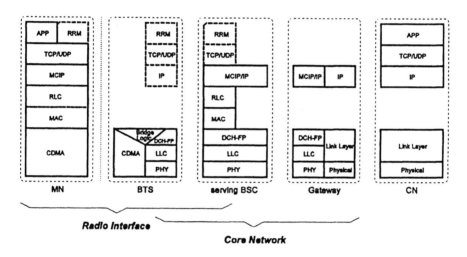

Figure 2-3. Protocol stacks in the MCIP access network.

2.4.1.1 Radio Interface Protocol

The radio interface protocol defines the operations of the MN, its associated BTS and the serving BSC, to make the wireless link transparent to IP traffic and capable of providing bandwidth-on-demand services to the IP layer and above. The radio interface comprises the CDMA layer [25, 1], the Layer 2 consisting of the MAC [25, 2] and RLC sublayers [25, 3], and the Radio Resource Management (RRM) layer [25, 4]. Radio interface is actually the most complex part of a wireless network, and the detailed discussion is out of the scope of this chapter. However, some conceptual description is necessary for presenting the proposed soft handoff scheme.

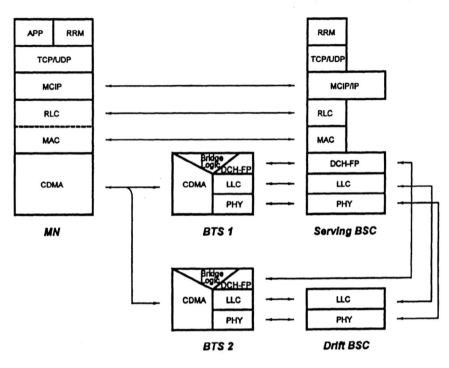

Figure 2-4. Space diversity in inter-cluster soft handoffs.

The CDMA layer is the physical layer over a wireless channel, which provides digital data transmission over the wireless air interface. The MAC sublayer controls how much wireless bandwidth is allocated to a certain user, which can be considered as a logic channel. The logic channel is mapped to a *dedicated channel* in the CDMA layer, and the dedicated channel can be identified by the scrambling code and the spreading code used for the channel [25]. The RLC has the functions similar to those of its wireline counterpart, which implements error detection, error correction and the ARQ (Automatic Repeat Request) control upon the radio frames forwarded by the MAC sublayer. Basically, the wireless layers replace their wireline counterparts to achieve an IP-based radio interface. In Fig. 2-3, we can see that the protocol stack in the MN is designed exactly in this way. The complexity is brought by the other end of the wireless link, as the physical channel layer and logical channel layer reside in two different devices, the former in the BTS and the latter in the serving BSC. Therefore the BTS needs two interfaces: one is the wireline interface to the BSC, which receives the logic channel assignment and data radio frames from the BSC in the downlink and forwards data radio frames to the BSC in the uplink; the other is the wireless interface, which establishes the physical dedicated CDMA channel for data transfer. The bridge logic is necessary to understand the MAC resource allocation parameters received via the wireline interface and to map the logical channel to the physical dedicated channel.

In the MCIP scheme, the wireline interface of the BTS is enhanced so that it can establish an IP connection with its serving BSC to obtain system parameters and to submit measurement reports. This function also brings great flexibility for future IP-based network management. Hence, there are two data paths between the BTS and its serving BSC. One is for the radio resource management information and user data transfer:

$$\underbrace{\text{Bridge Logic} \leftrightarrow \text{LLC (Link Layer Control)} \leftrightarrow \text{PHY (Physical Layer)}}_{BTS\,side}$$

$$\Leftrightarrow \underbrace{\text{PHY} \leftrightarrow \text{LLC} \leftrightarrow \text{MAC} \leftrightarrow \text{upper layers}}_{BSC\,side}$$

The other is for the IP connection between the BTS and its serving BSC:

$$\underbrace{upper\,layers \leftrightarrow IP \leftrightarrow LLC \leftrightarrow PHY}_{BTS\,side}$$

$$\Leftrightarrow \underbrace{PHY \leftrightarrow LLC \leftrightarrow IP \leftrightarrow upper\,layers}_{BSC\,side}$$

In this two-path situation, when the LLC layer receives data from the PHY layer, it should determine which data path to follow. LLC itself does not have this function, so we introduce the DCH-FP (Dedicated Channel - Frame Protocol) sublayer sitting on top of the LLC sublayer. The DCH-FP sublayer is to integrate the radio interface protocol stack and the core network protocol stack smoothly.

With the above protocol structure, all the radio resource management information, the user data, and other control messages can be transmitted as IP packets, which greatly simplify the network operation. The complex legacy network management model, where date plane and control plane are managed separately, is now unnecessary. Instead, we propose a *specific* application layer RRM protocol, which has control access to the RLC sublayer, the MAC sublayer and the CDMA layer (including receiving measurement reports from these layers), in addition to utilizing the transmission services offered by the TCP/UDP layer. On the other hand, a negative effect is that the MCIP air interface becomes a loosely layered protocol stack, due to the introduction of the RRM application layer.

Soft Handoff Support. The radio interface protocol stack is defined basically within a cell cluster, but can be extended to two or more clusters during handoffs. Fig. 2-4 shows the operation of the radio interface during an inter-cluster soft handoff. The *serving BSC* for an MN is the BSC that performs the Layer 2 (MAC and RLC sublayers) processing of the IP packets to/from the *Transport Blocks* (TBs) [25]. The radio resource management operations, and outer loop power control, etc., are also executed in the serving BSC. Each MN has one and only one serving BSC. An MN is also associated with *drift BSC(s)*, which is the BSC(s) controlling those target BTSs involved in the soft handoff. In Fig. 2-4, the drift BSC is the controlling BSC of BTS 2. During an inter-cluster soft handoff, the communication between the MN and its serving BSC takes place via multiple air channels from multiple BTSs. Drift BSCs participate in the macrodiversity combining/splitting by forwarding the LLC PDUs, each of which encapsulates a DCH-FP PDU (containing a second copy of a TB) as its payload, between new target BTSs and the old serving BSC. The drift BSCs do not perform any Layer 2 processing of the IP packets. The combining/splitting procedures in an intra-BSC handoff are basically the same except that the drift BSC does not exist and the target BTSs have direct connections with the serving BSC. Here we only consider the general inter-cluster soft handoff. Although there is no limit for the number of BTSs participating in the soft handoff, it has been shown that no additional benefit will be achieved by involving more than two BTSs [5]. As a result, for presentation clarity, in the following we only describe the situation of two BTSs in the protocol design, even though the protocol structure is suitable for multiple (more than two) BTS soft handoff.

In the downlink direction, as illustrated in Fig. 2-4, the RLC sublayer of the serving BSC segments and encapsulates the IP packets into RLC PDUs. One or more RLC PDUs are encapsulated into a TB at the MAC sublayer. Since both BTS 1 and BTS 2 are in the *active set* [29], the TB should be sent to both BTS 1 and BTS 2. The MAC sublayer at the serving BSC keeps the LLC addresses of both BTS 1 and BTS 2. The TB, the logic channel ID, and the LLC addresses of both BTS 1 and BTS 2 are passed down to the DCH-FP sublayer and encapsulated into a DCH-FP PDU at the serving BSC. The DCH-FP PDU is further encapsulated into two LLC PDUs each with a destination address of either BTS 1 or BTS 2. The two LLC PDUs are sent to BTS1 and BTS2, separately. In this way, the same TB together with its logical channel ID is delivered in Layer 2 to both the BTSs. At each of the two BTSs, the bridge logic receives the TB and the associated logical channel ID. The TB and the logical channel ID are passed down to the CDMA layer. After a series of physical layer processing procedures, two separate CDMA radio signals are sent to the MN by BTS 1 and BTS 2. At the MN, the two radio signals are combined by a *maximal ratio combining RAKE receiver* at the CDMA layer of the MN.

In the uplink direction, the radio frames sent from the MN are received by both the BTS 1 and BTS 2. The radio frames are combined into a TB and encapsulated with the CRC check result into a DCH-FP PDU at each BTS, and then routed to the DCH-FP sublayer at the serving BSC. The MAC layer of the BSC receives two versions of the same TB and CRC check result from the two BTSs, and selects the better one. The selected TB and CRC check result are submitted to the RLC layer for further processing.

Note that, during the soft handoff, IP layer processing is operated in the serving BSC. It is the LLC PDU copies of a packet that are sent to or collected from multiple BTSs. Therefore, the IP QoS processing has the same effect on all the paths involved in the soft handoff and does not affect the data content synchronization. Furthermore, implementing the LLC PDU transfer between the serving BSC and the drift BSC requires the BSCs to have the Layer 2 switching ability. It is not complicated to achieve in either hardware or software. We assume that the BSCs in the MCIP system indeed have the Layer 2 switching function.

2.4.1.2 Core Network Protocols

There exist two differences in the MCIP access system from the Cellular IP or the Internet. First, even though the network layer is based on the IP packet routing paradigm, two different routing algorithms are employed: the regular IP routing algorithm, and the MCIP routing algorithm. MCIP routing is used by user traffic and two signaling messages (*path-teardown*, *crossover-discovery*). All other signaling messages are routed by regular IP routing. Second, the data link layer comprises 2 sublayers: DCH-FP sublayer, and LLC sublayer which can be any regular link layer protocol supporting metropolitan area networking, such as ATM, Frame Relay, etc.. As shown in Fig. 2-3, the CDMA layer terminates at the MN and BTS, but the MAC and RLC sublayers terminate at the MN and the serving BSC. The DCH-FP sublayer is introduced to extend the transport channel, through which the CDMA layer provides services to the MAC sublayer, from the BTS to the serving BSC. Therefore, in addition to providing packet data transmission services to the network layer, the data link layer offers transparent data transmission services between the bridge logic at the BTS and the MAC sublayer at the serving BSC.

2.4.1.3 MCIP Routing

In MCIP, mobility management of active MNs is integrated with the routing and handoff schemes, as in the Cellular IP scheme, but explicit *path-setup* and *path-teardown* packets are employed to create and delete, respectively, an MN-specific *virtual path* [5] used by user traffic. A *path-setup* packet always originates from the MN. It is first sent to the MN's serving BSC over the air via the random access channel and then routed towards the gateway by regular IP routing. The path taken is recorded by the intermediate nodes (BSCs) as an MN-specific routing entry. Such an entry is a triplet (*MN IP address, uplink incoming interface, uplink outgoing interface*). The downlink packet can find its destination MN by taking the reverse direction of the routing entry. Fig. 2-2 illustrates the MN-specific routing entries maintained for paths from gateway to MN X and MN Y. MN X and MN Y can exchange data easily by MCIP routing. After a virtual path is set up, all the user packets will follow it. A *path-teardown* packet may originate from any node on the virtual path. When the teardown decision is made by an intermediate BSC, it may send two *path-teardown* packets towards both directions respectively. Details about MCIP routing are given in [31].

2.4.1.4 Other Functions

In MCIP, *IP header compression* is used to improve the utilization efficiency of wireless resources. When transmitted over the air, the large headers of the IP packets degrade the spectrum efficiency and cause further delay to user traffic. As there exists a high degree of redundancy in the headers of consecutive packets that belong to the same packet stream, an IP header compression algorithm compacts the information by maintaining a context, and attaching each packet only a compressed header which carries information of changes to the context. During an inter-cluster handoff, the context is transmitted to the new BSC by handoff messages originating from the old BSC, and the IP address of the new BSC is sent to the MN over the old CDMA channel, so that the compression algorithm can work continuously. Details of IP header compression are given in [31].

MCIP also deploys a paging process, similar to that used in Cellular IP, except that the process is implemented at the IP layer. A gateway domain can be partitioned into one or several paging areas. After an MN is turned on or migrates to a new paging area, it receives the paging area ID number from the broadcast channel of the best cell, and then sends a *paging-update* packet to the gateway. A paging cache at the gateway maintains an entry for each MN. A Mobile IP binding update is needed if the MN is new to the gateway domain. When a call to an MN arrives at the gateway, a *paging* message is sent to the known paging area, and broadcast over the air via all the BTSs in the paging area.

2.4.2 Inter-Cluster Handoff

Both soft and hard handoffs are supported by the access system. Intra-cluster handoffs are handled by the BSC locally without intervention of other system elements. Here, we focus on the inter-cluster soft and hard handoff.

2.4.2.1 Soft Handoff

The inter-cluster soft handoff comprises two phases: *add* and *drop*. To highlight the handoff process, a segment of the access network (Fig. 2-2) is redrawn in Fig. 2-5(a). MN X connects to the network via BTS 1, and is moving toward BTS 2. MN X periodically performs the measurements of received signal-to-interference plus noise ratio over the pilot channels from its surrounding BTSs, averages the measurements over an averaging window, and reports the results to its serving BSC, BSC 1 [25]. After MN X moves further into the overlapping area, the new BTS (BTS 2) is added to the *active set* (Fig. 2-5(b)); when MN X leaves the overlapping area, the old BTS (BTS 1) is dropped from the active set as shown in Fig. 2-5(c). Both the add and drop decisions are made by BSC 2.

The add phase has 4 signaling messages: (1) *add-request* from BSC 2 to BSC 3: it carries all parameters of the radio connection to MN X so that a second channel from BTS 3 to MN X can be established; (2) *ack-add-request* from BSC 3 to BSC 2: at BSC 3, upon receiving the request message, a radio resource allocation algorithm is executed. If available, the required radio resources at BSC 3 and BTS2 are reserved for MN X. The downlink scrambling code ID assigned to the new channel and BSC 3's IP address are carried by this message. If the required resources are not available, the handoff request is queued at BSC 3 until resources are available or the communication session is forced to terminate; (3) *add-newBTS* from BSC 3 to BTS 2: it is sent if the radio resource allocation is successful. Upon receiving this message, BTS 2 calculates all parameters used by the new channel for MN X at the bridge logic, and is ready to participate in both the downlink and uplink space diversity procedures. The uplink diversity procedure starts at this moment; (4) *add-new* from BSC 2 to MN X: it informs the MN to add fingers to its receiver. After receiving the *ack-add-request* message, BSC 2 is ready to receive the second copy of the uplink TBs forwarded by BTS 2 via the DCH-FP sublayer and execute the uplink macrodiversity combining. The *add-new* message carries the ID number of BTS 2, the downlink scrambling code ID used by BTS 2, and BSC 3's IP address. After sending this message, BSC 2 forwards every downlink TB to both BTS 1 and BTS 2 via the DCH-FP sublayer.

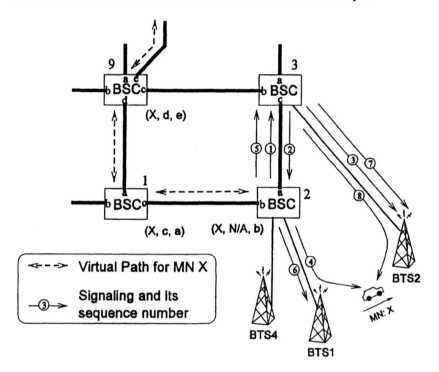

(a) MCIP soft handoff process

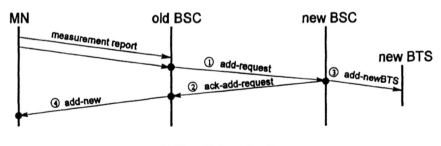

(b) The *add* phase signaling

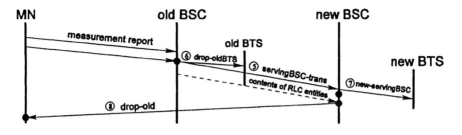

(c) The *drop* phase signaling

Figure 2-5. MCIP soft handoff.

The drop phase also has 4 signaling messages: (1) *drop-oldBTS* from BSC 2 to BTS 1: it informs BTS 1 to release the old channel once the drop decision is made. BSC 2 also stops functioning as the serving BSC, stops popping downlink TBs into the DCH-FP sublayer, and releases resources of the old connection; (2) *servingBSC-trans* from BSC 2 to BSC 3: it informs BSC 3 to take over the function of serving BSC. Following it, BSC 2 encapsulates the contents and parameters of the corresponding RLC sublayer entities into an IP packet, and sends it to BSC 3. IP packets awaiting for transmission in the IP layer are also forwarded to BSC 3; (3) *new-servingBSC* from BSC 3 to BTS 2: it informs BTS 2 to reconfigure the bridge logic so that the newly received TBs will be forwarded to BSC 3; (4) *drop-old* from BSC 3 to MN X: it informs the MN to drop BTS 1 from its active set. The ID number of BTS 1 is the only content of this message. MN X then removes from its receiver the fingers corresponding to the scrambling codes used by BTS 1.

2.4.2.2 Hard Handoff

The hard handoff signaling procedure is given in Fig.2-6. The details are omitted as the procedure is relatively straightforward and self-explanatory.

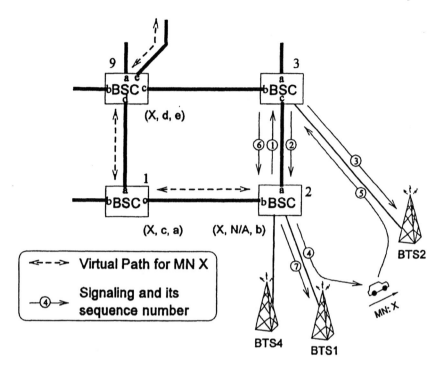

(a) MCIP hard handoff process

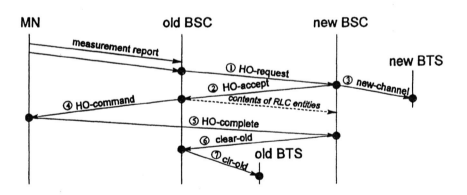

(b) MCIP hard handoff signaling

Figure 2-6. MCIP hard handoff.

2.4.2.3 Path Optimization

For the case shown in Fig. 2-5, after MN *X* hands off from BTS 1 to
BTS 2, its virtual path is extended from BSC 2 to BSC 3 as shown in Fig. 2-
7(a). The path is 2 hops longer than the shortest path from BSC 3 to the
gateway. A *path optimization* process is proposed to find the shortest path
for a maximal path reuse efficiency. The *crossover node* is the common
branch node of the two paths (the path from the old BSC to the gateway and
the path from the new BSC to the gateway) that is closest to the old BSC.
The handoffs can be classified into 3 cases in terms of the logical location of
the crossover node: (1) the old BSC is the crossover node; (2) the new BSC
is the crossover node; (3) neither the old nor the new BSC is the crossover
node as shown in Fig. 2-5. In the first two cases, the path after the handoff is
the shortest, and the path optimization is not necessary. The optimization
signaling process for case (3) is shown in Fig. 2-7(a), and the path for MN
X after the optimization is shown in Fig. 2-7(b). Details of the signaling are
given in [31].

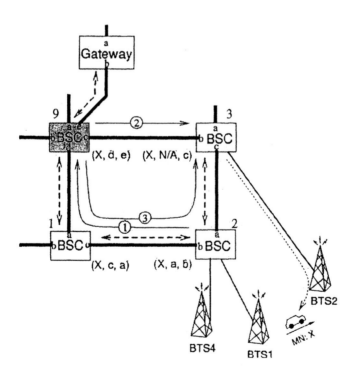

(a) Before the path optimization

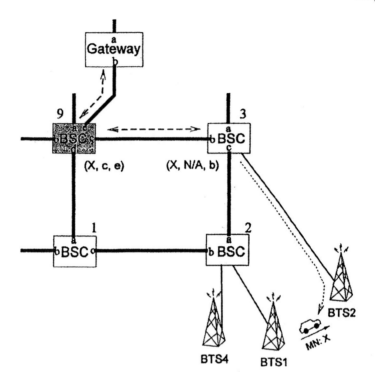

(b) After the path optimization

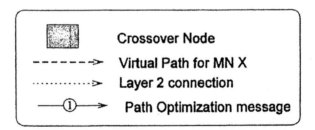

Figure 2-7. Path Optimization.

So far, we have presented the MCIP architecture, which inherits the advantages of existing micromobility protocols to support fast, seamless local handoff. Furthermore, in MCIP, the access network is properly structured to support soft handoff by utilizing space diversity in Layer 2. The protocol architecture is designed in such a way that the Layer 2 RAN and the IP core can cooperate well to provide wireless IP connection to any MN in the MCIP wireless domain. Next we will discuss the IP QoS provisioning over a wireless domain within the IP-based mobility management architecture.

2.5   DIFFSERV REGISTRATION DOMAIN

The main objective of the next generation wireless networks is to provide real-time multimedia services to mobile users, which require strict QoS guarantee on packet loss, delay, or delay jitter [44]. Thus call admission control (CAC) is necessary to ensure the QoS satisfaction of the accepted connections. Also the associated resource allocation procedure should be fast enough to achieve seamless handoff.

The IntServ approach uses per-flow resource reservation to achieve hard QoS guarantee and high efficient resource utilization compared to other QoS schemes [34]. Furthermore, in a wireless access network, the traffic load is not so heavy as that in an Internet backbone, and the per-flow processing is acceptable. But in a wireless domain, the IntServ has several obvious disadvantages as listed in the following.

- IntServ uses RSVP for resource reservation. RSVP involves hop by hop admission control and signaling processing. The RSVP signaling procedure may not be fast enough for a handoff, especially the soft handoff.
- In a wireless domain, normally the wireless links are the bottlenecks, while the bandwidths on the wireline links are sufficient. Using RSVP to reserve resources on those wireline links are not necessary.
- RSVP is receiver-oriented. In an "IntServ access, DiffServ backbone" environment [6], the gateway router of a wireless domain (see Fig. 2-2) acts as the receiver for all the MNs in the domain in resource reservation. The single hot spot leads to a long processing delay, even though we can use backup devices to avoid the potential reliability problem.
- When the IntServ domain interfaces with the DiffServ backbone, complex inter-operations are unavoidable to map the per-flow IntServ QoS specification to aggregate DiffServ QoS metrics [7].

Here we propose to implement the DiffServ QoS support over a wireless domain. The domain under consideration has a general architecture, and can deploy IP based micromobility protocols such as MCIP, Cellular IP, and HAWAII. The IP layer resource allocation is performed by a bandwidth broker in a centralized way. Therefore, the signaling overhead can be obviously decreased. Also, in the next section, we will propose a domain based call admission control algorithm to achieve fast resource allocation.

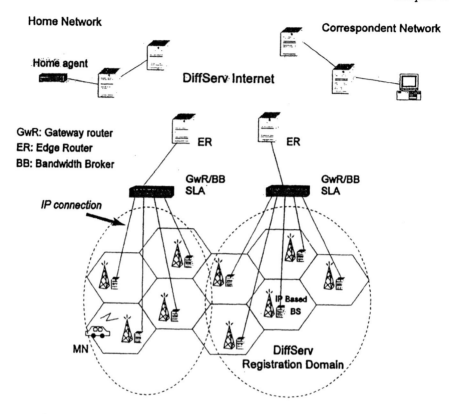

Figure 2-8. The DiffServ registration-domain-based wireless network architecture.

The DiffServ registration-domain-based wireless network architecture is shown in Fig. 2-8. In the system, all the registration domains are DiffServ administrative domains in which all the routers are DiffServ IP routers. The gateway and BSs are edge routers, and they are connected through core routers. Note that, in the system using MCIP, the BTS and the BSC cooperate to serve as one IP end of the wireless link. In discussing IP layer resource allocation, for simplicity, we can combine the two nodes into an *IP based base station*. The gateway is the interface to the DiffServ Internet backbone, where an SLA is negotiated to specify the resources allocated by the Internet service provider to serve the aggregate traffic flowing from/to the gateway. We consider wireless links as bottleneck links in the domain and the SLA is negotiated mainly based on the wireless resource availability. The gateway conditions the aggregate traffic for each class according to the SLA resource commitments. The BSs provide MNs the wireless access points to the Internet, and perform per-flow traffic conditioning and marking when data streams flow in the uplink direction. All the BSs in the same registration domain are connected to the same gateway router. All DiffServ routers use three separate queues to provide the premium service [26], the assured service [24], and the best-effort service, respectively. The three buffers are served under priority scheduling or weighted fair queue (WFQ) scheduling [46]. The traffic classes in the next-generation wireless networks can be mapped to these three DiffServ classes. For example, in a UMTS (Universal Mobile Telecommunications System) wireless network [20], the conversational class and the streaming class can be mapped to the premium service and the assured service, respectively, while the interactive class or the background traffic can be mapped to the best effort class. A bandwidth broker residing in the gateway router is responsible for the resource allocation and call admission control over the DiffServ registration domain.

2.6   DOMAIN-BASED CALL ADMISSION CONTROL

For simplicity, we assume that an *effective bandwidth* can be used to characterize both the traffic characteristics and the QoS requirements, and the calls belonging to the same service class have homogeneous traffic characteristics, and thus have the same effective bandwidths. For example, for a premium service call the peak rate can be used as the effective bandwidth, and for an assured service call the effective bandwidth can be determined using the approach given in [16]. The resource commitments specified in the SLA can then be represented in terms of how many calls for each class are allowed in the registration domain. As a result, the proposed admission control procedure is straightforward: whenever a new MN requests admission to a registration domain, the bandwidth broker determines whether to admit or reject the new call, based on the number of the calls currently in service and the SLA allocation for the service class to which the new call subscribes. The new call has to be dropped if all the SLA allocation has been occupied. This procedure requires very simple communications between the edge router (the BS) and the bandwidth broker (in the gateway router), and can be executed very fast. Furthermore, once an MN is admitted to a registration domain, it can hand off to other cells within the domain without the involvement of further call admission control in the bandwidth broker. During a soft handoff, a connection occupies bandwidth in two or more cells. However, as the handoff duration is very short compared to the connection's lifetime, the soft handoff's effect on bandwidth consumption can be ignored for steady-state analysis.

The above simple resource allocation scheme based on effective bandwidth in fact implies a very complicated design problem. The number of BSs in a domain, the resources allocated to each service class in each base station, and the resource commitments in the SLA should be determined carefully so that the new call blocking and handoff call dropping probabilities are reasonably low, while considering the traffic load in the registration domain, the mobility information and the call duration statistics. In [35], this design problem is solved for the situations that the interval between call arrivals, cell residence time and call duration are independently and exponentially distributed. An important conclusion is that a predetermined handoff call dropping probability can always be guaranteed by properly designing the admission controller. However, in the analysis, handoff calls and new calls are not differentiated. From users' point of view, it is better to be blocked at the beginning of a call than to be dropped in the middle of the call. As a result, handoff calls should be serviced with a higher priority than new calls. To further decrease the handoff call dropping probability, here we use the guard channel scheme [45] to reserve a fixed percentage of each BS's resources for handoff calls and extend the analysis given in [35] to include this situation.

Under the assumption that the complete partitioning scheduling mechanism among the service classes [36] is used, the admission control of each class can be considered separately. Consider a registration domain including M cells, where the new call arrivals of a certain service class are Poisson with a mean rate of λ calls per cell per unit time, and the call duration is exponentially distributed with mean $1/\mu$. Each cell can serve up to C calls of the class under consideration, and a percentage (α) of the cell capacity, αC is set as the guard channel to protect the handoff calls. The channel holding time in a BS (i.e., the time that a call spends with any BS before handing off to another BS) is exponentially distributed with mean $1/h$, i.e., the handoff rate is h. Similarly as in [35,36], we assume that the handoff rate from any cell to any other cell is such that all cells experience the same rate of handoff call arrivals. The handoff call dropping can be taken into account by considering the fact that the effective (actual) call departure rate μ_e is higher than the "natural" call departure rate μ. Then the approximation technique in [35, 36] can be used to calculate the new call blocking and handoff call dropping probabilities.

A new call can be blocked by the admission controller if the total number of calls in the domain exceeds the SLA resource allocation N ($C < N < MC$), and/or if the serving cell is full and can not accept any additional connections. Let P_{Badm} denote the probability of a new call being blocked by the admission controller, and P_H denote the handoff call dropping probability. P_{Badm} and P_H can be calculated by solving the following set of nonlinear equations:

$$\mu_e = \mu + hP_H \tag{1.1}$$

$$\lambda_e = \lambda(1 - P_{Badm}) \tag{1.2}$$

$$P_{Badm} = E(M\lambda / \mu_e, N) \tag{1.3}$$

$$P_H = E_{gH}\left(\frac{\lambda_e}{\mu_e}, \frac{h}{\mu + h}\frac{\lambda_e}{\mu_e}, \alpha C, C\right) \tag{1.4}$$

where (1.1)-(1.3) are the same as those used in [35, 36], and (1.4) is different because of the guard channel introduced. In the above equations, $E(\rho, C)$ represents the Erlang loss formula defined as

$$E(\rho, C) = \frac{\left(\dfrac{\rho^C}{C!}\right)}{\left(\sum_{i=0}^{C} \dfrac{\rho^i}{i!}\right)}.$$

$E_{gH}(\rho_{tot}, \rho_h, C_g, C)$ represents the handoff call dropping probability in a fixed guard channel system, where ρ_{tot} is the total Erlang load consisting of both the new calls and handoff calls, ρ_h is the Erlang load consisting of only the handoff calls, and C_g is the resources allocated for the guard channel. In the system, each call intends to hand off to a neighbor cell with a probability of $h/(\mu + h)$, so in the steady state the handoff Erlang load occupies $h/(\mu + h)$ of the total traffic load. Letting $E_{gB}(\rho_{tot}, \rho_h, C_g, C)$ denote the new call blocking probability in the guard channel system, we have

$$E_{gH}(\rho_{tot}, \rho_h, C_g, C) = P_0 \frac{\rho_{tot}^C}{C!}\left(\frac{\rho_h}{\rho_{tot}}\right)^{C_g} \tag{1.5}$$

$$E_{gB}(\rho_{tot}, \rho_h, C_g, C) = 1 - P_0 \sum_{i=0}^{C-C_g-1} \frac{\rho_{tot}^i}{i!} \tag{1.6}$$

where P_0 is the probability that all channels are unoccupied and is given by

$$P_0 = \left[\sum_{i=0}^{C} \frac{\rho_{tot}^i}{i!}\left(\frac{\rho_h}{\rho_{tot}}\right)^{\max(i-C+C_g, 0)}\right]^{-1} \tag{1.7}$$

Based on (1.5) and (1.7), equations (1.1)-(1.4) can be solved recursively [35, 36]. After obtaining P_{Badm} and P_H, we can calculate μ_e and λ_e from (1.1) and (1.2), respectively. Then based on μ_e and λ_e, we can compute the new call blocking probability in a cell from (1.6), given by

$$P_{Bcell} = E_{gB}\left(\frac{\lambda_e}{\mu_e}, \frac{h}{\mu + h}\frac{\lambda_e}{\mu_e}, \alpha C, C\right)$$

Then, the total blocking probability of a new call is given by

$$P_B = 1 - (1 - P_{Badm})(1 - P_{Bcell}) \approx P_{Badm} + P_{Bcell}.$$

2.7   ADAPTIVE ASSURED SERVICE

The streaming class defined in the 3G wireless networks can be supported by the assured service in a DiffServ architecture. A streaming class traffic, such as a streaming video, normally does not require very strict timely delivery but requires a guaranteed minimum delivery rate. Adaptive coding can be applied to this type of traffic to improve sustain probability when the network congests. A good example is the MPEG video coding format where a base layer contains basic and extension layer additional information. The video quality and bandwidth consumption can be scaled down to the bottom by only transmitting the base layer information. In a wireless network, because resource availability fluctuates frequently due to user mobility and channel quality variations, this type of adaptive service is very important to improve resource utilization efficiency. The adaptive framework to be discussed in this section only takes mobility into consideration. That is, the bandwidth allocated to an adaptive video changes only when there is a new call arrival, call completion, or handoff.

Here we propose to use a partitioned buffer [16] with size B to serve a layered video at each DiffServ router. Assume that a video traffic is coded to three layers, L_1, L_2, and L_3. The assured service buffer uses three configurations (1, 2, and 3) to provide three levels of QoS (high, medium and low) to the video traffic, as shown in Table 2-1, where X denotes the number of packets queued in the buffer, and ε_1, ε_2, and ε_3 ($\varepsilon_1 < \varepsilon_2 < \varepsilon_3$) denote different levels of loss probability provided by the different buffer configurations, respectively. In a homogeneous environment, all the video traffic for the assured service in the domain has homogeneous statistical characteristics and the buffer for the assured service in a DiffServ router is a homogeneous multiplexing system. We model each video traffic flow as a multiclass Markov-modulated fluid source (MMFS), where an underline Markov chain determines the traffic generation rate at each time instant. At each state of the Markov chain, three layers of traffic, L_1, L_2 and L_3, are generated. In such a scenario, an *optimal effective bandwidth* can be calculated by using the technique developed in [16], which is the minimal channel capacity required to guarantee the loss requirements for all the layers of the MMFS traffic when the buffer partition thresholds are optimally selected. For each QoS level listed in Table 2-1, the associated effective bandwidth can be calculated and is denoted by e_1, e_2 and e_3 ($e_1 > e_2 > e_3$), respectively.

Table 2-1. The buffer configurations used to provide the assured service

Buffer Configuration	Management Rule	QoS Level	Loss Probabilities	Effective Bandwidth
$[0, B]$	any packet accepted when $0 \le X < B$	high	ε_1 to L_1, L_2 and L_3	e_1
$[0, B_1, B]$ $(0 < B_1 < B)$	L_1 packets accepted when $0 \le X < B$ L_2 and L_3 packets accepted when $0 \le X < B_1$	medium	ε_1 to L_1, ε_2 to L_2 and L_3	e_2
$[0, B_1', B_2', B]$ $(0 < B_1' < B_2' < B)$	L_1 packets accepted when $0 \le X < B$ L_2 packets accepted when $0 \le X < B_2'$ L_3 packets accepted when $0 \le X < B_1'$	low	ε_1 to L_1, ε_2 to L_2, ε_3 to L_3	e_3

Under the assumption that the wireless link is always the bottleneck and the traffic will be served without loss in the wireline links, the adaptive mechanism is mainly used for the buffers in the MNs (for uplink traffic flows) or in the BSs (for downlink traffic flows). Each BS has a local admission/rate controller (LARC) to manage the admission control and the adaptive bandwidth allocation in the cell. In the downlink direction, the LARC can directly adjust the buffer configuration in the BS when bandwidth adaptation happens; in the uplink direction, the LARC should send messages to MNs to control the buffer configuration adjustment for bandwidth adaptation. The adaptive algorithm based on the effective bandwidth is straightforward. If we begin with a light load situation when sufficient resources are available, a new or handoff call arrival to the cell is admitted with bandwidth allocation of e_1, meaning that the high level of QoS is provided to the traffic. As the traffic load increases to a level where a new or handoff call can not be admitted with bandwidth e_1, the LARC then reduces the bandwidth allocation to each already accepted video call from e_1

to e_2, and adjusts the assured service buffer in the BS or MNs from configuration 1 to configuration 2 to fit the adaptive bandwidth allocation. The new request is accepted also with bandwidth allocation of e_2. In this situation, the medium level of QoS is to be provisioned. If the traffic load further increases to a certain degree, the LARC reduces the bandwidth allocation of both existing traffic and new requests to e_3 and changes to buffer configuration 3, trying to accept as many calls as possible by degrading the QoS to the low level. On the other hand, when traffic load decreases, the buffer can shift back to the higher level of configuration and the LARC can allocate more bandwidth to the calls for better QoS. Consider that at the current level each call is allocated an effective bandwidth of e_i $(i = 2,3)$. At the moment that a call completes or hands off to a neighbor cell, the LARC will check whether the available resources are enough to improve the bandwidth allocation of each call to e_{i-1}. If the available resources can be used to accept two more calls with allocation of e_{i-1} after the bandwidth increase, [6] the LARC will implement the adaptation and the buffer will shift up to the one-level higher configuration.

When the adaptation happens from one configuration to another providing higher QoS, the buffer is lightly loaded at that moment and has free queueing space left. The higher level configuration can use the space to allow more packets to be buffered to provide better QoS. The transition is always smooth. On the other hand, when the buffer needs to switch into a configuration providing lower QoS, more packets should be dropped. In [17], a simple and efficient mechanism is designed to achieve a smooth transition from the higher QoS level to the lower QoS level.

2.8   PERFORMANCE EVALUATION

In this section, we present some numerical examples to show the performance of the proposed MCIP micromobility management and the DiffServ resource allocation techniques.

2.8.1 MCIP Performance

In the following, we discuss the scalability of the MCIP access system through a numerical example, compare its handoff signaling cost with those of HAWAII and Cellular IP, and give a qualitative analysis why HAWAII and Cellular IP are not appropriate for 3G/IP interworking.

The HAWAII domain used in the comparison has the same topology and configuration as the MCIP domain shown in Fig. 1-2, except that the BSCs and BTSs are replaced by routers and BSs, respectively. The layout of clusters in an MCIP or HAWAII domain is shown in Fig. 2-9(a). Since Cellular IP does not support the topology shown in Fig. 2-2, the wireline connections between routers in its second tier are modified into a tree topology, as that shown in Fig. 2-9(b) while other configurations in the Cellular IP domain remain unchanged. Note that, in a HAWAII or Cellular IP access network, the radio interface is defined between an MN and its serving BS.

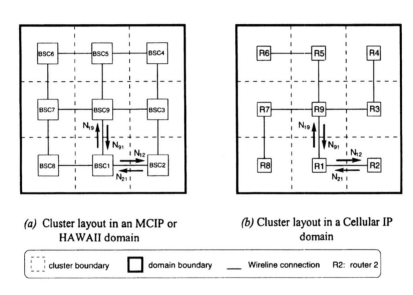

(a) Cluster layout in an MCIP or HAWAII domain

(b) Cluster layout in a Cellular IP domain

cluster boundary   domain boundary   Wireline connection   R2: router 2

Figure 2-9. Cluster layout.

Scalability. The configuration parameters of an MCIP domain are listed in Table 1-2, which are similar to those used in [41]. The coverage area of a BTS is a square. The scalability of a HAWAII domain with similar configurations is demonstrated in [41]. For MCIP, the number of mappings at the gateway (which is the same as the number of active users) is 39,438. It is well within the capability of modern routers [41]. Furthermore, a majority of these MNs are completely specified for a particular domain/subnet. In this case, perfect hashing is possible, resulting in $O(1)$ memory access for IP route lookup. Thus, route lookup for data forwarding can be done efficiently at the gateway [41].

Table 2-2. Domain Configuration Parameters

Symbol	Description	Value
n_1	BTSs per domain	144
n_2	BSCs per domain	9
n_3	BTSs per cluster	16
L_c	Perimeter of a cell	10.6 km [41]
ρ	User density (active users)	39 per km^2 [41]
L_R	Perimeter of a cluster $[4L_c]$	42.4 km
L_D	Perimeter of a domain $[3L_R]$	127.2 km
A	Coverage area of a domain $[(L_D/4)^2]$	1011.24 km^2
N	Number of active users in a domain $[A\rho]$	39438

Handoff Signaling Efficiency. Because the routing and handoff schemes used in MCIP, Cellular IP and HAWAII are different, their signaling overheads are also different. Here we focus on their mean handoff signaling costs. Since an MN in an MCIP domain keeps IP layer connection with its serving BSC, while an MN in a HAWAII or Cellular IP domain keeps IP connection with its serving BS, to make a fair comparison, we only compare the handoff signaling in the core network (at the second tier). Also, the metric we use in the comparison is "number of hops per second", but not the number of signaling messages per second. This is because a signaling message which travels several network nodes consumes more resources than a message which is delivered from a node to its next hop neighbor. In the following analysis, we use a fluid flow mobility model [43]. The model, also used in [41], assumes that MNs are moving at an average velocity of $v = 112$ km/hr, and their direction of movement is uniformly distributed over [0, 2π]. Assuming that MNs are uniformly populated with a density of $\rho = 39$ users per km^2, the rate of cluster boundary crossing, R, is given by $R = \rho v L_R / 3600\pi = 16.4$ (1/s). In terms of the cluster boundary crossing rate, letting N_{ij} denote the mean number of signalings from BSC i to BSC j per second, we calculate N_{12}, N_{21}, N_{19}, and N_{91} for the system shown in Fig. 2-9. The *path-setup* and *path-teardown* messages for the inter-domain handoffs are also included in the calculation, without considering the turn-ons and turn-offs inside each domain. With the symmetrical characteristic of the network topology, the total signaling cost due to inter-cluster (and inter-domain) handoffs in an MCIP domain is given by $N_{HO} = 4(N_{19} + N_{91}) + 8(N_{12} + N_{21})$. The number is the same for both soft and hard handoff schemes. The signaling cost for HAWAII *MSF hard handoff* scheme [41] can be calculated in the same way, where intra-cluster handoffs do not cause extra signaling in the core network. For Cellular IP *indirect semi-soft handoff* scheme [14] (which is actually a hard handoff scheme), because its network topology is different from MCIP and HAWAII, its signaling cost is given by $N_{HO} = 4(N_{19} + N_{91} + N_{12} + N_{21})$, where both inter-cluster and intra-cluster handoffs require signaling exchanges in the core network. Table 2-3 lists the numbers of the signaling messages in MCIP, HAWAII, and Cellular IP, respectively. For MCIP, both the numbers including (MCIP I) and excluding (MCIP E) the path optimization signalings are given.

Table 2-3. Comparison of Handoff Signaling Costs in the Core Network

	MCIP I	MCIP E	HAWAII	Cellular IP
N_{19}	20.5	18.4	10.2	98.3
N_{91}	10.2	6.1	4.1	49.1
N_{12}	7.2	6.1	4.1	24.6
N_{21}	12.3	10.2	6.1	49.1
N_{HO}	278.8	228.4	138.8	884.4

In MCIP, the signalings are not symmetric in the two directions of a link, i.e., $N_{19} \gg N_{91}$ and $N_{21} \gg N_{12}$. This is mainly due to the effect of the inter-domain handoffs (note that the gateway router is collocated with BSC 9). When an inter-domain handoff happens, for example, when an MN moves into the domain under consideration, a *path-setup* message is sent from the serving BSC to the gateway; when an MN moves out of the domain, a *path-teardown* message is sent to the gateway from the serving BSC. Thus, the number of signalings from the surrounding BSCs towards BSC 9 is more than that in the reverse direction.

The signaling cost in MCIP is approximately twice as much as that in HAWAII, due to three reasons. First, because HAWAII paths are soft-state, for inter-domain handoffs, when an MN moves out of the domain, it does not signal the intermediate nodes which cache entries for its soft-state path, but just leaves the soft-state path over there for time-out. By using explicit signaling messages to set up and tear down a virtual path for each MN, MCIP reduces the number of mappings cached by each BSC. This in turn reduces the route lookup time and improves the network scalability. Second, in HAWAII there is no path optimization process. However, the path optimization is necessary, because the cost of each optimization process is 6-hop signaling, but without it each data packet has to travel 2 hops more than necessary. This degrades the network resource utilization efficiency and causes a longer delay to user traffic, which is not shown in Table 2-3. Third, both MCIP soft and hard handoff schemes are mobile-assisted backward handoffs, which requires 3 signaling message exchanges between the old and new BSCs, while HAWAII MSF handoff scheme is a mobile-controlled forward handoff, which only needs 2 signaling message exchanges between the old and new BSs. MCFHO scheme may work well in random access radio networks, but it is not suitable for 3G/IP interworking.

Cellular IP is the least efficient in terms of the signaling cost for handoffs. Cellular IP handoff also suffers from a long handoff delay due to the MCFHO approach, similar to that used in HAWAII. Moreover, after a handoff, the old BS cannot quickly release the radio resources previously used by the MN, because there is no feedback mechanism from the MN or the new BS to inform the old BS that the MN has switched its transceiver to the new BS. This degrades the resource utilization efficiency.

2.8.2 DiffServ Resource Allocation Performance

In the following, we present two numerical examples to show the performance of the proposed resource allocation techniques. The first example shows that handoff call dropping probability can be decreased by setting fixed guard channel in each cell for handoff calls. The second one shows that the proposed adaptive scheme can decrease the new call blocking and handoff call dropping probabilities as compared to the non-adaptive scheme.

2.8.2.1 Call Admission With Guard Channel

Here, we use the parameter configuration given in [35] and make comparisons between call admission with guard channel and that without guard channel in terms of the   call blocking and dropping probabilities. With the complete partitioning scheduling mechanism, we focus on the admission control of assured service traffic. New calls arrive at each cell according to a Poisson process, and the exponentially distributed call duration has an average of $1/\mu = 0.5$ units of time. The channel holding time in each BS is also exponentially distributed with an average of $1/h = 0.1$ units of time. Each registration domain has $M = 20$ cells. Each cell can support up to $C = 20$ assured service calls. Bandwidth adaptation is not considered here. New call requests are rejected by the domain admission controller if there are $N = 320$ (80% of the total capacity of the domain) calls currently in service   in the domain.   When guard channel is used, $\alpha = 10\%$ of the cell   capacity (2 calls) is reserved for the handoff calls.

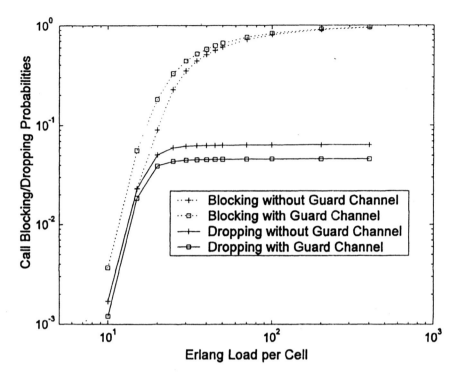

Figure 2-10. A comparison of the domain-based admission control systems with guard channel and without guard channel.

Fig. 2-10 shows the new call blocking and handoff call dropping probabilities versus the Erlang load of new calls per cell, ρ. For comparison, the performance curves of the admission control without guard channel [35] are also included. From the figure, we can see that the introduction of guard channel can decrease the handoff call dropping probability at the cost of an increase in the new call blocking probability. In the light load situation, it is highly possible that the number of on-going calls in the registration domain is less than 320 and no call is rejected by the domain admission controller. The mechanism used in each cell determines the call blocking and dropping probabilities. The tradeoff between the decrease of handoff call dropping and the increase of new call blocking is clearly observed. As the Erlang load increases, both the call dropping and blocking probabilities increase. When the traffic load is extremely high, most of the new requests will be rejected by the domain admission controller which limits the Erlang load in each cell to approximately N/M [35], and therefore limits the handoff call dropping probability to 0.0644 without guard channel and to 0.0463 with guard channel. We can see that the handoff call dropping probability can be guaranteed to a predetermined level by properly selecting N when other parameters are fixed. This analysis should be helpful in determining the resource requirement in the SLA negotiation between the registration domain and the Internet service provider.

2.8.2.2 Adaptive Bandwidth Allocation

Consider a registration domain with the same configuration as that in the above numerical analysis example with the guard channel. To show how new call blocking and handoff call dropping probabilities can be reduced by implementing the adaptive assured service, the bandwidth allocation to each call can now be adjusted between e_1, e_2 and e_3. The capacity allocated to the assured service in each cell is Ce_1 packets/second. If at some time the LARC adjusts the QoS to medium level and reaches the steady state, then each cell can admit $\lfloor Ce_1/e_2 \rfloor$ calls, and the domain admission controller will allow $0.8M\lfloor Ce_1/e_2 \rfloor$ calls to be accepted. The guard channel in each cell will be set to $\lceil \alpha Ce_1/e_2 \rceil$. If the LARC adjusts the QoS to the low level, then the cell capacity, the admission controller capacity and the guard channel will be adjusted to $\lfloor Ce_1/e_3 \rfloor$, $0.8M\lfloor Ce_1/e_3 \rfloor$ and $\lceil \alpha Ce_1/e_3 \rceil$ calls, respectively.

Table 2-4. The new call blocking and handoff call dropping probabilities achieved by the adaptive assured service.

Erlang Load	New Call Blocking Probability			Handoff Call Dropping Probability		
	High Config.	Medium Config.	Low Config.	High Config.	Medium Config.	Low Config.
15	0.0563	0.0246	0.0102	0.0185	0.0062	0.0026
50	0.6614	0.6286	0.5960	0.0458	0.0336	0.0303
400	0.9675	0.9535	0.9494	0.0462	0.0339	0.0307

For simplicity, we model a video flow as an on-off source as in [16] to illustrate the efficiency of the adaptive scheme. Each video source at the ``on'' state generates traffic with a total rate $R_p = 170.21$ packets/second, which can be coded into three layers L_1, L_2 and L_3 at rates of $R_p/8$, $R_p/8$ and $3R_p/4$, respectively. The buffer size in an MN or a BS is set to $B = 250$ packets. The three levels of loss probabilities, ε_1, ε_2 and ε_3, are set to 10^{-10}, 10^{-4} and 10^{-1}, respectively. For the high configuration listed in Table 2-1, a first-in-first-out buffer is used and the effective bandwidth (e_1) is 142.45 packets/second; for the medium configuration, the technique in [16] is used to calculate the effective bandwidth (e_2), equal to 119.82 packets/second, while the optimal partition threshold B_1 is 210 packets; for the low configuration, the effective bandwidth (e_3) is 112.65 packets/second and the optimal thresholds B_1' and B_2' are 64 and 171 packets respectively. By scaling down the QoS requirement, the bandwidth allocated to the ongoing calls can be reduced. The resources accumulated by bandwidth reduction are then used to accept more new calls and handoff calls. Fig. 2-11 shows the new call blocking and handoff call dropping probabilities for the three configurations. It is observed that both the new call blocking and handoff call dropping probabilities decrease obviously in the adaptive scheme as the QoS level is reduced. For clarity, Table 2-4 gives the new call blocking and handoff call dropping probabilities for the medium load ($\lambda = 15$), high load ($\lambda = 50$), and extremely high load ($\lambda = 400$) conditions, respectively.

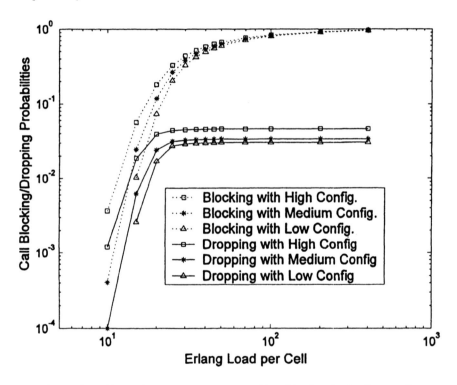

Figure 2-11. The performance improvement achieved by the adaptive framework

2.9   CONCLUSION AND FUTURE RESEARCH

This chapter addresses two important issues in establishing a mobile wireless Internet, i.e., support of fast soft handoff and provision of QoS over IP based wireless access networks.

First, the MCIP micromobility system is established. The MCIP architecture not only supports fast, seamless local hard handoff by inheriting the advantages of the existing micromobility protocols, but also supports soft handoff by properly selecting the access network structure and designing new handoff protocol. The scalability of the MCIP access system is illustrated by a numerical example, and the handoff signaling cost is compared with those of HAWAII and Cellular IP. The advantages of supporting soft handoff and QoS provisioning for real-time services are achieved at slightly increased system complexity.

Second, the DiffServ registration-domain-based resource allocation architecture is presented to support fast handoff with QoS guarantee. A novel adaptive assured service is proposed to improve wireless resource utilization efficiency. For the statistical environment with exponential distributions, the analysis demonstrates that (a) the handoff call dropping probability can be guaranteed to a predetermined level by properly allocating the resources to a certain class of traffic in a registration domain, (b) the guard channel scheme can be used to further reduce the handoff call dropping probability, and (c) the adaptive service can improve resource utilization while guaranteeing the call level QoS (i.e. the call blocking and dropping probabilities). This analysis can help to determine the resource requirement in the SLA negotiation between the registration domain and the Internet service provider.

The research results presented in this chapter is a first step towards developing the wireless mobile Internet. There are many open issues that need further investigation, some of which are described in the following.

MCIP supports fast soft handoff of mobile IP connections, mainly due to its adoption of the "Layer 2 RAN, IP core" architecture. However, there are several motivations for the use of IP transport within the RAN [20]: (a) IP is quickly becoming the widely accepted standard for packetization of voice, data, signaling, and operation, administration, and management (OAM) in the networking world; (b) IP as a network layer protocol is carefully designed to be independent of link/physical layers, so it allows easy cooperation of networks using different link/physical layer techniques; (c) the 3G core network is IP-based, and therefore an IP-based RAN will facilitate consistent backbone infrastructure, simple protocol stacks, operational efficiency, and industry standard OAM; (d) IP QoS management is approaching maturity.

With MNs' IP packets being routed directly in the RAN, IP over the air interface would be necessary. The current major impediment to IP over the air is the header size, but new work in header compression may eliminate this hurdle. Given that a spectrally efficient representation of IP on the radio medium is possible, IP packets can be sent out over the air by an MN, and the BS can handle the packets in the same way as it currently does with the specialized radio frames. However, there is still a problem with the multipath nature of macrodiversity. In the downlink direction, the routing from a single source at the BS to multiple BSs is similar to multicast, but with extremely tight transmission delay and delay jitter constraints for real-time traffic. In the uplink direction, however, traffic flows need to be combined (in the selective diversity mode) in the wired network after the BSs. In order to accommodate macrodiversity, the RAN should be a routing domain using a multipath routing protocol, which performs frame selection in some specific routers and features real time routing table convergence (for soft handoff). These characteristics involve a considerable change from existing IP routing algorithms (which do not require real time convergence and do not involve multipath nor selection of particular packets in the diversity combining) [27]. Therefore, although the prospects for moving IP into the RAN for transport of RAN protocols and user traffic to/from the MN are promising, a lot of research work remains to be done.

While various micromobility protocols have been proposed to support fast and seamless handoff of real time multimedia IP connections, there is very little work on a suitable QoS model for micromobility. Due to the obvious disadvantages of the IntServ approach listed in Section 5, extending the DiffServ model to micromobility seems to be a better choice. The DiffServ resource allocation techniques proposed in this chapter are just a start point for this topic.

The domain based call admission control technique clearly illustrates that DiffServ QoS model can integrate micromobility naturally, and the resource allocation based on SLA and effective bandwidth should be fast enough for seamless handoff while guaranteeing QoS at both packet level and call level, but the model is over-simplified to facilitate mathematical analysis. First, the registration domain is assumed to be an exponentially distributed statistical environment, so that the new call blocking and handoff call dropping probabilities can be calculated by using the Erlang formula. However, the statistical characteristics of the arriving process, the connection lifetime, and the channel holding time of future wireless IP multimedia connections are unknown at this moment. To acquire accurate knowledge of the above random process/random variables, further investigation is necessary. Second, it is assumed that an effective bandwidth can be calculated for each connection of any service class. Although we present a technique to calculate effective bandwidth for the assured service provisioning packet loss guarantee, how to include the delay requirement (if any) in the effective bandwidth calculation is still unknown. Third, complete resource partition among different service classes is assumed; however, dynamic resource sharing among service classes is an important way to improve wireless bandwidth utilization [37]. As a result, dynamic resource sharing techniques should be investigated, and the call admission control for a service class should consider the impacts of other service classes on the resource availability.

NOTES

1. In different context, the mobile IP host may also be called mobile host, mobile station, or mobile user. Here we use the mobile IP terminology.
2. There are two mechanisms identified for handoff: backward and forward handoff. In general, backward handoff is initiated by the serving BS, whereas forward handoff is initiated via the target (new) BS.
3. As the 3G wireless wireless networks will be extensively deployed in the near future, we try to make our scheme compatible with the 3G standard as much as possible.
4. BTS is the 3G terminology for base station.
5. When the mobile-specific routing scheme is used, there is only one path between the domain root router and the MN. All packets belonging to the same connection will follow this path in the domain so as to keep the in-sequence packet order. Therefore, we call a mobile-specific path as a virtual path.
6. Here the capacity for two calls with allocation of e_{i-1} serves as a threshold to detect whether the traffic load really decreases. If the threshold is set to 0, the fluctuation of available capacity due to the new arrival, handoff or call completion will lead to too frequent adaptation of the buffer configuration, but with little improvement of the QoS. How to optimally set the threshold needs further investigation.

REFERENCES

[1] 3GPP TS 25.211 v3.3.0 (2000-06). Physical Channels and Mapping of Transport Channels onto Physical Channels (FDD) (Release 1999).
[2] 3GPP TS 25.321 v3.8.0 (2001-06). MAC Protocol Specification (Release 1999).
[3] 3GPP TS 25.322 v3.7.0 (2001-06). RLC Protocol Specification (Release 1999).
[4] 3GPP TS 25.922 v4.0.0 (2001-03). Radio Resource Management Strategies (Release 4).
[5] A.H. Aghvami and P. Smyth. Forward or Backward Handover for W-CDMA? *First Int'l. 3G Mobile Communication Technologies*, (Conf. Publ. No. 471), pp. 235-239, 2000.
[6] Y. Bernet. The Complementary Roles of RSVP and Differentiated Services in the Full-Service QoS Network. *IEEE Commun. Mag.*, vol. 38, no. 2, pp. 154-162, Feb. 2000.
[7] Y. Bernet et al. A Framework for Integrated Services Operation over DiffServ Networks. IETF RFC 2998, Nov. 2000.
[8] S. Blake, D. Black, M Carlson, E. Davies, Z. Wang, and W. Weiss. An Architecture for Differentiated Services IETF RFC 2475, Dec. 1998.
[9] L. Bos and S. Leroy. Toward an All-IP-Based UMTS System Architecture. *IEEE Net.*, vol. 15, no. 1, pp. 36-45, Jan.-Feb. 2001.

[10] R. Braden, D. Clark, and S. Shenker Integrated Services in the Internet Architecture: An Overview. Internet RFC 1633, June 1999.

[11] R. Braden, L. Zhang, S. Berson, S. Herzog, and S. Jamin. Resource Reservation Protocol (RSVP)--Version 1 Functional Specification. Internet RFC 2205, Sept. 1997.

[12] A.T. Campbell et al. Comparison of IP Micromobility Protocols *IEEE Wireless Commun. Mag.*, vol. 9, no. 1, pp. 2-12, Feb. 2002.

[13] A.T. Campbell and J. Gomez. IP Micro-Mobility Protocols. *ACM SIGMOBILE, Mobile Comp. and Commun. Rev.*, vol. 4, no. 4, pp. 45-54, Oct. 2001.

[14] A.T. Campbell, J. Gomez, S. Kim, A.G. Valko, C-Y Wan, and Z.R. Turanyi. Design, Implementation and Evaluation of Cellular IP. *IEEE Pers. Commun.*, vol. 7, no. 4, pp. 42-49, Aug. 2000.

[15] A.T. Cambell, S. Kim, J. Gomez, and C-Y. Wan. Cellular IP Performance. Internet draft, draft-gomez-cellularip-perf-00.txt, work in progress, Oct. 1999.

[16] Y. Cheng and W. Zhuang. Optimal Buffer Partitioning for Multiclass Markovian Traffic Sources. *Proc. IEEE GLOBECOM'01*, vol. 3, 2001, pp. 1852-1856.

[17] Y. Cheng and W. Zhuang. DiffServ Resource Allocation for Fast Handoff in Wireless Mobile Internet. *IEEE Commun. Mag.*, vol. 40, no. 5, pp. 130-136, May 2002.

[18] D.D. Clark and W. Fang. Explicit Allocation of Best-Effort Packet Delivery Service. *IEEE/ACM Trans. Networking*, Vol. 6, no. 4, pp. 362-373, Aug. 1998.

[19] S. Das, A. Misra, P. Agrawal, and S.K. Das. TeleMIP: Telecommunications-Enhanced Mobile IP Architecture for Fast Intradomain Mobility. *IEEE Pers. Commun.*, vol. 7, no. 4, pp. 50-58, Aug. 2000.

[20] S. Dixit, Y. Guo, and Z. Antoniou. Resource Management and Quality of Service in Third-Generation Wireless Networks. *IEEE Commun. Mag.*, vol. 39, no. 2, pp. 125-133, Feb. 2001.

[21] S.F. Foo and K.C. Chua. Regional Aware Foreign Agent (RAFA) for Fast Local Handoffs. Internet draft, draft-chuafoo-mobileip-rafa-00.txt, work in progress, Nov. 1998.

[22] V.K. Garg. *IS-95 CDMA and CDMA 2000*. Prentice Hall, 2000.

[23] E. Gustafsson, A. Jonsson, and C. Perkins. Mobile IP Regional Registration. Internet draft, draft-ietf-mobileip-reg-tunnel-03, work in progress, July 2000.

[24] J. Heinanen, F. Baker, W. Weiss, and J. Wroclawski. Assured Forwarding PHB Group. IETF RFC 2597, June 1999.

[25] H. Homa and A. Toskala. *WCDMA for UMTS: Radio Access for Third Generation Mobile Communications*. John Wiley \& Sons, 2000.

[26] V. Jacobson, K. Nichols, and K. Poduri. An Expedited Forwarding PHB. IETF RFC 2598, June 1999.

[27] J. Kempf, P. McCann, and P. Roberts. IP Mobility and the CDMA Radio Access Network: Applicability Statement for Soft Handoff. Internet draft, draft-kempf-cdma-appl-02.txt, work in progress, Sept. 2001.

[28] J. Kim and A. Jamalipour. Traffic Management and QoS Provisioning in Future Wireless IP Networks. *IEEE Pers. Commun.*, vol. 8, no. 5, pp. 46-55, Oct. 2001.

[29] J. Laiho-Steffens, M. Jasberg, K. Sipila, A. Wacker, and A. Kangas. Comparison of Three Diversity Handover Algorithms by Using Measured Propagation Data. *Proc. IEEE VTC*, vol. 2, pp. 1370-1374, 1999.

[30] J-H Lee, T-H Jung, S-U Yoon, S-K Youm, and C-H Kang. An Adaptive Resource Allocation Mechanism Including Fast and Reliable Handoff in IP-Based 3Gwireless Networks. *IEEE Pers. Commun.*, vol. 7, no. 6, pp. 42-47, Dec. 2000.

[31] X. Liu and W. Zhuang. Inter-Cluster Soft Handoff in 3G/IP Interworking. *Proc. 3Gwireless'2002*, pp. 778-783, San Francisco, May 2002.

[32] J.W. Lockwood. Implementation of Campus-wide Wireless Network Services Using ATM, Virtual LANs and Wireless Base Stations. *Proc. IEEE WCNC*, vol. 2, pp. 603-605, New Orleans, LA, Sept. 1999.

[33] A. Misra, S. Das, A. Mcauley, A. Dutta, and S.K. Das. Integrating QoS Support in TeleMIP's Mobility Architecture. *Proc. IEEE Int'l Conf. Personal Wireless Commun.*, pp. 57-64, 2000.

[34] B. Moon and H. Aghvami. RSVP Extensions for Real-Time Services in Wireless Mobile Networks. *IEEE Commun. Mag.*, vol. 39, no. 12, pp. 52-59, Dec. 2001.

[35] M. Naghshineh and A.S. Acampora. Design and Control of Micro-Cellular Networks with QoS Provisioning for Real-Time Traffic. *Proc. IEEE 3rd Int'l Conf. Universal Personal Commun.*, 1994, pp. 376-381.

[36] M. Naghshineh and A.S. Acampora. QoS Provisioning in Micro-Cellular Networks Supporting Multimedia Traffic. *Proc. IEEE INFOCOM'95*, 1995, pp. 1075-1084.

[37] D. Partain, G. Karagiannis, P. Wallentin, and L. Westberg. Resource Reservation Issues in Cellular Access Networks Internet draft, draft-partain-wireless-issues-00.txt, work in progress, April 2001.

[38] C. Perkins. IP Mobility Support IETF RFC 2002, Oct. 1996.

[39] C. Perkins and D.B. Johnson. Route Optimization in Mobile IP. Internet draft, draft-ietf-mobileip-optim-10.txt, work in progress, Nov. 2000.

[40] R. Ramjee et al. IP Micro-Mobility Support Using HAWAII. Internet draft, draft-ietf-mobileip-hawaii-00.txt, work in progress, June 1999.

[41] R. Ramjee, T. La Porta, S. Thuel, K. Varadhan, and S.Y. Wang. HAWAII: A Domain-based Approach for Supporting Mobility in Wide-area Wireless Networks. *IEEE/ACM Trans. Networking*, Vol. 10, No. 3, pp. 396-410, June 2002.

[42] A. Terzis, L. Wang, J. Ogawa, and L. Zhuang. A Two-Tier Resource Management Model for the Internet. *Proc. of IEEE GLOBECOM'99*, vol.3, pp. 1779-1791, 1999.

[43] R. Thomas, H. Gilbert, and G. Mazziotto. Influence of the Mobile Station On the Performance of a Radio Mobile Cellular Network. *Proc. 3rd Nordic Seminar, Paper 9.4,*penhagen, Denmark, 1988.

[44] X. Xiao and L.M. Ni. Internet QoS: A Big Picture. {\it IEEE Net.}, vol. 13, no. 2, pp. 8-18, March/April 1999.

[45] O.T.W. Yu and V.C.M. Leung. Adaptive Resource Allocation for Prioritized Call Admission over an ATM-Based Wireless PCN. *IEEE J. Select. Areas Commun.*, vol. 15, no. 7, pp. 1208-1225, Sept. 1997.

[46] H. Zhang. Service Disciplines for Guaranteed Performance Service in Packet-Switching Networks. *Proc. IEEE*, vol. 83, no. 10, pp. 1374-1396, Oct. 1995.

[47] T. Zhang, P. Agrawal, and J-C Chen. IP-Based Base Stations and Soft Handoff in All-IP Wireless Networks *IEEE Pers. Commun.*, vol. 8, no. 5, pp. 24-30, Oct. 2001.

Chapter 3

Wireless Access of Internet Using TCP/IP
A Survey of Issues and Recommendations

Sridhar Komandur, Spencer Dawkins and Jogen Pathak
Cyneta Networks, Inc

3.1   INTRODUCTION

The Internet has many different kinds of applications (e.g., audio, email, chat). The applications interact with a few transport protocols like TCP, UDP and RTP. There is only one widely-deployed network protocol, the Internet Protocol (IP). The IP protocol facilitates routing and subnetwork specific details from the transport protocols. While this allows transport protocols to run over many different kinds of subnet (e.g., Ethernet, wireless network, FDDI), they are not optimized for any specific subnetwork.

The basic functionality provided by IP is a best effort delivery of packets to their destination. Packets flowing between a source and a destination are not acknowledged by intermediate gateways and need not follow the same path through the network. Thus it is acceptable in the Internet for the packets to be delivered out-of-order, duplicated or even lost in transit. Applications requiring reliability and in-order delivery use TCP over IP.

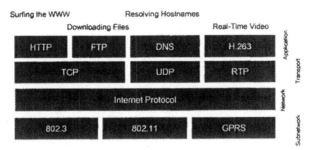

Figure 3.1. Internet Protocol (IP) Interaction

In the next section we delve into the details of TCP. In Section 3.3 we discuss wireless networks. Section 3.4 describes the issues related to wireless access of Internet using TCP and some recommendations to improve performance. We summarize current standards-track TCP recommendations and describe some current research topics in Section 3.5.

3.2 TRANSMISSION CONTROL PROTOCOL (TCP)

TCP provides an in-order and reliable byte-stream across the network between two endpoint hosts. The underlying services are:
- Use all available bandwidth between the end points,
- Retransmit lost datagrams,
- Present out-of-order datagrams to applications in-order and
- Absorb duplicated datagrams.

The main objective of TCP is to avoid sustained network congestion in the network through congestion avoidance and congestion control.

3.2.1 Window Based Flow Control

TCP uses a windows based mechanism for flow control. As Figure 3.2 shows, a windows based flow control has better performance than a sender that sends a new packet only upon receiving acknowledgement of the previous packet.

Don't want to stop and wait for each ack

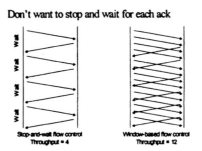

Figure 3.2. Windows Based Flow Control

The receiver notifies the sender of the available buffer space referred to as the *receiver's advertised window*. The sender determines the optimum amount of unacknowledged data that can be in-flight at a given time and is referred to as *congestion window.* The *window size* used by TCP at a given time is the minimum of advertised and congestion windows. Figure 3.3 shows the sliding window mechanisms:

Window = 5 segments

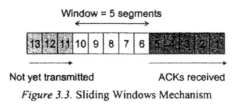

Figure 3.3. Sliding Windows Mechanism

In the figure above, the client has acknowledged packets 1 through 5, while packets *6* through *10* are in-flight.

TCP uses an ACK-clocking window control mechanism to maintain equilibrium. Initially it is in Slow Start mode, where the congestion window increases exponentially. After a certain threshold on the congestion window is reached, the sender's TCP stack enters a congestion avoidance phase as illustrated in the figures below.

TCP Slow Start

• Probe for bandwidth aggressively

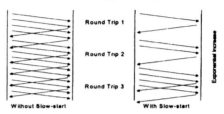

Figure 3.4. TCP Slow Start

During slow start phase sender congestion window increases by 1 MSS for every ACK received and by 1/MSS during congestion avoidance phase. The slow start phase ends when the congestion window reaches *slow start threshold (ssthresh)*.

TCP Congestion Avoidance

• Probe for bandwidth gently

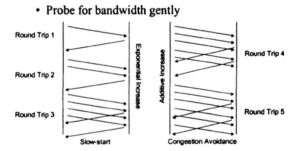

Figure 3.5. TCP Congestion Avoidance

The ideal congestion window size (cwnd) is $\Delta = (RTT * bandwidth)$. If sender's cwnd $> \Delta$, then it will increase RTT, causing potential packet loss and retransmission timeout (RTO) at the sender. The following figure illustrates the congestion window growth:

Slow Start and Congestion Avoidance

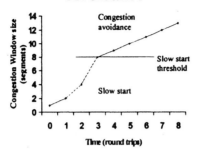

Figure 3.6. Slow Start and Congestion Avoidance

3.2.2 Detecting Packet Loss

TCP detects packet loss using retransmission timeout (RTO) and duplicate acknowledgements (DupAcks).

At a given time, the sender times the earliest unacknowledged packet. If the ACK is not received within this time, TCP assumes loss of packet due to network congestion.

If the sender receives some (typically 3) duplicate ACKs requesting the same packet, TCP assumes loss of that packet and resends it again (Fast Retransmit & Fast Recovery).

In the subsections below we discuss the consequences of RTO and Fast Retransmit.

3.2.2.1 Retransmission Timeout (RTO)

The sender dynamically calculates RTO as (mean RTT + 4 * mean RTT deviation). Implications of this approach are:

- Fewer false "positives": Responds well to large variations in RTT
- Loss of data leads to increased recovery time and as the RTO is increased.
- RTO time is doubled for each subsequent timeout. So the losses need to be infrequent.
- Congestion window is reduced to initial congestion window (typically 1 or 2 MSS)

3.2.2.2 Fast Retransmit and Fast Recovery

Fast Retransmit: Since RTOs take a long time to recover lost packet, duplicate ACKs are used to signal the need to retransmit the lost packet.

Fast Recovery: DupAcks indicates that packets are actually reaching the client, even though a packet has been declared lost. So instead of drastically reducing the cwnd to one MSS, a more gentle approach is followed:

- ssthresh =   min(cwnd, receiver's advertised window)/2 (at least 2 MSS)
- retransmit the missing segment (fast retransmit)
- cwnd = ssthresh + number of dupacks

- When a new ACK arrives, enter congestion avoidance
- Congestion window is cut in half but is still larger than RTO case with cwnd = 1

3.3   WIRELESS NETWORKS

We highlight characteristics of wireless networks that are not in tune with typical network congestion assumptions made while designing TCP for the Internet.

Low bandwidth: The available bandwidth is low, depending on the number of timeslots allocated over air interface channel for the TCP session. In the presence of contention for the air interface bandwidth, a session's bandwidth could be in the low hundreds of bits per second.

High latency: The round trip times are high and at times highly variable due to deterioration in air interface bandwidth. The minimum round-trip delay close to one second is not exceptional.

High Bit Error Rate (BER): This results in frame or packet losses, or long variable delays due to local link-layer error recovery.

Internal Buffering: Some wireless network elements have **a lot of internal buffer space** and tend to accumulate data, thus resulting in increased and variable queuing delays, increasing retransmission due to timeouts.

Variable Latency: The wireless network users (or simultaneous TCP sessions for same user) may **share common channels** for their data packet delivery which, in turn, may cause **unexpected delays** to the packet delivery of a user due to simultaneous use of the same channel resources by the other users. This could potentially lead to higher RTO and retransmissions.

Access Disruptions: Unexpected link disconnections (or intermittent link outages) may occur frequently and the period of disconnection may last a very long time.

Link Recovery: (Re)setting the link-connection up may take a long time (several tens of seconds or even minutes)

Wireless Links are Expensive. The use of most **wireless links is expensive.** Many of the service providers apply time-based or volume (bytes)-based charging.

Most of the above characteristics are applicable to both 2.5G (e.g., GPRS) and 3G (e.g., UMTS) networks.

3.4 TCP AND WIRELESS NETWORKS

The characteristics of wireless network make it difficult for TCP to perform appropriately. The desired behavior is:

- Retransmit a packet lost due to errors immediately
- Take congestion control actions if and when there is congestion in the network

Fast Retransmit is followed by Fast Recovery, which reduces the congestion window. However, this will result in unnecessarily reduced throughput if the packet loss was due to corruption over air interface, and not to congestion at all.

On a CDMA channel, errors may occur due to interference from other users or due to noise or fading. Interference due to other users is an indication of congestion. If such interference causes transmission errors, it is appropriate to reduce congestion window. However, if noise causes errors, it is not appropriate to reduce window.

When a channel is in a degraded state for an extended period of time, it might be better to let TCP back off, so that it does not unnecessarily attempt retransmissions while the channel remains degraded.

With its end-to-end feedback mechanisms, TCP is unable to distinguish between packet losses due to congestion and transmission errors. The cwnd is reduced when TCP encounters a transmission error loss and reacts as if the loss was due to network congestion. This mismatch of design results in poor throughput over wireless access networks.

The following recommendations have been put forward by IETF's Performance Implications of Link Characteristics (PILC) working group to improve TCP performance over 2.5G and 3G wireless networks, and are under IETF review as of this writing:

Appropriate Advertised Window Size (RFC 3150, Sec 2.5): The window size should reflect (bandwidth & delay) product.

Window-scaling Option – 3G networks (RFC 1323): This allows for congestion window greater than 64K bytes.

Large Initial Window (RFC 2581): Allows for an initial congestion window of at least 2MSS/4K bytes and an aggressive slow start behavior. A proposal to increase this value to 4MSS/4K bytes is currently under review in IETF, as of this writing. Note that an RTO timeout still sets cwnd = 1 MSS.

Limited Transmit (RFC 3042): It has been observed in at least one reported case that over 50% of web retransmits occur after RTO, following this scenario:

- Lost packet sequence number near cwnd
- Sender doesn't send 3 more packets
- Receiver doesn't generate 3 dupacks

This results in

- RTO, which slows throughput more than Fast Retransmit
- Slow Start, which slows throughput more than Fast Recovery

"Limited Transmit" was proposed to allow recovery even when three additional packets do not follow a lost packet.

When duplicate acknowledgements begin to arrive at the congestion window edge, the congestion window is extended and new packets are injected until either a new acknowledgement is received, indicating that the missing segment did arrive, or three duplicate acknowledgements are received, triggering Fast Retransmit/Fast Recovery instead of RTO/Slow Start.

MTU Larger than Default IP MTU (RFC 791 and RFC 1191): According to RFC 791, IP's minimum Message Transfer Unit is 576, which results in an MSS value of 536 bytes. A larger MTU is recommended especially for 3G networks because the higher bitrates allow larger MTUs without taking away from the application's interactive feel as long as serialization delays are less than 200ms. For example, 3G networks with bandwidth of 48KBps could use MSS = 48KB * 200ms = 960bytes. However, it is less beneficial for 2.5G networks MSS = 1.2KB * 200ms = 240bytes (considering a low bandwidth of 1.2KBps).

Path MTU can be determined using probe based methods (RFC 1191).

TCP SACK Option (RFC 2018 and RFC 2883): Allows the retransmissions to be more effective by identifying "holes" in the receive window of the receiver.

Explicit Congestion Notification option (RFC 3168): This allows the sender to explicitly identify congestion – avoiding packet retransmission due to network congestion.

Timestamp Option (RFC 1323): Allows better estimation of RTT and avoid spurious retransmissions. Spurious retransmissions can also be detected using Eiffel proposal.

PILC's recommendations form the basis of the Wireless Application Protocol (WAP) Forum's "Wireless Profiled TCP" recommendation as well.

WAP 1.0/1.1/1.2 was intended to support limited number of applications/devices using a vertical model (modify every layer to be in tune with wireless access networks). However, this added significant complexity as there is no end-to-end transparency – TCP, SSL, HTTP, and HTML all need to be translated. WAP 2.0 has adopted Wireless Profiled TCP as its transport protocol.

3.5   CONCLUSION

This chapter addressed the challenges and proposed solutions for wireless access of Internet using the ubiquitous TCP/IP protocol. This area is in its infancy, with researchers focusing on the issue of *seamless Internet access* – the user/application is unaware of the particular access networks (WLAN/ cellular/ DSL/ Ethernet).

TCP's reliance on end-to-end mechanisms for loss recovery have served the Internet very well in avoiding sustained congestion losses, but this has come at the expense of TCP's ability to recover from transmission errors without significant reductions in bandwidth utilization. Limited Transmit is the current "state of the art" in standards-track recovery from transmission error losses.

Additional research is underway to detect unnecessary Fast Recovery procedures, using TCP Timestamp options (Eifel) or duplicate selective acknowledgements (D-SACK) to detect that a segment that was retransmitted unnecessarily so that the sending TCP can reclaim an unnecessarily-reduced congestion window, and may be less sensitive to reordering in the future.

"Forward RTO Recovery" is a proposal for detecting unnecessary retransmissions with sender-side modifications only, without requiring use of either TCP timestamps or selective acknowledgements.

Additional interesting challenges for wireless Internet access include billing, routing, mobility and security.

REFERENCES

The following RFCs ("Requests For Comments") are published by the Internet Society, and are available from the Internet Engineering Task Force.

"RFC 791 Internet Protocol", J. Postel, September 1981

"RFC 793 Transmission Control Protocol", J. Postel, September 1981
"RFC 1191 Path MTU discovery", J.C. Mogul, S.E. Deering, November 1990

"RFC 1323 TCP Extensions for High Performance", V. Jacobson, R. Braden, D. Borman, May 1992

"RFC 2018 TCP Selective Acknowledgement Options", M. Mathis, J. Mahdavi, S. Floyd, A. Romanow, October 1996

"RFC 2414 Increasing TCP's Initial Window", M. Allman, S. Floyd, C. Partridge, September 1998

"RFC 2581 TCP Congestion Control", M. Allman, V. Paxson, W. Stevens, April 1999

"RFC 2883 An Extension to the Selective Acknowledgement (SACK) Option for TCP", S. Floyd, J. Mahdavi, M. Mathis, M. Podolsky, July 2000

"RFC 3042 Enhancing TCP's Loss Recovery Using Limited Transmit", M. Allman, H. Balakrishnan, S. Floyd, January 2001

"RFC 3150 End-to-end Performance Implications of Slow Links", S. Dawkins, G. Montenegro, M. Kojo, V. Magret, July 2001

"RFC 3155 End-to-end Performance Implications of Links with Errors", S. Dawkins, G. Montenegro, M. Kojo, V. Magret, N. Vaidya August 2001

"RFC 3168 The Addition of Explicit Congestion Notification (ECN) to IP", K. Ramakrishnan, S. Floyd, D. Black, September 2001

"TCP/IP Illustrated Volume 1", W. Richard Stevens, Addison-Wesley, 1994

"Wireless Profiled TCP", WAP Forum, WAP-225-TCP-20010331-a, March 2001

Eifel is described in "The Eifel Retransmission Timer", Reiner Ludwig, Keith Sklower, Computer Communication Review, July 2000, and "The Eifel Algorithm: Making TCP Robust Against Spurious Retransmissions", Reiner Ludwig and Randy Katz, Computer Communication Review, January 2000.

Forward RTO Recovery is described in "F-RTO: A TCP RTO Recovery Algorithm for Avoiding Unnecessary Retransmissions", P. Sarolahti, M. Kojo, June 2002 (Work in Progress, available from Internet-Draft repositories)

Chapter 4

MOBILITY PREDICTION FOR QOS PROVISIONING

Hyong S. Kim and Wee-Seng Soh
Carnegie Mellon University

4.1   INTRODUCTION

In recent years, there has been a rapid increase in cellular network deployment and mobile device market penetration. With vigorous research that promises higher data rates, future wireless networks will likely become an integral part of the global communication infrastructure. Ultimately, wireless users will demand the same reliable service as of today's wireline telecommunications and data networks. However, there are some unique problems in cellular networks that challenge their service reliability. In addition to problems introduced by fading, user mobility place stringent requirements on network resources. Whenever an active mobile terminal (MT) moves from one cell to another, the call needs to be handed off to the new base station (BS), and network resources must be reallocated. Resource demands could fluctuate abruptly due to the movement of high data rate users. Quality of Service (QoS) degradation or even forced termination may occur when there are insufficient resources to accommodate these handoffs.

From a mobile user's point of view, forced termination of an ongoing call is more objectionable than the blocking of a new call request. Therefore, handoff requests are generally prioritized over new call requests when they compete for radio resources. In the classic handoff prioritization problem, each BS prioritizes handoff requests by setting aside some wireless resources that could only be utilized by incoming handoffs. Since any such resource reservation would inevitably increase the blocking probability of new calls, and reduce the overall system resource utilization, it is therefore extremely important that the reservations are made as sparingly as possible while meeting the desired degree of handoff prioritization. In this way, wireless service providers would be able to provision high quality services without compromising their revenues unnecessarily.

Early work in handoff prioritization proposed static reservation at each BS as a solution [1], in which a fixed portion of the radio capacity is permanently reserved for handoffs. However, such a static approach is unable to handle variable traffic load and mobility; it might underutilize precious radio resources when handoffs are less frequent, and could experience unacceptably large number of forced terminations when mobility is high. In order to meet the desired forced termination probability without over-reserving precious radio resources, the amount of reservation at each BS should be dynamically adjusted according to the requirements of anticipated incoming handoffs.

If the system has prior knowledge of the exact trajectory of every MT, it could take appropriate steps to reserve resources, so that QoS may be guaranteed during the MT's connection lifetime. However, such an ideal scenario is very unlikely to occur in real life. The next best available option is to predict the mobility of MTs, and perform resource reservations based on these predictions. Many such predictive schemes have been proposed in the literature, but they differ in their reservation efficiencies. In the process of meeting the same *forced termination probability* (P_{FT}) requirement, a more efficient scheme will be able to accomplish the task with a lower *new call blocking probability* (P_{NC}) than a less efficient scheme. The efficiency of the scheme depends on whether the reservations are made at the right place and time. Therefore, the efficiency of a predictive scheme is closely associated with its prediction accuracy. Since reservation efficiency has a direct impact on operators' revenues, there are strong incentives to design more accurate prediction schemes.

In the United States, the FCC recently mandates that cellular-service providers must be able to pinpoint a wireless emergency call's originating location to within 125 m. This spurs intensive research in mobile-tracking techniques. One promising approach is the integration of a global positioning system (GPS) receiver in each MT. According to [2], it is very reasonable to expect assisted GPS positioning methods to be able to yield an accuracy of under 20 m during 67 percent of the time. During 2003-2009, a new batch of GPS satellites will be launched to include two additional civilian carrier frequencies that could potentially yield positioning accuracy within 1 m for civilian users, even without the use of ground-based augmentation system [3]. As more breakthroughs in positioning techniques take place, fuelled by the strong interest in location-based services from the industry, MTs are likely equipped with reasonably accurate location-tracking capability in the near future. The time is thus ripe for active research into how such inherent capability may be harnessed for QoS provisioning in cellular networks.

In this chapter, we present two mobility prediction schemes that utilize real-time mobile positioning information, and show how they can be used for dynamic resource reservation to improve mobile users' call experience. The first scheme, known as the *linear extrapolation* (LE) scheme, predicts the future direction of travel by using linear extrapolation over the MT's recent positions. The algorithm is simple but yet it is capable of handling irregular cell boundaries. The second scheme, known as the *road topology based* (RTB) scheme, utilizes road topology information in its predictions. The scheme takes advantage of the fact that high-mobility MTs are usually those that are carried in vehicles, and vehicular movements are usually restricted to roads. This achieves more accurate predictions than the LE scheme. The improved accuracy comes at the cost of increased complexity, but the resulting gain in reservation efficiency may justify this cost.

The remaining of this chapter is organized as follows. Section 2 reviews some related work in the literature, and discusses the motivation for designing prediction schemes based on real-time mobile positioning information. Section 3 describes the two mobility prediction schemes that have been developed. In Section 4, we shall see how these mobility prediction techniques are to be used for dynamic resource reservation in cellular networks to achieve efficient tradeoff between P_{NC} and P_{FT}. In Section 5, we evaluate the performance of the schemes via simulations. Simulation results are presented to compare the efficiencies of these techniques with several other methods. Finally, Section 6 concludes the main findings in this chapter.

4.2   RELATED WORK AND MOTIVATION

As discussed above, static reservation schemes are only suitable for stationary traffic conditions. Since traffic demands are expected to fluctuate from time to time, various schemes have been proposed to dynamically reserve bandwidth resources so as to adapt to such changes. In the following, we review some of the previous work, and identify the areas in which improvements need to be made.

In [4], Oliveira *et al.* propose an admission control scheme in which a new/handoff call is accepted only if it can successfully reserve bandwidth in all adjacent cells, in addition to the availability of bandwidth in the local cell. The redundant reservation of radio resources in all adjacent cells is inevitable, since there is no consideration for the individual trends of the users in the network, e.g. position, speed, and direction. Reservations are made even when the MT is stationary. In addition, each MT has to reserve bandwidth in its neighboring cells the moment it enters a cell. This inflexibility in reservation location and timing could lead to under-utilization of bandwidth resources.

Levine *et al.* [5] propose the concept of a *shadow cluster* – a set of BSs to which a MT is likely to attach in the near future. The scheme partitions time into equal intervals, and estimates the probability of each MT being in any wireless cell within the shadow cluster for future time intervals. During each time interval, the BSs exchange information about the predicted bandwidth demands for future time intervals so as to determine the feasibility of admitting new call requests. The effectiveness of this scheme, however, is closely associated with the amount of knowledge about individual MT's dynamics and call holding patterns in the form of probability density functions (pdfs), as well as the variances of these pdfs. Near-optimum solutions could be achieved if these pdfs could be accurately obtained, provided that their variances are sufficiently small. However, this remains a difficult task because mobile user behavior could depend on a large number of factors, such as weather, time of the day, traffic conditions, and emergencies. It is very difficult to define pdfs that could account for all the above factors. In addition, their variances could become too large for the pdfs to be useful.

In [6], Choi *et al.* propose to reserve resources in neighboring cells according to the estimated probability that a MT would hand off into these cells within an estimation time window, based upon the previous cell of the MT, and its extant sojourn time (the time that the MT has already spent in the current cell). It requires the use of a knowledge base containing the time spent by previous MTs in the cell, the previous cells that they came from, and their corresponding target handoff cells. However, it may be insufficient to predict the mobility of a MT based on its previous cell information and its extant sojourn time. Moreover, calls that are newly generated in the cell do not have previous cell information. This may hinder the scheme's prediction accuracy.

With the vision that future MTs are likely equipped with reasonably accurate positioning capability, it is time for active research into how it may be used for mobility prediction in wireless networks. The use of real-time positioning information for mobility prediction could potentially give rise to better accuracy and greater adaptability to time-varying conditions than previous methods. While there has been previous work in the literature that attempted to perform mobility prediction based on mobile positioning information [7]–[9], none of the work has addressed the fact that the cell boundary is normally fuzzy and irregularly shaped due to terrain characteristics and the existence of obstacles that interfere with radio wave propagation. Instead, either hexagonal or circular cell boundaries have been assumed for simplicity. The applicability and performance of these techniques in actual networks are therefore unclear.

Another observation is that none of the previous work has integrated road topology information into its prediction algorithm. Since MTs that are carried in vehicles are the ones that would encounter the most frequent handoffs, the integration of road information in mobility prediction algorithms could potentially yield better accuracy, which is crucial for more timely and efficient reservations. With real-time location information of MTs, it is now possible to take advantage of knowledge about road layouts.

Having seen the potential benefits of using real-time mobile positioning information for mobility predictions, the following section describes two such schemes that have been developed.

4.3 MOBILITY PREDICTION SCHEMES

In this section, we present two mobility prediction schemes that utilize real-time mobile positioning information. Both schemes require the serving BS to receive regular updates about each active MT's position every ΔT, say 1 sec. This will consume a small amount of wireless bandwidth (several bytes per update for each MT), which might be negligible for future broadband services. While the positioning accuracy of current commercially available GPS-based MTs is still poor (> 100 m), these schemes are built upon the assumption that future MTs could achieve much better accuracy than today (say < 10 m).

4.3.1 Linear Extrapolation (LE) Scheme

This is the simpler of the two schemes to be presented. The key idea is to predict a MT's future direction of travel using linear extrapolation over its most recent Q positions. The objective is to predict whether a handoff would occur within a threshold time $T_{threshold}$, and if so, predict the target handoff cell as well as the remaining time from handoff. In order to perform the predictions, we need to estimate the "boundary" of each radio cell. Here, the "cell boundary" is associated with the average distance from the antenna site at which *handoff-requests* usually take place. Due to terrain characteristics and the existence of obstacles that interfere with radio wave propagation, the boundary of a cell is normally fuzzy and irregularly shaped [10]. This is in contrast to many existing work that derived models for trajectory prediction and cell residence time distribution based on either hexagonal or circular cell assumptions.

4.3.1.1 Approximating The Irregular Cell Boundary

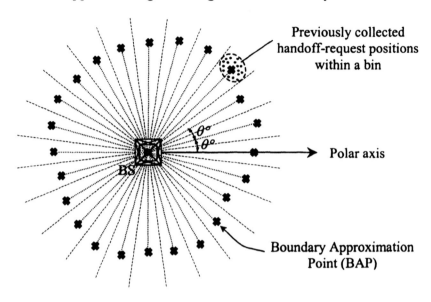

Figure 4.1. Approximating the cell boundary.

The cell boundary is approximated using M points surrounding a BS, arranged in such a way that the radial lines drawn from the BS to these points are spaced at regular angles $\theta° = 360°/M$ apart (see Figure 4.1). We shall refer to these points as *boundary approximation points* (BAPs). These BAPs are represented in polar coordinates as $(R_n, n\theta°)$, $n = 0, 1, ..., M - 1$, where R_n is the radial distance of the nth point from the BS. The polar axis could be chosen to point towards any arbitrary magnetic direction. In order to compute the BAPs, the BS has to collect data about the locations where previous handoff-requests were made, as well as the corresponding target handoff cell for each such request. Note that a MT's target handoff cell has already been chosen by the time it makes a handoff-request. The region around each BS is divided into M sectors, where the nth sector spans from azimuth $(n - ½)\theta°$ to $(n + ½)\theta°$, $n = 0, 1, ..., M - 1$. The collected handoff-request positions are divided into M bins according to the sector in which they are found. The collected positions within each bin are in turn reduced to a single BAP $(R_n, n\theta°)$. R_n is obtained as the average distance of the collected positions from the BS among those found within the nth bin. Specifically, let $D_{i,n}$ be the radial distance of the ith collected position within bin n, and let H_n be the total number of collected positions in bin n. R_n is calculated as follows:

$$R_n = \frac{\sum_{i=1}^{H_n} D_{i,n}}{H_n}. \tag{1}$$

A most likely target handoff cell T_n is also associated with each BAP, if it happens to be the most commonly chosen target handoff cell among the handoff-requests that occurred within the bin.

The BAPs (R_n, $n\theta^o$) and their corresponding most likely target handoff cells T_n are stored in a table located at each BS. This table could be updated, say, once a day, based on handoff-request positions collected over the previous day. In this way, they would be robust against changes in terrain characteristics or man-made structures that could affect radio propagation, as well as modifications in handoff parameters (e.g., hysteresis margin, handoff thresholds [10]) by the network operator. The number of entries in the table is exactly equal to the number of BAPs M. The parameter M could differ for each BS; it should be chosen experimentally to ensure that they are indeed good approximations of the cell boundary. The choice of M should depend on the size of actual BS coverage area, as well as topography; a larger BS coverage area or more irregular topographical features would require a larger M to increase the resolution of the approximated cell boundary.

4.3.1.2   Prediction Algorithm

The algorithm presented here is a variant of the one proposed in [11]. However, instead of delegating the prediction responsibility to individual MTs, here we assign the task to the BS. This reduces the computational power requirement at the MTs, which could be more attractive since battery power limitation and component cost are major concerns for mobile device manufacturers. Also, the BS should be able to handle this additional computational requirement without any difficulty.

For every active MT in the cell, the BS performs a series of simple calculations to estimate whether a handoff may occur within $T_{\text{threshold}}$, based on the approximated cell boundary information. Specifically, a line segment is extrapolated from the MT's current location in its estimated direction of travel by a distance equal to the product of $T_{\text{threshold}}$ and the MT's current speed (see Figure 4.2). This represents the path that could be traveled by the MT if it were to maintain its current speed and direction. The line segment is to be tested for any intersection with the approximated cell boundary. If an intersection is found, the BAP that is closest to this intersection point is determined, which we shall refer to it as the *closest BAP* (CBAP). The remaining time before handoff is then estimated as the time taken to reach the intersection point, while the corresponding target handoff cell is predicted to be the T_n associated with the CBAP.

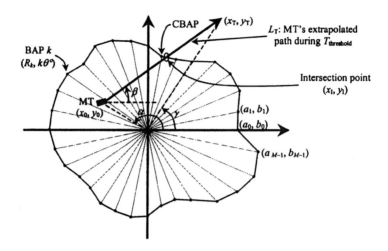

Figure 4.2. Searching for cell boundary intersection.

As mentioned earlier, the BS obtains the position information of active MTs at regular time intervals ΔT (typically 1sec [12]). It shall also store the MT's previous positions over the last $Q - 1$ intervals. Using a rectangular coordinate system with the current BS as the origin, the current and previous $Q - 1$ points are labeled as (x_0, y_0), (x_1, y_1), ..., (x_{Q-1}, y_{Q-1}), where (x_0, y_0) is the current position of the MT. The MT's direction of travel is estimated using linear regression over these Q points. Note that $Q > 2$ is required to mitigate the effects of GPS positioning errors. However, Q should not be too large, else the method could become insensitive to real changes in direction. Since we are only interested in small-scale predictions of whether the MT would hand off in the next few tens of seconds (typically less than 30 sec), it is reasonable to assume that a MT's direction and speed would remain almost unchanged during this period. This is especially true for vehicular movements, as roads are normally straight and sudden acceleration or deceleration is less frequent. Moreover, sharp turns are usually accompanied by a significant reduction in the vehicle's speed. Since the BS could perform prediction as frequent as every ΔT, the reduction in speed caused by sharp turns would give more time allowance for alternative reservations to be made if the previous reservation decision has become invalid.

A fast search algorithm has been developed to determine the intersection point between the extrapolated path and the approximated cell boundary. An important point to stress here is that not all the M BAPs are expected to be available. There might exist areas around the BS that are inaccessible to MTs, hence no handoff has ever occurred in such areas, and consequently no BAP is defined in its vicinity. The algorithm must be able to work even if some of the BAPs are missing. Another important point to note is that the

MT could have already exceeded the approximated cell boundary but has not yet made a handoff request. The algorithm must be able to detect such a situation. In this case, a handoff may be imminent. The target handoff cell could be predicted to be the most commonly chosen target handoff cell associated with the BAP that is nearest to the MT's current position. Another option to deal with this scenario is to predict the target handoff cell based on actual signal strength measurements, such as predicting that the MT will hand off to the neighboring BS with the strongest received signal strength. Note that the algorithm to be described below is based on the first option.

The following notations are used in the search algorithm:

B'	Set of indexes of valid BAPs in the cell, where $|B'| \le M$.
α	Azimuth of MT's position w.r.t. BS, $0° \le \alpha < 360°$.
k	Index of BAP whose azimuth w.r.t. BS is closest to α, $k \in B'$.
(a_i, b_i)	Rectangular coordinates of BAP i; polar coordinates $(R_i, i\theta°)$.
L_{ij}	Line segment connecting BAP i and BAP j, with endpoints (a_i, b_i) and (a_j, b_j).
β	MT's estimated direction of travel, where $0° \le \beta < 360°$.
γ	Azimuth of MT's projected position w.r.t. BS at time $T_{threshold}$, where $0° \le \gamma < 360°$.
n	Index of BAP whose azimuth w.r.t. BS is closest to γ, $n \in B'$.
s	Average speed of MT over last Q positions.
D_T	Estimated distance traveled by MT within the next $T_{threshold}$.
(x_T, y_T)	Estimated position of the MT at time $T_{threshold}$.
L_T	Line segment representing the extrapolated path of MT within the next $T_{threshold}$, with endpoints (x_0, y_0) and (x_T, y_T).
(x_I, y_I)	Rectangular coordinates of the intersection point.
T_i	Most commonly chosen target handoff cell recorded at BAP i.
t_{remain}	Estimated remaining time from handoff.

First, we determine the index k of the BAP whose azimuth is closest to the MT's azimuth w.r.t. the BS, α. Only valid BAPs are admissible, i.e., BAP $k \in B'$. We utilize a simple function called $Diff(\phi_1, \phi_2)$ that will return the polar angle difference in the range $[0°, 180°]$ for any two given polar angles ϕ_1 and ϕ_2. Specifically,

$$Diff(\phi_1, \phi_2) = \min\left\{ |\phi_1 - \phi_2|, 360° - |\phi_1 - \phi_2| \right\} \qquad (2)$$

k is then defined as follows:

$$k = \arg \min_{i \in B'}\left\{ Diff(i \cdot \theta, \alpha) \right\} \qquad (3)$$

Instead of performing an exhaustive search to test every value from the entire set B', Eq. (3) can be solved by searching in the vicinity of an initial value k_{init}, where

$$k_{init} = \left\lfloor \frac{\alpha}{360°} \times M + \frac{1}{2} \right\rfloor \bmod M . \qquad (4)$$

Note that k_{init} does not necessarily belong to B'. If $k_{init} \in B'$, then $k = k_{init}$ and Eq. (3) is solved. Otherwise, the test for membership in B' continues for $[(k_{init} \pm 1) \bmod M]$, $[(k_{init} \pm 2) \bmod M]$, ..., $[(k_{init} \pm N) \bmod M]$. The test stops at the first instant when the test value is found in B'. The value found this way is guaranteed to satisfy Eq. (3). Here, N is the maximum allowed deviation from k_{init} to be searched, where $N \leq M/2$. It may be necessary to choose a small value for N, because the resulting value of k may not be meaningful if the deviation is too large. This could arise when the BAPs are sparse and the MT is still very far away from any of these BAPs. If none of the test values belongs to B' upon finishing the tests, the search algorithm is terminated and no prediction is made. When this happens, the system will not reserve any resources for this particular MT.

Suppose a valid k can be found above. Before proceeding to search for any boundary intersection, we need to determine whether the MT has exceeded the approximated cell boundary using the following test:

$$x_0^2 + y_0^2 \geq R_k^2 . \qquad (5)$$

If the above inequality test in Eq. (5) is true, the MT is deemed to have exceeded the approximated cell boundary. In this case, the target handoff cell is predicted to be T_k, and the estimated remaining time to handoff t_{remain} is 0, i.e., the handoff is imminent.

If the inequality test in Eq. (5) is false, we proceed to compute D_T:

$$D_T = s \cdot T_{threshold} . \qquad (6)$$

The coordinates (x_T, y_T) can then be computed as:

$$x_T = x_0 + D_T \cdot \cos \beta , \text{ and} \qquad (7a)$$

$$y_T = y_0 + D_T \cdot \sin \beta . \qquad (7b)$$

After obtaining (x_T, y_T), it is straightforward to compute γ. Next, the index n is defined in a similar way as Eq. (3):

$$n = \arg \min_{i \in B'} \{ \mathit{Diff}(i \cdot \theta, \gamma) \}. \tag{8}$$

Eq.(8) can also be solved efficiently using a similar approach as the one previously presented for k.

Next, a series of line intersection tests are performed to determine if the line segment L_T intersects with the approximated cell boundary (formed by line segments linking neighboring BAPs taken from B'). We require the use of a simple function called $\mathit{Intersect}(l_1, l_2)$ that will test for intersection between any two given line segments l_1 and l_2. Suppose (x_{11}, y_{11}) and (x_{12}, y_{12}) are the coordinates of the two endpoints of l_1, while (x_{21}, y_{21}) and (x_{22}, y_{22}) are the coordinates of the two endpoints of l_2. Also, let m_1 and m_2 be the gradients of l_1 and l_2, respectively. The function $\mathit{Intersect}(l_1, l_2)$ will return TRUE if the following inequalities shown in Eqs. (9a) and (9b) are satisfied, and will return FALSE otherwise:

$$\left[y_{21} - m_1(x_{21} - x_{11}) - y_{11} \right] \cdot \left[y_{22} - m_1(x_{22} - x_{11}) - y_{11} \right] \le 0, \text{ and} \tag{9a}$$

$$\left[y_{11} - m_2(x_{11} - x_{21}) - y_{21} \right] \cdot \left[y_{12} - m_2(x_{12} - x_{21}) - y_{21} \right] \le 0. \tag{9b}$$

The number of line intersection tests to be performed is limited to a maximum of $|B'|/2 + 1$. The search direction is first determined as follows. The search will be performed in a counterclockwise manner along the cell boundary if either one of the following conditions shown in Eqs. (10a) and (10b) holds true; otherwise, it will be performed in a clockwise manner.

$$\text{If} \quad (\gamma \ge \alpha) \quad \text{and} \quad (\gamma - \alpha \le 180°), \tag{10a}$$

$$\text{If} \quad (\alpha > \gamma) \quad \text{and} \quad (\alpha - \gamma > 180°). \tag{10b}$$

Let i_{start} and i_{end} be the indexes of the BAPs where the line intersection tests start and end, respectively. We define two additional functions as follows (their implementations are straightforward and will be skipped here):

$\mathit{Clockwise}(i)$: Returns the index of the neighboring BAP next to BAP i in clockwise direction (result must be modulo M and a member of B').

$\mathit{Counterclockwise}(i)$: Similar to above, but in counterclockwise direction.

The values of i_{start} and i_{end} are assigned according to the search directions as follows:

$$i_{start} = \begin{cases} Clockwise(k) & \text{if counterclockwise search,} \\ Counterclockwise(k) & \text{if clockwise search.} \end{cases} \tag{11a}$$

$$i_{end} = \begin{cases} Counterclockwise(n) & \text{if counterclockwise search,} \\ Clockwise(n) & \text{if clockwise search.} \end{cases} \tag{11b}$$

For a counterclockwise search, the algorithm is given below:

i, j: integers
found: Boolean

```
1    i ← i_start
2    j ← Counterclockwise(i_start)
3    found ← Intersect(L_T, L_ij)
4    while ((j ≠ i_end) and (found = FALSE))
5        i ← j
6        j ← Counterclockwise(j)
7        found ← Intersect(L_T, L_ij)
8    return found
```

If the above returns a TRUE, then an intersection exists between L_T and the most recently tested L_{ij}. For the case of a clockwise search, the algorithm is almost identical to the one presented above, except that function *Counterclockwise*() is replaced with function *Clockwise*() in Lines 2 and 6.

Using the resulting values of i and j obtained from the algorithm above, the coordinates of the boundary intersection point can be easily determined. Let m_T and m_{ij} be the gradients of the line segments L_T and L_{ij}, respectively. x_1 and y_1 can be determined as follows:

$$x_1 = \frac{b_i - y_0 + m_T x_0 - m_{ij} a_i}{m_T - m_{ij}}, \tag{12a}$$

$$y_1 = m_T(x_1 - x_0) + y_0. \tag{12b}$$

Next, CBAP is chosen from BAPs i and j depending on which one of them is closer to the intersection point (x_1, y_1):

$$\text{CBAP} = \begin{cases} i & \text{if } \|(x_1,y_1)-(a_i,b_i)\| < \|(x_1,y_1)-(a_j,b_j)\|, \\ j & \text{otherwise.} \end{cases} \tag{13}$$

If CBAP $= i$, the predicted target handoff cell is taken to be T_i. Otherwise, it is taken to be T_j. The estimated remaining time from handoff is obtained as follows:

$$t_{remain} = \frac{\|(x_1,y_1)-(x_0,y_0)\|}{s}. \tag{14}$$

An important point to emphasize here is that the location information of the MTs are only used to estimate t_{remain} and to predict the target handoff cell, so as to make dynamic adjustment of resource reservations; actual handoff-requests are still initiated based on received signal strength measurements, error rates, interference, as well as handoff protocols used [13].

Another important point to stress is that the scheme has reasonable tolerance for prediction errors. The key objectives are to predict the correct target handoff cell, and to have sufficient time to set aside spare resources for the handoff by rejecting new calls when necessary. While a MT could change its direction and speed as it approaches the cell boundary, the target handoff cell prediction is erroneous only if the trajectory of the MT changes so much that it enters a different cell from the one previously predicted. In addition, we do not require precise knowledge of the exact time at which a handoff-request will be made, which is an intractable task since a handoff could occur anywhere within the handoff region [13], and a MT's speed and direction could also change randomly. The scheme would have achieved its objectives so long as the system has sufficient time to set aside the required spare resources for the handoff. Moreover, predictions are made periodically and alternative reservations would be attempted if the previous decision has become invalid.

The novelty of the scheme lies in its measurement-based approach. No assumption is made about the distribution of call holding times and cell dwell times, unlike many other approaches, which could become less effective if the assumptions were invalid. By regularly updating the BAPs, the scheme is robust against changes in handoff policies, as well as in terrain and manmade features that could affect radio propagation.

4.3.2 Road Topology Based (RTB) Scheme

This scheme is slightly more complex than the LE scheme. It utilizes road topology information for its predictions. Since MTs that are carried in vehicles are the ones that exhibit the greatest mobility, and knowing that vehicles are generally confined' to roads, the use of road topology information could potentially yield even better prediction accuracy than the LE scheme. The outputs of this scheme are similar to the LE scheme. We predict whether an active MT would encounter a handoff within a threshold time $T_{threshold}$, and if so, estimate the remaining time from handoff. Unlike the LE scheme that predicts only a single target handoff cell, the RTB algorithm could return more than one candidate target handoff cell, along with the probability that the MT would hand off to each of these candidate cells within $T_{threshold}$.

4.3.2.1   Prediction Database

Similar to the LE scheme, the prediction tasks are also assigned to individual BSs, which should have sufficient computational and storage resources to implement the scheme. In order to incorporate road information into the mobility predictions, each BS needs to maintain a database of the roads within its coverage area. We shall treat the road between two neighboring junctions as a road segment, and identify each segment using a junction pair (J_1, J_2), where a junction can be interpreted as the intersection of roads, e.g. T-junction or cross-junction. The approximate coordinates of each junction pair are to be stored in the database. Since a road segment may contain bends, it can be broken down further into piecewise-linear line segments. The coordinates defining these line segments within each road segment are also recorded. All the above coordinates could be easily extracted from existing digital maps previously designed for GPS-based navigational devices. Infrequent updates to these maps are foreseen because new roads are not constructed very often, while existing road layouts are seldom modified.

The database also stores some important information about each road segment. Since two-way roads would probably have different characteristics for each direction, the database shall store information corresponding to opposite directions separately. Information stored in the database includes the average time taken to transit the segment, the neighboring segments at each junction, and the corresponding probability that a MT traveling along the segment would select each of these neighboring segments as its next segment. These transition probabilities could be automatically computed from the previous paths of other MTs. The database will be updated periodically every $T_{database}$ since many of its elements are highly dependent on current traffic conditions.

In reality, the transition probabilities between road segments would probably vary with time and traffic conditions. For stochastic processes whose statistics vary slowly with time, it is often appropriate to treat the problem as a succession of stationary problems. We shall model the transition between road segments as a second-order Markov process, and we assume that it is stationary between database update instances so as to simplify the computations. Based on this model, the conditional distribution of a MT choosing a neighboring segment given all its past segments is assumed to be dependent only on the current segment and the immediate prior segment. Using the road topology shown in Figure 4.3 as an illustration, consider two MTs (MT1 and MT2) that are currently traveling from junction B towards junction E. MT1 came from segment CB previously, while MT2 came from segment AB previously. Based on the assumed model, the conditional probability of MT1 going to segment EF will be computed differently from that of MT2. The conditional probability of MT1 going to segment EF is

$$P(S_{k+1} = \text{EF} \mid S_k = \text{BE}, S_{k-1} = \text{CB}),$$
(15a)

while that of MT2 is

$$P(S_{k+1} = \text{EF} \mid S_k = \text{BE}, S_{k-1} = \text{AB}),$$
(15b)

where S_k is the current segment that the MT transits. Note that our stationarity assumption implies that the above conditional probabilities are independent of the value of k.

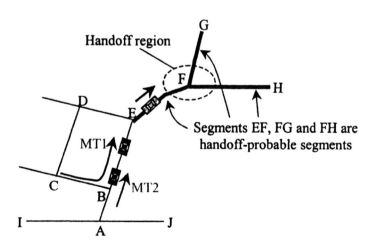

Figure 4.3. Utilizing road topology information for mobility prediction.

At the beginning of a new call, the previous segment of a MT is unknown, because it was not tracked previously. Therefore, we also need to store first-order conditional distribution in each segment, which are estimated from a subset of the data that were used to estimate the second-order conditional distribution. For instance, if we do not have any information about the previous segment of MT1 and MT2 in Figure 4.3, their conditional probabilities of going to segment EF are both

$$P\bigl(S_{k+1} = \mathrm{EF} \mid S_k = \mathrm{BE}\bigr). \tag{16}$$

If previous handoffs have occurred along a road segment, the probability of a handoff occurring in that segment is computed from previous data observed. The handoff probability, the target handoff cell, as well as the average time at which handoffs occur after entering the segment, are recorded. We shall refer to a segment that has experienced previous handoffs as a "handoff-probable segment" (HPS). An assumption made here is that MTs traveling along the same road segment in the same direction as previous MTs that have encountered handoffs are likely to encounter handoffs themselves.

Using the model described above, we could determine via chain rule the conditional probabilities of reaching and handing off at each of the HPSs from segments that are several hops away. We could also estimate the average time required to reach them, using current position and speed information, as well as previously collected statistics corresponding to each segment along the paths. The target handoff cell corresponding to each HPS is also available from the database. We could in turn estimate the probability that a MT would hand off to each neighboring BS within any specified threshold time $T_{\text{threshold}}$.

Before we proceed to describe in Section 3.2.2 how the prediction algorithm is implemented, we shall first explain how the prediction database is maintained. The following are the notations used:

$T_{\text{thres\_max}}$	The maximum $T_{\text{threshold}}$ that will be specified.
S	Set of road segments within the BS's coverage area.
$S(J_a, J_b)$	Directional segment from junction J_a to junction J_b.
$N(J_a)$	Set of junctions that are neighbors of J_a.
N_{Cells}	Set of neighboring cells adjacent to the BS of interest.
S_{HPS}	Set of handoff-probable segments (HPSs) in S.
S_{RSV}	Set of segments in which MTs may make reservations.
$P[S_{k+1} \mid S_k]$	1st order conditional transition probability.
$P[S_{k+1} \mid S_k, S_{k-1}]$	2nd order conditional transition probability, i.e., P[transiting to S_{k+1} | currently at S_k, previously at S_{k-1}].

$C_{HO}[S(J_a, J_b)]$	Most probable target handoff cell if a handoff were to occur along $S(J_a, J_b)$, where $C_{HO}[S(J_a, J_b)] \in N_{Cells}$.
$P_{HO}[S(J_a, J_b)]$	P[handoff along $S(J_a, J_b)$ | MT is in $S(J_a, J_b)$].
$T_{avg}[S(J_a, J_b)]$	Average time taken to transit segment $S(J_a, J_b)$.
$T_{HO,avg}[S(J_a, J_b)]$	Average time spent in $S(J_a, J_b)$ before handoff (only defined if $S(J_a, J_b) \in S_{HPS}$).
$D_{HO,avg}[S(J_a, J_b)]$	Average distance from J_b in $S(J_a, J_b)$ where handoffs occur (only defined if $S(J_a, J_b) \in S_{HPS}$).
X	Hop limit of routes that are considered.
$H(R)$	Number of hops in route R.
$S_{initial}(R)$	Initial segment of route R. This is the segment from which the route originates.
$S_{last}(R)$	Last segment of route R.
R'	Route R without the initial and the last segments, i.e., $\{R\} = \{R'\} \cup \{S_{initial}(R)\} \cup \{S_{last}(R)\}$.
$T'_{HO,avg}(R)$	Sum of the average time taken to transit R' and the average time spent in last segment before handoff, i.e., $\{\Sigma \, T_{avg}[S(J_i, J_j)] \, \forall \, S(J_i, J_j) \in R'\} + T_{HO,avg}[S_{last}(R)]$.
$R_X[S(J_a, J_b)]$	Set of all possible routes that are $\leq X$ hops from $S(J_a, J_b)$. Note: the routes must only contain segments in S.
$R_{X,HPS}[S(J_a, J_b)]$	Set of all routes in $R_X[S(J_a, J_b)]$ that terminate with HPS, i.e., if $R \in R_X[S(J_a, J_b)]$ and $S_{last}(R) \in S_{HPS}$, then $R \in R_{X,HPS}[S(J_a, J_b)]$. May have ≥ 1 HPS along each R.
$P_{HO}[R \mid S_k]$	1st order conditional probability that MTs in S_k would follow R and hand off at $S_{last}(R)$, where $R \in R_{X,HPS}[S_k]$.
$P_{HO}[R \mid S_k, S_{k-1}]$	2nd order conditional probability that MTs in S_k would follow R and hand off at $S_{last}(R)$, where $R \in R_{X,HPS}[S_k]$.

As mentioned earlier, many of the entries in the prediction database are highly dependent on the traffic conditions, which could depend on time of the day, weather conditions, road constructions, etc. In order to ensure that the entries are current, the database will be updated every $T_{database}$. The following procedure is performed during each database update:

1    $S_{HPS} \leftarrow \varnothing$
2    $S_{RSV} \leftarrow \varnothing$
3   **for** each $S(J_a, J_b) \in S$
4   compute $P[S_{k+1} = S(J_b, J_x) \mid S_k = S(J_a, J_b)]$
           $\forall \, J_x \in [N(J_b) - \{J_a\}]$
5   compute $P[S_{k+1} = S(J_b, J_x) \mid S_k = S(J_a, J_b), S_{k-1} = S(J_y, J_a)]$
           $\forall \, J_x \in [N(J_b) - \{J_a\}], \forall \, J_y \in [N(J_a) - \{J_b\}]$
6   compute $T_{avg}[S(J_a, J_b)]$
7   compute $P_{HO}[S(J_a, J_b)]$

```
8        if P_HO[S(J_a, J_b)] > 0
9          then S_HPS ← S_HPS ∪ {S(J_a, J_b)}
10              S_RSV ← S_RSV ∪ {S(J_a, J_b)}
11              compute C_HO[S(J_a, J_b)]
12              compute T_HO,avg[S(J_a, J_b)]
13              compute D_HO,avg[S(J_a, J_b)]
14   for each S(J_a, J_b) ∈ S
15     R_X,HPS[S(J_a, J_b)] ← ∅
16     for each R ∈ R_X[S(J_a, J_b)]
17       if S_last(R) ∈ S_HPS
18         then R_X,HPS[S(J_a, J_b)] ← R_X,HPS[S(J_a, J_b)]∪{R}
19              compute T'_HO,avg(R)
20              if T'_HO,avg(R) ≤ T_thres_max
21                then S_RSV ← S_RSV ∪ {S(J_a, J_b)}
22              compute P_HO[R | S_k = S(J_a, J_b)]
23              compute P_HO[R | S_k = S(J_a, J_b), S_k-1 = S(J_y, J_a)]
                    ∀ J_y ∈ [N(J_a) − {J_b}]
```

We assume that in between the database updates, the BS shall collect all the relevant data required for the subsequent database update. The above procedure begins by emptying both S_{HPS} and S_{RSV} (Lines 1 and 2) so that they can be regenerated based on the newly collected data. From Lines 3 to 13, we sequentially deal with all the road segments one at a time. Lines 4 and 5 compute the first and second order conditional distribution for the transition from the segment concerned to its neighboring segments. These can be computed based on the paths of MTs previously served by the BS. Lines 6 and 7 compute the average time spent by previous MTs in the segment, as well as the probability that a MT would request a handoff while transiting the segment. If handoffs have occurred along this segment previously, then the segment is identified as a HPS, and is entered into both S_{HPS} and S_{RSV} (Lines 9 and 10). Its membership in S_{RSV} signifies that MTs traveling in this segment are potential candidates for resource reservation. Lines 11 to 13 simply compute the database entries describing the handoff behavior of MTs traveling in this segment.

From Lines 14 to 23, we pass a second time through all the road segments, again dealing with them sequentially. For each segment $S(J_a, J_b)$, we reset the set $R_{X,HPS}[S(J_a, J_b)]$ so that it will be regenerated using newly computed database entries (Line 15). For each hop-limited route that originates from segment $S(J_a, J_b)$, we test whether its last segment is a HPS (Lines 16 and 17). Note that a route must include the origin segment $S(J_a, J_b)$, and at least one other segment. A hop limit is specified so as to reduce the computational load required. Also, note that $R_X[S(J_a, J_b)]$ is pretty much static, and does not need to be recomputed during each database update. It is modified only when there are changes to the road topology. If the examined route is found to have a last segment that is a HPS, it is added to the set $R_{X,\text{HPS}}[S(J_a, J_b)]$ (Line 18). In addition, the total average time for a MT to follow the route and eventually hand off at the last segment of the route is computed (Line 19). Note that this time does not include the estimated time taken to complete the current segment $S(J_a, J_b)$. The latter will be added during the prediction phase because we wish to utilize the dynamics of individual MTs for its computation. If the computed time is found to be less than $T_{\text{thres\_max}}$, the segment will be added to S_{RSV} if this has not been done previously (Lines 20 and 21). Finally, we proceed to compute the conditional probabilities that MTs currently in segment $S(J_a, J_b)$ would follow this route and hand off at its last segment (Lines 22 and 23).

Note that all the above database entries only need to be calculated once during each database update. Therefore, they should be well within the computational capability of a dedicated, average processor at the BS.

In the following, we give examples of how the database entries $P_{\text{HO}}[R \mid S_k = S(J_a, J_b)]$ and $P_{\text{HO}}[R \mid S_k = S(J_a, J_b), S_{k-1} = S(J_y, J_a)]$ can be computed. Using the sample road topology shown in Figure 4.3 again, suppose we are computing the above entries for segment $S_k = \text{BE}$, and we are interested in the route $R = \{\text{BE, EF, FH}\}$. The above database entries represent the conditional probabilities that a MT currently traveling along segment BE would go through segments EF and FH, and finally make a handoff request while traveling along segment FH. Note that segment EF is also a HPS. Therefore, these conditional probabilities must account for the probability that the MT does not make a handoff request while traveling along segment EF. For the first-order conditional probability that does not assume knowledge about the previous segment, it is obtained as

$$
\begin{aligned}
P_{\text{HO}}&[R = \{\text{BE, EF, FH}\} \mid S_k = \text{BE}] \\
&= P[S_{k+1} = \text{FH} \mid S_k = \text{EF}, S_{k-1} = \text{BE}] \cdot P[S_{k+1} = \text{EF} \mid S_k = \text{BE}] \\
&\quad \cdot (1 - P_{\text{HO}}[\text{EF}]) \cdot P_{\text{HO}}[\text{FH}].
\end{aligned} \tag{17}
$$

Since segment BE has two neighboring segments (segments CB and AB), there are two second-order conditional probabilities to be computed:

$$
\begin{aligned}
P_{\text{HO}}&[R = \{\text{BE, EF, FH}\} \mid S_k = \text{BE}, S_{k-1} = \text{CB}] \\
&= P[S_{k+1} = \text{FH} \mid S_k = \text{EF}, S_{k-1} = \text{BE}] \\
&\quad \cdot P[S_{k+1} = \text{EF} \mid S_k = \text{BE}, S_{k-1} = \text{CB}] \\
&\quad \cdot (1 - P_{\text{HO}}[\text{EF}]) \cdot P_{\text{HO}}[\text{FH}].
\end{aligned} \tag{18a}
$$

$$P_{\text{HO}}[R = \{\text{BE, EF, FH}\} \mid S_k = \text{BE}, S_{k-1} = \text{AB}]$$
$$= P[S_{k+1} = \text{FH} \mid S_k = \text{EF}, S_{k-1} = \text{BE}]$$
$$\cdot P[S_{k+1} = \text{EF} \mid S_k = \text{BE}, S_{k-1} = \text{AB}] \tag{18b}$$
$$\cdot (1 - P_{\text{HO}}[\text{EF}]) \cdot P_{\text{HO}}[\text{FH}].$$

Having seen the prediction database update procedure, we shall proceed to describe the mobility prediction algorithm in the next section.

4.3.2.2 Prediction Algorithm

As mentioned earlier, predictions are performed only for MTs that are currently traveling in segments that belong to the set S_{RSV}. These are the MTs that have the greatest potential of making handoff requests within $T_{\text{threshold}}$. Similar to the LE scheme, we assume here that the BS is able to obtain the position information of all active MTs every ΔT. The BS shall also maintain a record of the most recent Q positions of each MT. These will be used to determine its average speed, as well as to map its reported position onto the road topology database (a process known as map-matching [14]). In the prediction algorithm to be presented next, we do not describe how the map-matching is performed. Instead, we assume that the MT's position can be translated onto the correct position on the road topology database for simplicity. Interested readers can refer to relevant literature from Intelligent Transportation Systems (ITS) research for additional information, such as [14]. The algorithms for the actual map-matching process are not expected to be too tedious, especially if reasonably accurate positioning information were to be available (say < 10 m).

The follow notations are used to present the prediction algorithm:

s^i	Estimated speed of MT i, updated every ΔT.
$S^i(J_a, J_b)$	Current road segment in which MT i is traveling, updated every ΔT using map-matching.
S^i_{previous}	Previous segment from which MT i came from (may or may not be known).
$D^i_{\text{remain}}[S^i(J_a, J_b)]$	MT i's estimated remaining distance from J_b.
$T^i_{\text{remain}}[S^i(J_a, J_b)]$	MT i's estimated remaining time from reaching J_b.
$T_{\text{thres}}[C_j]$	$T_{\text{threshold}}$ of neighboring cell C_j.
$P^i_{\text{HO,Thres}}[C_j]$	Probability of MT i handing off into neighboring cell C_j within $T_{\text{thres}}[C_j]$.
$T^i_{\text{HO,remain}}[R]$	MT i's estimated remaining time from handoff if it follows route R and hands off at $S_{\text{last}}(R)$.
$T^i_{\text{HO,remain}}[S^i(J_a, J_b)]$	MT i's estimated remaining time from handoff in segment $S^i(J_a, J_b)$, only defined if $S^i(J_a, J_b) \in S_{\text{HPS}}$.

The prediction algorithm for any MT i of interest is given below:

1    $P^i_{\text{HO,Thres}}[C_j] \leftarrow 0$ for all $C_j \in N_{\text{Cells}}$

2   compute $D^i_{\text{remain}}[S^i(J_a, J_b)]$

3   if $s^i > 0$

4   then $T^i_{\text{remain}}[S^i(J_a, J_b)] \leftarrow D^i_{\text{remain}}[S^i(J_a, J_b)] / s^i$

5   else $T^i_{\text{remain}}[S^i(J_a, J_b)] \leftarrow \infty$

6   if $S^i(J_a, J_b) \in S_{\text{HPS}}$

7   then $T^i_{\text{HO,remain}}[S^i(J_a, J_b)]$
$$\leftarrow \max(D^i_{\text{remain}}[S^i(J_a, J_b)] - D_{\text{HO,avg}}[S^i(J_a, J_b)], 0) / s^i$$

8   if $T^i_{\text{HO,remain}}[S^i(J_a, J_b)] \leq T_{\text{thres}}[C_{\text{HO}}[S^i(J_a, J_b)]]$

9   then $P^i_{\text{HO,Thres}}[C_{\text{HO}}[S^i(J_a, J_b)]]$
$$\leftarrow P^i_{\text{HO,Thres}}[C_{\text{HO}}[S^i(J_a, J_b)]] + P_{\text{HO}}[S^i(J_a, J_b)]$$

10   for each $R \in R_{X,\text{HPS}}[S^i(J_a, J_b)]$

11     $T^i_{\text{HO,remain}}[R] \leftarrow T^i_{\text{remain}}[S^i(J_a, J_b)] + T'_{\text{HO,avg}}(R)$

12   if $T^i_{\text{HO,remain}}[R] \leq T_{\text{thres}}[C_{\text{HO}}[S_{\text{last}}(R)]]$

13   then if S^i_{previous} is known

14   then $P^i_{\text{HO,Thres}}[C_{\text{HO}}[S_{\text{last}}(R)]]$
$$\leftarrow P^i_{\text{HO,Thres}}[C_{\text{HO}}[S_{\text{last}}(R)]]$$
$$+ P_{\text{HO}}[R \mid S_k = S^i(J_a, J_b), S_{k-1} = S^i_{\text{previous}}]$$

15   else $P^i_{\text{HO,Thres}}[C_{\text{HO}}[S_{\text{last}}(R)]]$
$$\leftarrow P^i_{\text{HO,Thres}}[C_{\text{HO}}[S_{\text{last}}(R)]]$$
$$+ P_{\text{HO}}[R \mid S_k = S^i(J_a, J_b)]$$

In Line 1, we reset the MT's probability of handing off to cell C_j (within cell C_j's $T_{\text{threshold}}$), for all C_j that are neighbors of the current cell. Next, in Line 2, we estimate the MT's remaining distance from the end of its current segment. If the MT is determined to be moving, we compute its estimated time from completing its current segment. Otherwise, we simply state that this time is infinite (Lines 3 to 5). If the MT's current segment is a HPS, then there is a chance that a handoff could occur while it is traveling along this segment. Therefore, we estimate the remaining time from handoff, using its current estimated speed as well as its current distance from the average handoff position (Lines 6 and 7). The estimated time will be set to 0 if the average handoff position along the segment has already been exceeded, indicating that a handoff may be imminent. If this estimated time is within $T_{\text{threshold}}$ of the corresponding C_j, where C_j is the most commonly chosen target handoff cell in this segment, we add the segment's handoff probability to the MT's probability of handing off to C_j (Lines 8 and 9). Next, we proceed to examine previously recorded candidate routes that would lead to HPSs. For each of these routes, we compute its estimated remaining time from handoff if the MT were to follow the route in its entirety and hand off at the very last segment (Lines 10 and 11). If this time is found to be within

$T_{threshold}$ of C_j, where C_j is the most commonly chosen target handoff cell from the last segment of this route, we will update the MT's probability of handing off to C_j. We shall utilize either first or second order conditional probability for the update, depending on whether we know the previous segment of the MT (Lines 12 to 15). Upon completing the search through all the possible routes, we would have accumulated the estimated handoff probability (within the time $T_{threshold}$ of each neighboring cell) into each neighboring cell for the MT examined. The estimated remaining time from handoff for each possible route has also been computed.

4.4. DYNAMIC RESOURCE RESERVATION BASED ON MOBILITY PREDICTIONS

In order to perform handoff prioritization, the usual approach is to require each BS to set aside some wireless resources that could only be utilized by incoming handoff calls. By predicting when and where handoffs would occur, the system could dynamically adjust the amount of resource reservations at each BS, so as to meet the desired forced termination probability requirement in an efficient manner. In this way, more resources would be reserved only when mobility is high, while little or no resources would be reserved when mobility is low. This could achieve more efficient resource utilization than static or non-predictive approaches.

While the two mobility prediction schemes previously described in Section 3 perform predictions via different mechanisms, once these predictions are translated into resource reservation requirements, the way they are used for admission control are essentially the same. In both schemes, each BS shall have a "reservation target" (R_{target}) that is being updated periodically by the prediction algorithm. A new call is accepted if the remaining resource after its acceptance is at least R_{target}, i.e.,

$$R_{total} - \left(R_{used} + R_{new}\right) \geq R_{target} , \tag{19}$$

where R_{total} is the total resource capacity of the BS, R_{used} is the amount of resources currently in use, and R_{new} is the resource requirement of the new call. For a handoff request, the admission control rule is more lenient – it is admitted so long as there is sufficient remaining capacity to accommodate the handoff, regardless of the value of R_{target}:

$$R_{total} - R_{used} \geq R_{handoff} , \tag{20}$$

where $R_{handoff}$ is the resource requirement of the handoff request.

As we have seen previously in Section 3, both the LE and the RTB prediction schemes require the use of a threshold time $T_{threshold}$. Resources are to be reserved at a neighboring cell only for MTs that are predicted to hand off to this neighboring cell within the $T_{threshold}$ of the neighboring cell. Note that each BS should maintain its own $T_{threshold}$, as the value required could be characteristic of the BS's configuration, such as its total capacity, coverage area, subscriber density, etc. The threshold time $T_{threshold}$ could be interpreted as the time given to the BS to set aside the requested amount of spare resources for the anticipated incoming handoffs. During this time, spare resources are accumulated as they are released by active MTs that either end their calls or hand off to other cells; new calls are blocked so long as R_{target} is compromised. Thus, the value of $T_{threshold}$ could indirectly affect the forced termination probability (P_{FT}) experienced by handoff calls entering the cell. Since the required value of $T_{threshold}$ for the same target P_{FT} could vary over time when there are changes in dynamic factors such as system load, traffic conditions, user mobility, etc., $T_{threshold}$ should be dynamically adjusted to keep P_{FT} at the desired target value. We utilize an adaptive algorithm used in [6] to control its value. The algorithm counts the number of forced terminations among a number of observed handoffs. It increases $T_{threshold}$ by 1 sec if the measured forced termination ratio exceeds a preset value, and decreases it by 1 sec otherwise. In the algorithm, the value of $T_{threshold}$ is limited to the range [0, T_{thres_max}].

Next, we shall describe how the reservation target R_{target} at each BS is computed. The following are the additional notations introduced:

T_{RSV}	Reservation target modification interval.
$MT[S(J_a, J_b)]$	Set of active MTs currently traveling in $S(J_a, J_b)$.
$R_{MT}(x)$	Resource requirement of MT(x).
$R_{RSV\_RQ}[C_i{\rightarrow}C_j]$	Reservations to be requested at cell C_j on behalf of MTs currently in cell C_i that are predicted to hand off to C_j within C_j's $T_{threshold}$.

The reservation target R_{target} at each BS is to be updated every T_{RSV}. When it is time for an update, the BS of cell C_j sends a message carrying its current value of $T_{threshold}$ to each of its neighboring BSs, and instructs these neighbors to estimate the amount of resources that their active MTs would need to reserve in C_j. Suppose C_i is a neighboring cell of C_j. Upon receiving the instruction from C_j, the BS of C_i will perform either one of the following routines, depending on whether the LE scheme or the RTB scheme is implemented:

LE scheme:
1     $R_{RSV\_RQ}[C_i{\rightarrow}C_j] \leftarrow 0$
2   **for** each active MT(x) in cell C_i

    3   predict whether handoff would occur within $T_{\text{thres}}[C_j]$,
                and determine the CBAP if so.
    4   **if** $(T_{\text{CBAP}} = C_j)$
    5   **then** $R_{\text{RSV\_RQ}}[C_i{\rightarrow}C_j] \leftarrow R_{\text{RSV\_RQ}}[C_i{\rightarrow}C_j] + R_{\text{MT}}(x)$

RTB scheme:
    1    $R_{\text{RSV\_RQ}}[C_i{\rightarrow}C_j] \leftarrow 0$
    2   **for each** $S(J_a, J_b) \in S_{\text{RSV}}$ of C_i
    3   **for each** $\text{MT}(x) \in \text{MT}[S(J_a, J_b)]$
    4   compute $P^x_{\text{HO,Thres}}[C_j]$ based on the given $T_{\text{thres}}[C_j]$
    5            $R_{\text{RSV\_RQ}}[C_i{\rightarrow}C_j] \leftarrow R_{\text{RSV\_RQ}}[C_i{\rightarrow}C_j] + P^x_{\text{HO,Thres}}[C_j] \times R_{\text{MT}}(x)$

For the LE scheme, the BS of C_i simply examines all its active MTs and predicts for each of them whether its remaining time from handoff is within the given $T_{\text{threshold}}$ of C_j. If this is the case, and the target handoff cell is indeed predicted to be C_j, the BS will add the MT's resource requirement to the total requirement to be requested at C_j for MTs currently in C_i. The accumulated requirement will only be conveyed to C_j after all the active MTs have been examined. R_{target} in C_j will be updated as the sum of the reservation requirements from each of its neighboring cells, after it has received all the computed values from its neighbors.

For the RTB scheme, the BS of C_i shall examine each of its active MTs that are currently traveling on a segment that belongs to the set of segments eligible for reservation consideration. If the MT is found to have a non-zero probability of handing off into C_j within C_j's $T_{\text{threshold}}$, the BS will need to reserve resources in C_j by an amount equal to the resource requirement of this MT, weighted by the above probability. Similar to the LE scheme, R_{target} in C_j will be updated as the sum of the reservation requirements from each of its neighboring cells upon receiving the computed values from them.

As we have seen, the dynamic reservation schemes described above are suitable for heterogeneous resource requirements, because the reservation at each BS is adjusted according to the needs of individual MTs. Also, since reservations are dynamically adjusted based on the projected demands obtained via mobility predictions, the schemes could potentially achieve higher resource utilization than static or non-predictive approaches.

4.5 PERFORMANCE EVALUATION

Handoff prioritization schemes are commonly evaluated in terms of two QoS metrics, namely new call blocking probability (P_{NC}) and forced termination probability (P_{FT}). As mentioned previously, P_{FT} may be reduced at the expense of increasing P_{NC}. However, in the process of meeting the same P_{FT} requirement, a more efficient scheme will be able to accomplish the task with a lower P_{NC} than a less efficient scheme. The efficiency of the scheme depends on whether the reservations are made at the right place and time. Therefore, a predictive scheme should outperform a non-predictive scheme. Similarly, the efficiency of a predictive scheme should improve with its prediction accuracy.

4.5.1 Simulation Model

To facilitate the evaluation of the schemes presented, a novel simulation model was designed. Previous work in the literature either assumes that MTs travel in straight lines for long periods of time, or assumes that MTs follow random movements that do not resemble vehicular motion on roads. The simulation model used here incorporates road layouts that place constraints on MTs' paths. This establishes a more realistic platform to evaluate the performance of any positioning-based prediction algorithm.

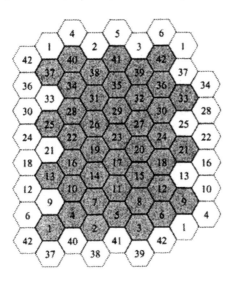

Figure 4.4. Simulation network with wrap-around at network boundary.

The simulation network consists of 42 wireless cells. In order to eliminate boundary effects that could make it very difficult to comprehend the performance evaluation results, a common approach found in the literature is used [6]; cells at the boundary wrap around as shown in Fig. 4.4. In this way, whenever a MT travels out of the network boundary, it is re-injected into the network again via the appropriate wrap-around cell as though a handoff has occurred from outside the simulation network. This compensates for any traffic loss at the network boundary. The simulation model also consists of arbitrary road layouts that are randomly generated based on some heuristic rules; real maps are not used because we require the roads to wrap around at the network boundary. The road layouts are designed to imitate those found in city areas. Figure 4.5 shows an example of the road topology that was randomly generated for simulation purpose.

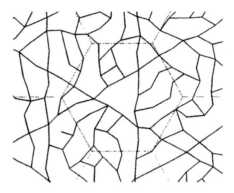

Figure 4.5. A sample road layout randomly generated based on heuristic rules.

Although the cell layout shown in Figure 4.4 adopts the hexagonal cell model, the simulation model does not assume that handoffs occur at the hexagonal boundary. In the simulation network, the hexagonal model is merely used to determine the BS locations. In contrast to previously mentioned work in which handoffs are assumed to occur at either circular or hexagonal cell boundaries, the simulation model used here does not have well-defined cell boundaries. Instead, we randomly generate $P = 100$ points around each BS that influence the positions at which handoffs occur. We shall call these points as *handoff influence points* (HIPs). Suppose R is the cell radius (assumed to be 1000 m in the simulations), which is typically defined as the distance from the BS to the vertex of the hexagonal cell model. When a MT comes within $0.075R$ from one or more of these HIPs, we assume that a handoff will occur during its transit through this region. The time at which the handoff shall occur is assumed to follow a uniform distribution within the time spent in the region. The target BS is assumed to be the nearest neighboring BS at the time when the handoff occurs, although this may not be the case in real life. The HIPs are created around the BS at regular angles $\theta^\circ = 360^\circ/P$ apart. The distance between each point and the BS is first generated using truncated Gaussian distribution, with a mean of $1.15R$ and a standard deviation of $0.2R$. All the distances are truncated to the range [0.95R, 1.35R]. Next, we perform smoothing by averaging the distance of each point with those of its immediate neighboring points. The smoothing step is necessary so that the entire handoff region would be continuous. Otherwise, some MTs might wander through any existing gap without handing off during the simulations.

We do not claim that the above model resembles the actual handoff position distribution in a real cellular network. However, it should be sufficient for the purpose of creating an irregular handoff region with some uncertainty, so as to evaluate the performance of different handoff prioritization schemes. To our knowledge, no work has modeled the 2-D distribution of handoff positions in real cellular networks. Therefore, we are unable to make use of any previously known model in the simulations.

To make the problem more interesting, traffic lights are introduced into the simulation model. Two sets of traffic lights are assumed. When one set is GREEN, the other set is RED. At a T-junction, one traffic light set is chosen randomly and assigned to the two roads that make the largest angle. The other remaining road will be assigned the opposite set. At a cross-junction, the roads are assigned alternate traffic light sets. Each GREEN and RED signal shall last for 60 sec. A speed limit is also assigned to each road segment, chosen from the set 40 km/h, 50 km/h, and 60 km/h with equal probability. Each MT will be randomly assigned a speed as it enters a new road segment, using truncated Gaussian distribution. The mean speed will be the speed limit of that particular road segment. The standard deviation is assumed to be 5 km/h, and the speed is truncated to a limit of three standard deviations from its mean.

For the sake of simplicity, we do not model the effects of positioning errors in the simulations. As mentioned earlier, the mobility prediction techniques presented here are based on the assumption that the positioning errors of future MTs are relatively small (say < 10 m). Therefore, we do not foresee any drastic effect on the simulation results if positioning errors were to be included, but we shall examine their effects in future studies.

In the simulations, the unit of bandwidth is called bandwidth unit (BU), which is assumed to be the required bandwidth to support a voice connection [6]. Each cell is assumed to have a fixed link capacity C of 100 BUs. For simplicity, the bandwidth requirement of each MT is assumed to be symmetric, meaning that they have the same requirement in both uplink and downlink. However, it is straightforward to modify the scheme to handle asymmetric requirements.

The traffic model used here is similar to the one used by [6]. Connection requests are generated according to Poisson distribution with rate λ (connections/sec/cell) in each cell. The initial position of a new call and its destination can be on any road with equal probability. The path chosen by the MT is assumed to follow the shortest path algorithm. For each call request, we assume that it is either of type "voice" (requires 1 BU), or of type "video" (requires 4 BUs) with probabilities R_{vo} and $1-R_{vo}$ respectively, where R_{vo} is also called the "voice ratio". In the simulations, R_{vo} is set to 0.5. The lifetime for each connection is exponentially distributed with mean 180 sec. We define the *normalized offered load* per cell as

$$L = \frac{[1 \cdot R_{vo} + 4 \cdot (1 - R_{vo})] \cdot \lambda \cdot 180}{C} . \tag{21}$$

Three additional schemes have been simulated for comparison purposes. They are described as follows:

a) <u>Reactive scheme</u>: This method is purely reactive with no prediction. It serves as a lower bound for the efficiency of the schemes considered. It measures the forced termination ratio among a number of handoffs recently observed, and increases the reservation in the cell when the P_{FT} target is not achieved, or decreases it otherwise.

b) <u>Choi's AC1 scheme</u>: This is one of the three methods proposed in [6]. In their simulations based on 1-D cell layout, their AC3 method performed best among the three methods, namely AC1, AC2 and AC3. However, in our simulations based on 2-D cell layout, we discover that AC1 has the best performance, whereas AC2 and AC3 are over-conservative and has much worse efficiency than the Reactive scheme. Therefore, we shall only present the results for AC1 here. This method works by estimating the probability that a MT would hand off into a neighboring cell within an estimation time window T_{est}, based upon its previous cell, and its extant sojourn time (i.e., the time it has already spent in the current cell). It requires the use of a knowledge base containing the time spent by previous MTs in the cell, the previous cells that they came from, and their corresponding target handoff cells. T_{est} is dynamically adjusted based on the measured forced termination ratio among a number of handoffs recently observed, and it indirectly controls the amount of resource reservations.

c) <u>Benchmark scheme</u>: This method serves as a benchmark indicating the best achievable results if we were to have perfect knowledge regarding when and where handoff requests will occur. It is impossible to achieve this in real-life. Reservations are computed for each active MT at regular time intervals (1 sec). If a MT's handoff time is within $T_{threshold}$, the MT's resource requirement will be reserved in the target handoff cell. $T_{threshold}$ is dynamically adjusted using a mechanism similar to the one used for adjusting T_{est} in [6], so as to meet any specified P_{FT} target.

4.5.2   Simulation Results

In the following, we present the results obtained from the simulations. Note that all results are the averages over 42 cells in the simulation network. In the simulation with no handoff prioritization, both P_{NC} and P_{FT} are 7.6%. This is unacceptably high for P_{FT}, thus explaining the need for handoff prioritization. Figure 4.6 shows the plots of P_{NC} versus P_{FT} for the five schemes considered. For each scheme, the target P_{FT} is varied so as to illustrate its tradeoff with P_{NC}. For any fixed L (set to 1.0 in the simulations), the relative positions of such tradeoff curves could demonstrate the relative efficiencies among the different schemes. A curve that is closer to the origin represents a more efficient scheme. It means that the scheme is able to achieve the same P_{FT} target while trading off a smaller increase in P_{NC}.

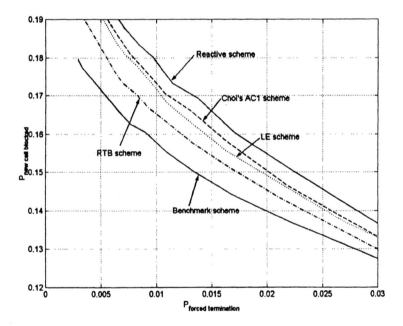

Figure 4.6. Plots of P_{NC} vs. P_{FT} comparing the efficiencies of different schemes.

Among the five schemes, the Reactive scheme has the worst efficiency since it does not make use of any prediction. Choi's AC1 scheme has better efficiency than the Reactive scheme because it possesses some intelligence in where and when the resources should be reserved. However, it has lower efficiency than the next three schemes. This is probably because it may be insufficient to predict the mobility of a MT based on its previous cell information, and its extant sojourn time. Moreover, calls that are newly generated in the cell do not have previous cell information. This hinders the scheme's prediction accuracy, thus lowering its efficiency. The LE scheme has slightly better efficiency over Choi's AC1 scheme. The improvement is even more significant in the RTB scheme. These demonstrate that mobility prediction schemes based on mobile positioning information are more accurate, thus leading to more efficient reservations. The most efficient scheme among the five schemes considered is the Benchmark scheme. As mentioned earlier, this is an idealized scheme that possesses complete knowledge of when and where the next handoff of each MT will occur. It merely serves as a bound to the best efficiency that could be achieved by other schemes. For a target P_{FT} of 1%, the Reactive scheme has a P_{NC} of 17.9%, while the lower bound set by the Benchmark scheme is 15.8%. The RTB scheme is able to achieve a P_{NC} of 16.5%.

As we have seen from the simulation results, the plots agree with intuition that handoff prioritization efficiency improves as the amount of knowledge incorporated into the schemes increases. With the additional knowledge of real-time mobile positioning information, the LE scheme is able to outperform Choi's AC1 scheme, even though it is based on a simple linear extrapolation approach. For the RTB scheme, the use of both real-time mobile positioning information and road topology knowledge further reduces the uncertainty in predicting the MTs' future movements. As a result, its performance is even closer to the limit set by the Benchmark scheme.

4.6   CONCLUSION

In this chapter, we have seen two mobility prediction techniques that are built upon the assumption that future MTs would be equipped with reasonably accurate positioning capability. The predictions are to be performed by each individual BS for all the active MTs under its service. Unlike previous attempts to perform mobility predictions using mobile positioning information, which have either assumed hexagonal or circular cell geometries, the presented schemes do not require any cell geometry assumption. Instead, the handoff regions are approximated based on previously recorded handoff locations.

The utilization of road topology information by the RTB scheme could potentially yield better prediction accuracy for MTs that are carried in vehicles. Since such MTs are expected to exhibit the greatest mobility and urgency for advance resource reservation, they are the main focus of the prediction techniques presented.

Among the many possible applications for which mobility prediction could prove useful, this chapter focuses on its use for dynamic resource reservation so as to prioritize handoff calls over new calls. This improves the user's perception of QoS by reducing the probability of forced call termination or call quality degradation during handoffs. With mobility prediction, the reservations at each BS could be dynamically adjusted according to the resource demands of MTs that are anticipated to hand off into the cell from its neighboring cells. By comparing the plots featuring the tradeoffs between new call blocking probability and forced termination probability obtained from several schemes, we demonstrate that reservation efficiency improves as the amount of knowledge incorporated into the scheme increases, and the RTB scheme has the potential to achieve performance that is closest to the limit set by the idealized Benchmark scheme. The use of the RTB scheme could therefore reduce the operational costs of cellular service providers, while offering the subscribers with improved QoS experience throughout their call duration.

With the emergence of telematics systems in vehicles, motorists may receive dynamic route guidance based on real-time traffic information. If this routing information were to be made available to the wireless network, it could help to further diminish the uncertainty in mobility predictions, and realize even more efficient resource reservation schemes. This shall constitute an interesting research area for the years to come.

REFERENCES

[1] D. Hong and S. S. Rappaport, "Traffic model and performance analysis for cellular mobile radio telephone systems with prioritized and non-prioritized handoff procedures," *IEEE Trans. Veh. Technol.*, vol. VT-35, no. 3, Aug. 1986, pp. 77-92.

[2] Y. Zhao, "Standardization of mobile phone positioning for 3G systems," *IEEE Commun. Mag.*, Jul. 2002, pp. 108–116.

[3] E. A. Bretz, "X marks the spot, maybe," *IEEE Spectrum*, Apr. 2000, pp. 26-36.

[4] C. Oliveira, J. B. Kim, and T. Suda, "An adaptive bandwidth reservation scheme for high-speed multimedia wireless networks," *IEEE J. Select. Areas Commun.*, vol. 16, no. 6, Aug. 1998, pp. 858–74.

[5] D. A. Levine, I. F. Akyildiz, and M. Naghshineh, "A resource estimation and call admission algorithm for wireless multimedia networks using the shadow cluster concept," *IEEE/ACM Trans. Networking*, vol. 5, no.1, Feb. 1997, pp. 1–12.

[6] S. Choi and K. G. Shin, "Predictive and adaptive bandwidth reservation for handoffs in QoS-sensitive cellular networks," in *Proc. ACM SIGCOMM'98*, Vancouver, British Columbia, Sep. 1998, pp. 155–66.

[7] T. Liu, P. Bahl, and I. Chlamtac, "Mobility modeling, location tracking, and trajectory prediction in wireless ATM networks," *IEEE J. Select. Areas Commun.*, vol. 16, no. 6, Aug. 1998, pp. 922–36.

[8]   A. Aljadhai and T. Znati, "Predictive mobility support for QoS provisioning in mobile wireless environments," *IEEE J. Select. Areas Commun.*, vol. 19, no. 10, Oct. 2001, pp. 1915–1930.

[9]   M.-H. Chiu and M. A. Bassiouni, "Predictive schemes for handoff prioritization in cellular networks based on mobile positioning," *IEEE J. Select. Areas Commun.*, vol. 18, no. 3, Mar. 2000, pp. 510–522.

[10] D. Grillo, R. A. Skoog, S. Chia and K. K. Leung, "Teletraffic engineering for mobile personal communications in ITU-T work: the need to match practice and theory," *IEEE Pers. Commun.*, pp. 38–58, Dec. 1998.

[11] W.S. Soh and H.S. Kim, "Dynamic guard bandwidth scheme for wireless broadband networks", in *Proc. IEEE INFOCOM'01*, Anchorage, Alaska, USA, Apr. 2001, pp. 572-81.

[12] T. Kobayashi, N. Shinagawa, and Y. Watanabe, "Vehicle mobility characterization based on measurements and its application to cellular communication systems," *IEICE Trans. Commun.*, vol. E82-B, no. 12, Dec. 1999, pp. 2055–60.

[13] N. D. Tripathi, J. H. Reed, and H. F. Vanlandingham, "Handoff in cellular systems," *IEEE Pers. Commun.*, Dec. 1998, pp. 26–37.

[14] Y. Zhao, *"Vehicle Location and Navigation Systems,"* Chapter 4, Artech House, 1997.

Chapter 5

SEAMLESS MOBILITY

James Kempf
DoCoMo USA Communications Laboratories

5.1 INTRODUCTION

Maintaining IP service to a mobile host in the Internet is largely a matter of negotiating boundaries. As a host moves, it crosses various natural boundaries in the Internet. When the host is inside one of these boundaries, both the host and the network contain state associated with maintaining the host's IP service. The state is in certain cases specific for the network within the boundary. Some aspects of this state change when the mobile host moves across a boundary, so that the old state is either no longer valid or no longer accessible. As a consequence of that primary change, secondary effects may occur.

Without some action on the part of the mobile host, the network, or both, the mobile host's IP service may be cut off, or certain characteristics of that service may be modified so that the IP service no longer meets the user's preferences, as perhaps expressed in the user's contract with their Internet service provider. The user naturally prefers not to deal with such boundaries, since the boundaries are often not particularly obvious nor do they relate much to the kinds of tasks the user wants to accomplish. The user would rather have IP service continue without pause or interruption as the boundary is crossed. Seamless mobility provides users with the illusion that there is no boundary, by automatically smoothing the change to maintain the host's IP service in accordance with the user's preferences.

Table 5.1. Natural Internet Boundaries

Boundary	Primary Change	Secondary Effects
Wireless cell	Link layer addressing	Link security
Subnet	IP routing	Transition packet loss, Access router location, Authentication, Authorization, Dormant host location, IP service characteristics, Security
Wireless Internet service provider	Network access authentication	Authentication, Authorization, Accounting, Security
Wireless medium technology	Link layer addressing, IP routing, possibly authentication, authorization, and accounting	Same as for link layer and IP routing change, possibly also for wireless service provider change

In Table 5.1, some natural boundaries in the Internet are listed in the left column. The central column lists the primary change when one of those boundaries is crossed, and the right column lists the secondary effects caused by the primary change. In the subsections below, we examine the effects of crossing the various boundaries in a bit more detail.

5.1.1   The Cell Boundary

The most basic boundary is between one wireless cell and another. A wireless access point typically has a limited geographic area in which the mobile host can obtain adequate coverage. In the absence of any topographical obstructions, the area would be a circle, but hills, cars, buildings and the like often modify this. Around the core coverage area, a thin ring exists in which wireless coverage for the access point falls of quickly, as the signal power and link quality from the access point decrease. When the mobile host transitions between one wireless access point and another, it crosses this ring, and the wireless link layer arranges for the mobile host to switch its link to an access point with better signal power and link quality properties.

The wireless link layer may also maintain security on the link between the mobile host and the wireless access point. For example, on 802.11 wireless LAN, a security protocol called Wired Equivalent Protocol (WEP) can be used to encrypt the frames sent between the access point and the mobile host. The encryption doesn't extend beyond the wireless link, however. When the mobile host moves from one access point to another, the link encryption state (such as session keys, etc.) on the old access point is lost, and it becomes necessary to re-initialize link security on the new access point, or otherwise arrange for the new access point to obtain the link security state. If the link layer depends on network layer protocols to implement some aspects of link security, then solutions for moving network layer security state may be reusable for link security as well.

5.1.2 The Subnet Boundary

The next most basic boundary is between two IP subnets. The wired backbone is typically partitioned into subnets, with access points and their surrounding cells providing wireless service. The mapping between the subnet topology and the geographic coverage area of the cells determines where the subnet boundary is located. The boundary may not be firmly located, as it may change when moving vehicles or other obstacles modify wireless reception. When the mobile host crosses from one cell to another, it may also cross a subnet boundary. Crossing a subnet boundary causes the host's IP mobility management protocol, typically Mobile IP, to rearrange routing to the host. Packets routed to the old subnet will not reach the host in the new subnet unless the routing is updated. Since a subnet boundary always follows one or several cell boundaries, part of the task of Mobile IP is to determine whether the crossing of a cell boundary also constitutes the crossing of a subnet boundary.

Routing is the primary change that occurs when a subnet boundary is crossed. There are also secondary changes that are consequences of the routing change. If the mobile host or the network is able to detect the subnet change before it occurs, the change can be optimized in various ways. For example, potential new access routers can be detected on the new link prior to the link switch removing the need to solicit a router advertisement on the new link, or packets can continue to be routed to the mobile host after the link switches but before the completion of the Mobile IP routing change.

The subnet change affects other aspects of the mobile host's basic IP service. If the host is operating in a public access network, the host's authentication and authorization credentials must be available in the new subnet, so the new subnet's Network Access Server (NAS) can check whether the mobile host is authorized for IP service. The host may have specific service characteristics, like wireless link bandwidth or other quality of service parameters, which must be re-established on the new subnet. If the host does not have an active traffic session up and is in power saving, or dormant, mode, the host or network must take action to allow the host to be found should packets arrive for the host. Finally, the host may be utilizing a security gateway that becomes inaccessible when the host moves to the new subnet. Arrangements must be made to re-establish or transfer the security associations from the old gateway to the new.

5.1.3   The Internet Service Provider Boundary

While moving around a large geographic area, a mobile host may move across the boundary between the coverage area of one Wireless Internet Service Provider (WISP) and another. Alternatively, the user may travel to some remote location and start using a mobile host in a geographic area where the user's home WISP does not provide service. The user must either have an account with the new WISP or the two WISPs must have a roaming agreement allowing their users to operate on each other's networks. Such roaming agreements are common for wired ISPs that only cover a small geographic area.

The boundary between two WISPs may be difficult to hide, because the new WISP may require the user to redo the basic authentication exchange that permits network access. The mobile host may also be required to re-initialize any security gateway state it is utilizing, for example, to obtain VPN service with its home network. If the two WISPs deploy compatible authentication, authorization, and accounting software and protocols, and they have the proper interoperability procedures in place, it may be possible to provide the user with single sign on, so that a network access procedure executed with one WISP is sufficient to provide the user with access in the other WISP's network as well, but transferring this state when the host moves from one network to another may be difficult.

5.1.4   Boundaries Between Wireless Media Technologies

The boundary between wireless media technologies does not necessarily involve physical movement of the mobile device, though movement is required to transition between the geographical coverage area of one wireless medium and another. A mobile host may have access to more than one wireless medium at a particular location. The best example is hotspot service. A hotspot is a small, constrained area of broadband, local area wireless coverage embedded in a broader area of wide area, narrowband wireless coverage. A mobile host with a wide area interface and a hotspot interface has the option of sending and receiving traffic on either interface. Outside the hotspot, the broadband coverage does not exist. When the host enters the hotspot, it must arrange for handover of IP service from the wide area interface to the hotspot interface if hotspot service is desired. When the host leaves the hotspot, IP service must be moved back to the wide area interface. Such intertechnology handover allows the mobile host to optimize its use of wireless resources depending on the available media options.

One of the first requirements for smoothing the boundary between wireless media technologies is some way for the mobile host to detect that another wireless coverage option exists. For reasons of power conservation, it is desirable for the host to keep only one interface active at a time, and if no hotspot service is available, that is the wide area interface. In order to utilize the hotspot, the mobile must be able to determine through the wide area interface whether hotspot service is available. If service is available, and the user's preferred policy is to use hotspot service, handover of IP service from the wide area interface to the hotspot interface is required. If the mobile host moves beyond the geographical coverage area of the hotspot service, the host's IP service must be moved back onto the wide area interface. In addition, the hotspot coverage may be provided by a different WISP than that for the wide area coverage. If so, the boundary between WISPs must be crossed in as smooth a manner as possible.

5.1.5   Seamless Mobility Solutions

Cell boundary seamless mobility solutions depend critically on the amount of support available from the link layer, and the details of the solution depend on the particular link technology. Mobility between wireless Internet service providers is in its infancy, and support for seamless inter-WISP mobility has yet to be developed beyond such basic aspects as seamless subnet mobility and inter-provider roaming, the latter of which is well developed for the wired network. Therefore, we confine the discussion in the rest of this chapter to seamless subnet mobility and intertechnology handover. We discuss proposed seamless mobility solutions for those aspects of subnet handover not covered by basic Mobile IP: maintaining seamless packet delivery while Mobile IP routing changes occur, seamlessly re-establishing various characteristics of a host's IP service, such as quality of service, on the new subnet, intertechnology handover, and dormant mode location tracking.

5.2 SEAMLESS IP ROUTING CHANGE

Mobile IP provides a complete solution for managing routing to a mobile host in the Internet. As a mobile host moves across a collection of wireless access points, Mobile IP manages the routing transition when the mobile host crosses a subnet boundary. At a subnet boundary, the host's local address, the care of address, is no longer valid. The host must obtain a new care of address in the new subnet. The host's Mobile IP stack performs a subnet handover algorithm called movement detection to determine whether the link switch, between the old wireless access point and the new, also constitutes a switch in subnet. If a subnet switch occurred, the mobile host obtains a new care of address. The mobile host informs its Home Agent about the change by requesting a change in binding between the mobile host's home address and its care of address. This allows correspondent hosts that send traffic to the home address to obtain routing to the mobile host at its new care of address. When a switch in binding occurs, the Home Agent simply switches the routing to the new care of address.

This relatively straightforward solution to mobility management works well if the amount of time required for a handover is not an important factor in meeting the customer's expectation of service quality. The default Mobile IP movement detection algorithm depends on side effects of the link switch that manifest at the IP layer in order to determine whether the subnet has changed. For example, the access router[1] may periodically multicast or broadcast router advertisement messages. With movement detection, a mobile host detects that it has changed subnets by receiving a router advertisement message for a router on a different subnet. The amount of time required to detect the movement, and thus to begin the routing change, depends on the period of the network layer router advertisement message beacon, and not on the mobile host directly detecting that a link switch has occurred. While the link layer is switching the link, the host is waiting to detect the subnet change, and the routing change at the Home Agent is under way, packets are sent to the old access router; where they are dropped.

In order to smooth the routing transition, a number of algorithms have been proposed to replace movement detection. They are:

- Buffering incoming packets during the routing change and deliver them to the mobile host when the subnet handover is complete,
- Multicast all packets to a small set of wireless access points around the cell in which the mobile host is located so that any packets sent during

---

[1] We use the Mobile IPv6 term *access router* generically for the last hop router before the wireless link. Mobile IPv4 calls the specialized last hop router it requires a *Foreign Agent*.

the link switch and routing change will be available on the new access point,
- If the new access router is known before the link switch, begin the routing change prior to moving to reduce the amount of time required after the link switch,
- Establish a tunnel between the new access point or access router and the old prior to the routing change completion, so that any packets delivered to the old access point are forwarded to the mobile host in the new subnet.

We discuss these in the following subsections.

5.2.1   Buffered Handover

Buffered handover techniques propose to reduce packet loss by buffering packets while the mobile host's IP service is down due to the subnet handover. Buffering is done during the actual link switch and while the host is performing movement detection and the while the change in routing at the Home Agent is happening. Once the routing change has completed, packets are delivered to the host at its new care of address. Buffered handover can be combined with any of the other techniques, because it is really a technique for smoothing handover, not for accelerating it. Although buffering works well for nonreal time traffic, it can introduce "ringing" artifacts into real time traffic, like voice. If the buffer is too large, the additional traffic must be buffered while the original buffer is downloaded, etc., until the transient from the handover has subsided.

Buffering can be done in a variety of network elements: the access point, the access router, or a routing proxy located some distance away from the wireless subnet. A routing proxy is typically used as a localized mobility management agent, to hide mobility in the local administrative domain from correspondents on the Internet. Buffering for handover is an additional service that the routing proxy can offer.

The network element performing the buffering must either know when to start buffering, or it must use a sliding buffer for all traffic. In both cases, the buffering network element must be prepared to download the buffer into the mobile host when the mobile host provides signaling saying that it has moved. If a sliding buffer is used, no signaling is required to start buffering, but if not, some signal must be provided before the mobile host loses IP service on the old link. If the buffering is done in the access point or access router, the access point or access router can utilize signaling from the mobile host to determine when to start buffering, or it can utilize signaling from the link layer, if available. If buffering is done in a routing proxy, signaling is required from the mobile host, the access point, or the access router to the routing proxy to indicate when buffering should start.

5.2.2   Multicast Traffic Delivery

Another proposed handover accelerating technique is multicast distribution. Instead of tunneling packets to the mobile host, the mobile host's Home Agent multicasts packets to the mobile host's current wireless access point and a small set of wireless access points around the current point. The wireless access points maintain a sliding buffer of the multicast traffic. When the mobile host moves, it signals the new access point to download the buffer. No multicast is done to the mobile host, but the wireless access points are responsible for maintaining the multicast group through which the mobile host receives its packets. On the uplink from the mobile host, no multicast is required.

While multicast distribution can help to smooth traffic flow across the subnet handover boundary, it does break the end-to-end transport connection between the mobile host and its correspondent. The Home Agent must encapsulate packets and send them with UDP in order to multicast. The Home Agent cannot send packets using TCP or any other transport protocol because UDP is the only transport protocol used for multicast. TCP depends on the end-to-end connection to enforce congestion control on the Internet, so multicast distribution could impact congestion management. Also, UDP is unreliable, with no retransmission, but applications on the mobile host that start a conversation with a correspondent utilizing a reliable protocol would end up having unreliable transport for a part of the routing, so they would not compensate for packet loss. Finally, since the wireless access points are required to maintain multicast groups with a Home Agent that might be on the other side of the Internet, multicast routing protocols would have to be widely deployed on the Internet. Multicast is not widely deployed on the Internet today.

5.2.3   Anticipated Care of Address Change

If the mobile host or access router has information on the new access router prior to the link switch, it can arrange for the process of Mobile IP routing update to start prior to movement to the new subnet. The host or access router does this by anticipating the care of address change, and performing signaling to cause a change in the care of address prior to the link switch. Anticipated care of address change accelerates handover by speeding up movement detection allowing the Mobile IP routing change to proceed in parallel with the link switch.

Anticipated care of address change requires the mobile host to obtain a link layer identifier for the new access point or access router from the link layer. If it is capable, the host can resolve the identifier to the IP address of the new access router. The host sends either the link layer identifier or the IP address of the new access router to its current access router. The current access router returns a proxy router advertisement for the new access router, allowing the host form a new care of address. The current access router can also send an unsolicited proxy router advertisement to the mobile host, if the current access router receives the link layer information on the new router rather than the mobile host.

In Mobile IPv4, after the mobile host receives the proxy router advertisement, it sends a Home Agent registration message through the new Foreign Agent to start the change in routing at the Home Agent. Unless the IP layer can force the link layer to hand over, this change may occur prior to switching to the new link. The new Foreign Agent buffers the Home Agent's reply until the mobile host shows up on the new link.

In Mobile IPv6, when the current access router obtains the identity of the new access router, the current access router contacts the new access router about the handover. The new access router inserts the mobile host's care of address in its neighbor table. When the mobile host receives the proxy router advertisement from its current access router, it sends a binding update to the current (now old) access router telling the old access router that it is now addressable under the new care of address. Logically, it has moved to the new subnet, regardless of whether it has physically moved by switching the link. The old access router replies with a binding acknowledgement, which it sends through both the old and new links. When the mobile host shows up on the new link, it gets the binding acknowledgement.

Anticipated care of address change can reduce packet loss due to handover, but the reliability is very sensitive to the amount of time available for the prehandover wireless signaling. If the mobile host does not have enough time for completion of the prehandover wireless signaling prior to the old link going down, the signaling does not complete, and the mobile host must perform standard Mobile IP movement detection on the new subnet. As a result, packets are lost during movement detection and routing change on the new subnet, just as with standard Mobile IP handover, and they are additionally lost prior to the handover if the link bandwidth is so slow that the prehandover wireless signaling blocks the link. If enough time is available for completion of prehandover signaling, however, anticipated care of address change accelerates the subnet handover so that the mobile host obtains packets on the new subnet faster than with standard Mobile IP handover.

5.2.4   Tunnel-based Handover

Tunnel-based handover techniques smooth the handover boundary by temporarily fixing up the routing failure due to the link switch. The fix is in the form of a source route-based tunnel between the old access router and the new. If the old access router has information on where the mobile host is moving prior to the link switch, it uses that to contact the new access router. The routers establish a tunnel for the mobile host's traffic when it moves. If the information on the new access point isn't available until after the mobile host moves, the tunnel is set up after the link switch. The mobile host ends up losing packets during the time the link is down and while the tunnel is set up, but packets start flowing again prior to the Home Agent binding update completion, which may take some time.

Tunnel-based handover is a very flexible technique that can be adapted and used with other techniques, such as anticipated handover and buffering. However, it does require a lot of help from the link layer. For best results, the old access router should be able to put the tunnel in place prior to the link switch, and it should receive a trigger from the link layer notifying it that the mobile host has moved so it can begin routing through the tunnel. The tunnel can also be set up after the mobile host has moved, but packets are dropped if they are routed to the old access router while the link switch and tunnel setup signaling are underway.

5.3   SEAMLESS SERVICE TRANSITION

The typical wireless link handover protocol determines when to start handover by detecting some degradation in the quality of the link to the old wireless access point: a decrease in power level, an increase in bit error rate, etc. The link layer then determines a new access point and switches the mobile host's link to that. The IP layer may be informed either before the link switch or after, or the link layer provide no indication and thus require Mobile IP movement detection to detect a subnet change.

The following are additional factors that can smooth the establishment of the mobile host on the new subnet:

-   If the mobile host can discover the access router prior to switching the link, it can use anticipated care of address change or tunnel-based handover to accelerate the handover,
-   In addition to just switching routing, there are a variety of IP service characteristics that need to be re-established on the new subnet. These

characteristics include quality of service, authorization and accounting, and security.

Candidate Access Router Discovery (CARD) and Context Transfer (CT) cover these factors. We discuss CARD and CT in the following subsections.

5.3.1   Candidate Access Router Discovery

If the mobile host doesn't find out about the new link until after the link layer has made the switch, then the host has no choice but to use standard router discovery to discover the router, either by directly soliciting or by waiting for a routing advertisement beacon. Packet loss occurs during the period in which the host's old router is no longer valid. If, however, the mobile host discovers the new access point prior to the link switch, and it has a link layer identifier for the new access point, the mobile host may use this information to accelerate the subnet handover. To do that, the mobile host requires a mechanism for mapping the link layer identifier of the next wireless access point to an access router. This mechanism is CARD.

The CARD has two basic functions:

- Allow an IP network element such as a mobile host or access router to map from the link layer identifier of an access point to the IP address of an access router, and to obtain subnet prefix (in IPv6), and other subnet information essential for switching a host to the new router,
- Provide information on additional service characteristics, including link medium technology, available bandwidth, etc. so that the mobile host can make an informed choice about its next link, if the link layer allows the host to choose.

The mapping from the wireless access point identifier to the access router IP address allows the mobile host or access router to accelerate the handover by using one of the techniques discussed in Section 5.2. This mapping is similar in function to that performed by the Reverse Address Translation (RARP) protocol in IPv4 or inverse neighbor discovery in IPv6 on a single link. The difference in this case is that the link identifier of the wireless access point is not on the same link as the host or router performing the identifying.

Locating an access router based on its service characteristics has much in common with locating any service based on its characteristics. SLPv2 is an example of a protocol that allows services to be discovered by characteristics, and it has been used in a protocol called GAARD for doing candidate access router discovery. The service characteristics that may be of interest to a mobile host looking for a new link are:

- Wireless media technology,
- Bandwidth on the wireless link,
- Bandwidth on the wired link,
- Whether differentiated services quality of service categories, such as Expedited Forwarding, are available,
- IPsec gateways reachable via the link.

The wireless media technology is an especially important characteristic. Without CARD, a mobile host with multiple interfaces is forced to keep all interfaces active and consuming power in order to discover new access points for interfaces that are currently not carrying IP traffic. By using CARD, the mobile host can discover which media are currently available through a single interface, and shut down the interfaces for media that aren't, thereby saving power. Of course, the geographical granularity of the information available on the other media may affect the result. For example, consider a wide area wireless network in Tokyo with a subnet that covers the entire city, advertising the availability of hotspots. If a host in Asakusa gets information on a hotspot router in Shibuya, the hotspot interface would be activated even though it is a long subway ride before the hotspot service actually becomes available.

The other characteristics are important if the link layer allows the mobile host to control handover. For example, suppose the mobile host uses CARD to discover an access router having a particular set of bandwidth and differentiated service characteristics that support multimedia. The mobile host is only capable of using these characteristics if it can force a handover from its current access point to a new access point that supports these characteristics. Current link layer technologies typically only allow handover based on the power and link quality characteristics of the current link, and they often don't allow the mobile host to actually specify which wireless access point should be used after the handover. But if the access router is available through a different wireless interface, the IP layer can force the IP traffic onto the broadband interface through intertechnology handover.

5.3.2   Context Transfer

After the mobile host has switched to a new subnet, the Mobile IP protocol handles changing routing, and the accelerated handover techniques described in Section 5.2 help smooth packet loss during the routing change at the Home Agent. However, the mobile host and the network may require some additional steps before the IP service characteristics at the new router match those at the old. The host may be entitled to particular quality of service classifications for its packets. The new access router may require the host to authenticate with the network, in order to obtain authorization for network access and to arrange accounting. If session keys are in use at the IP layer for authenticating between the access router and mobile host on the link, these may need to be re-established. If header compression is in use on the link, the header compressor must be restarted. Finally, the new access router may offer an IPsec gateway having a topological location that offers more optimal routing than the old.

All of these service characteristics must be re-established when the host moves to the new subnet. This involves identifying those services that need to be re-established and performing the protocol transactions with the appropriate network elements to re-establish the service from scratch. The disadvantage of this approach is that it can be extremely time consuming, especially if multiple service characteristics are involved.

CT presents another alternative. The idea behind CT is to transfer exactly enough context, or state, describing an IP service characteristic in order to re-establish the characteristic on the new access router as if the mobile host had performed protocol transactions from scratch. For example, CT of authorization information from the Network Access Server (NAS) on one access router to another can re-establish the host's right to network access without requiring a lengthy re-authentication phase.

For CT to work, both the sending and receiving router must agree on the format of the context to be transferred. Each individual service characteristic has its own set of context variables that are required to re-initialize the service for a host, so a standardized feature context specification must be provided for each service characteristic. If for some reason CT fails, the host must be informed so that it can re-initialize the service characteristic for the failed transfer.

CT is of particular interest between two subnets belonging to the same network operator, since the likelihood of compatibility between the two routers is higher. CT between routers on two different network operators' networks is more problematic, since there is no guarantee that the two routers will implement the characteristics in compatible ways. Of particular concern are service characteristics, in which network operators may utilize their own proprietary attributes to describe the characteristic, such as with network access authentication, or where the interpretation of particular context may differ, such as with the type of service bits for quality of service. Even if the two operators have a business relationship allowing their users to roam into each others' networks, the existence of proprietary extensions or differing interpretations of various fields and flags might reduce the utility of CT for inter-operator handovers.

5.4   SEAMLESS INTERTECHNOLOGY HANDOVER

The boundary involved in intertechnology handover is not a result of the technical requirements for routing or business considerations, but rather, the boundary exists within the host itself. At its base level, the boundary is between two hardware devices and their software drivers, supporting two different wireless media types. An example is a host with an IEEE 802.11b wireless LAN card and a Bluetooth card. Exactly how intertechnology handover is performed depends on the configuration of the wireless access network.

5.4.1   Multitechnology Subnets

With some wireless technologies, it is possible for both wireless technologies to co-exist on the same IP subnet. This scenario typically involves two local area media types, since wide area media, such as cellular, typically require deep radio access networks that shut out other media and form a natural subnet. One possibility is that the wireless access points act as transparent bridges reflecting the wireless link layer onto the wired link layer. An example is the 802.11 family of wireless LAN protocols. They can support a single IP subnet containing both 802.11a and 802.11b access points. Another possibility is that the wireless link is accessed through an intermediate "Layer 2.5" protocol such PPP. In this case, multiple access technologies that all support PPP can co-exist on one subnet. An example of an intertechnology subnet is Bluetooth, which allows IP access over a serial-like link using PPP, and 802.11b access via PPP over Ethernet.

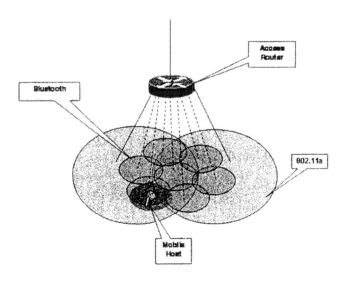

Figure 5.1. Intertechnology Handover on the Same Subnet

With two wireless technologies on the same subnet, Mobile IP need not become involved because there is no change in care of address. However, performing a handover between the two technologies may require a certain amount of ingenuity. Naturally, the first step is for the host to bring up the second interface to the point where solid link layer service exists. If the wired medium type is a shared, multi-access link type such as IEEE 802 series Ethernet and the wireless technology types are also shared access media in same MAC address space, like 802.11a and 802.11b, the router's ARP or neighbor cache must be flushed to remove the mapping between the IP address and the MAC address. Then, the new mapping must be established, or the host can simply wait until the ARP or neighbor cache mapping is re-established when traffic to or from the host transits the router.

If the access between the wired medium type and the wireless access points is through PPP, a PPP connection must be brought up on the new access point, a time consuming process, before the PPP connection to the old access point is torn down. Terminating the PPP connection on the router instead of the access point can save the expense of bringing up the new PPP connection, but the wireless driver on the mobile host may not allow this option.

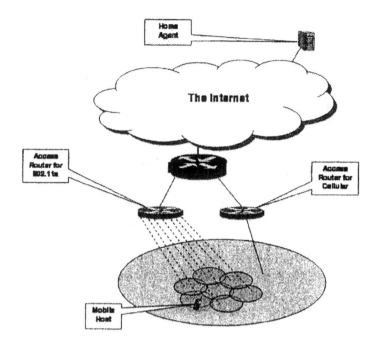

Figure 5.2. Intertechnology Handover Between Subnets

5.4.2 Different Technology Subnets

An easier case is when the different technologies are configured into different subnets. The handover then becomes primarily one moving the Mobile IP care of address between interfaces.

When the mobile host detects a link layer beacon on the second interface, it configures the interface and driver to receive traffic. The IP stack should then be configured onto the interface, but no IP address should be obtained. Instead, the Mobile IP stack solicits a Foreign Agent advertisement (Mobile IPv4) or a router advertisement (Mobile IPv6) from which the host can obtain a care of address. If the host is using co-located care of addresses in Mobile IPv4 or stateful address autoconfiguration in Mobile IPv6, it uses DHCP to obtain a care of address on the new interface.

At this point, the host can send an update to the Home Agent containing the new care of address to change routing to the new interface, but the mobile host should not bring down the old interface just yet. Some packets may be in transit to the old interface, and the host should keep the old interface up until traffic stops arriving on the old interface. At that point, the old interface can be brought down.

If the wireless medium requires PPP, the mobile host faces the same problem as with handover between two access points in a subnet. The Mobile IPv4 specification states that a mobile host connecting via PPP to a wireless network should use the PPP assigned address as a co-located care of address. RFC 2290 also provides a way for a mobile host to indicate to the PPP termination point that a Foreign Agent can be used if available. Since Mobile IPv6 only uses co-located care of addresses, this option is not necessary for Mobile IPv6; the host just uses the PPP assigned address as its care of address. Once the PPP connection is up, the mobile host can send an update to the Home Agent to change the routing to the new care of address.

5.4.3   Network Access Authentication

If different network operators own the two different technology subnets, the mobile host must negotiate a network access authentication exchange prior to changing the Home Agent routing. Unlike the case of a handover between two subnets owned by the same network operator, the probability is low that the access routers will be able to exchange authentication context between them, and thus avoid a delay-producing re-initialization from scratch. The authentication exchange is typically run over a link layer specific protocol, for example, 802.1x for 802.11a and 802.11b, or PPP. Cellular systems also have specific authentication protocols, such as that used for GSM SIM authentication. In any case, the handover is not likely to be seamless from the user's perspective if the user must re-authenticate from scratch in order to enter the new network; in particular, real time applications are likely to experience much delay and transferring them on handover to a new network will likely require some kind of verbal negotiation on the part of the user with the other party to the real time transaction, so that the delay is acknowledged by both parties.

5.5   SEAMLESS DORMANT MODE LOCATION MANAGEMENT

Most mobile hosts have the ability to enter a power saving, or dormant, mode in which power use is reduced by doing no wireless transmission and only sporadic and limited reception. Cellular telephony networks additionally support locating hosts in dormant mode with paging. Paging allows the circuit-switched telephone network to avoid allocating and moving costly dedicated circuits to cellular devices that currently have no active call.

In cellular networks with paging, the wireless access network is divided up into paging areas. A cellular device in dormant mode monitors a beacon that contains a paging area identifier for the paging area in which it is located. When the device moves from one paging area to another, it registers the new location with the network. If a call for the cellular device comes in, all cells in the paging area where the device is located are contacted to send out a paging message on a special channel to locate the device. The device responds by activating and establishing its location more precisely, so a circuit can be allocated for the call. Since cellular devices usually don't have a session active, dormant mode location management saves much power on the device, and reduces the amount of network traffic associated with location management of dormant mode devices. In effect, paging trades a reduced location update cost for an increased paging cost, under the assumption that movement from cell to cell is more frequent than an incoming call.

5.5.1   IP Paging

The success of paging for reducing power consumption in cellular devices and for reducing location update signaling in the wireless access network has suggested that paging might be an appropriate technology to apply to wireless IP networks. The architecture proposed for IP paging introduces a Paging Agent in the foreign network to act as a proxy for the mobile host while it is in dormant mode. In designs based on Mobile IPv4, the Foreign Agent doubles as the Paging Agent. In designs based on Mobile IPv6, a new network element is introduced to support paging.

Rather than aggregating cells, as in cellular networks, paging areas in wireless IP networks aggregate subnets. In most studies, the assumed cell to subnet mapping is 1:1, i.e. one cell per subnet, so the paging area actually serves to aggregate cells too. If the wireless link technology does not support paging, the router advertisement beacon multicast by the access router or Foreign Agent contains a paging area identifier. If the wireless link technology does support paging, the link layer cell beacon being monitored by the mobile host contains the paging area identifier.

When the mobile host detects that it has moved from one paging area to another, it sends a location update message to the Paging Agent. In Mobile IPv6, the mobile host must additionally send a binding update to the Home Agent, in order to transfer its care of address handling from one Paging Agent to another. In Mobile IPv4, the Foreign Agents acting as Paging Agents arrange for the care of address update to happen among themselves.

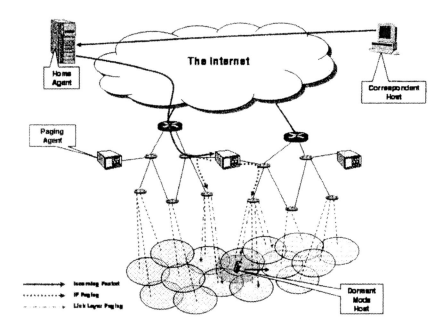

Figure 5.3. IP Paging

When a packet for the dormant mode mobile host arrives, the Home Agent routes the packet to the Paging Agent that is handling the mobile host's care of address. The Paging Agent sends an IP paging message to all access routers in subnets contained within the paging area, and the access routers either initiate a wireless link page, if the link technology supports paging, or multicast the IP paging message to the mobile hosts in the subnet. The dormant mode mobile host activates and responds to the page. The Paging Agent delivers the first packet to the mobile host. The host then changes its care of address binding at the Home Agent so that the care of address is in the mobile host's current subnet, and the Home Agent starts routing packets to the host.

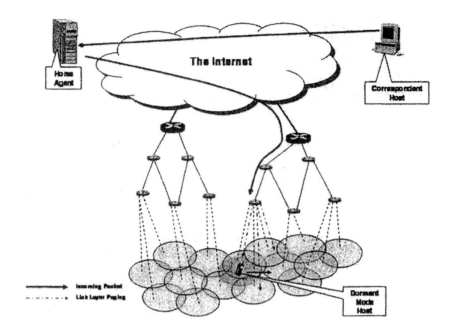

Figure 5.4. Mobile IPv6 Dormant Mode Location Management

5.5.2   Mobile IP for Dormant Mode Management

The mapping between the Paging Agent and Foreign Agent is natural for Mobile IPv4, but the introduction of a Paging Agent into Mobile IPv6 networks is not. Mobile IPv6 was specifically designed to remove the Foreign Agent. A Paging Agent, like the Foreign Agent, is a stateful, single point of failure in the foreign network. Mitigating the effects of single point of failure network elements usually requires replicated hardware, an expensive proposition, so introducing paging into Mobile IPv6 could result in substantially increasing the cost of deployment.

Most of the studies on IP paging have looked at network deployment scenarios based on Mobile IPv4, where a subnet is mapped into a single cell, and paging areas to aggregate subnet-cells into larger groups. These scenarios exactly match the cellular systems, where active mode location management is on a per cell basis and paging areas aggregate cells to cover larger geographical areas, thereby reducing location management signaling. However, in cdma2000, where Mobile IPv4 is used for mobility management, a different mapping between cells, subnets and paging areas is used. The cdma2000 IS695 packet data support system allows the operator to deploy multiple cells per subnet, and confines the paging area to a single subnet.

For IPv4, the single cell per subnet scenario makes sense if mobile hosts are assigned globally routable care of addresses. The 32-bit address length for IPv4 addresses limits the number of hosts with globally routable addresses that can be supported in a single subnet, since only that part of the 32-bit address not covered by the subnet mask is available for identifying hosts. The network operator has the choice of either using nonglobally routable ("net 10") addresses and Network Address Translation (NAT), thereby increasing the number of hosts that can be supported in a subnet, or limiting the geographical extent of the subnet by restricting it to single cell, thereby limiting the number of hosts that can use the subnet to those that will fit into a single cell.

None of these considerations apply to Mobile IPv6, however. Mobile IPv6 addresses have a fixed, 64-bit host identifier and a fixed, 64-bit subnet prefix. Theoretically, Mobile IPv6 could support 2^{64} hosts per subnet, far more than the total number of cellular hosts on the planet today. Broad geographical coverage at a modest location update cost can equally be achieved in Mobile IPv6 by having multiple cells per subnet, and using Mobile IPv6 for dormant mode location update.

To achieve best, the cell's link layer beacon should contain the subnet prefix in addition to the link layer paging area identifier, if the link layer supports paging. The link layer software uses the paging area identifier to determine when the mobile host has moved from one paging area to another. The link layer activates the Mobile IPv6 stack when a paging area transition occurs, supplying it with the subnet prefix necessary to construct a new care of address.

If a paging area transition occurs, the Mobile IPv6 stack causes a binding update to be sent to the Home Agent with the mobile host's new care of address in the new subnet. This avoids requiring the mobile host to solicit for a router advertisement when it moves between subnets, reducing power consumption. In addition, duplicate address detection or stateful address configuration using DHCP must be avoided; otherwise, the mobile host must engage in extensive signaling with the network before it can drop back into dormant mode, wasting power.

If the link layer does not support paging but does support dormant mode, the multicast router advertisement for the subnet serves as the beacon. The host must maintain an active Mobile IPv6 stack and monitor the router advertisements to determine when it has moved between subnets, perhaps in addition to a link layer beacon. This case results in a higher energy cost than if paging support is available from the link layer.

When the Home Agent receives the binding update from the dormant mode host, it changes the binding exactly as for an active mode host. Any incoming packets are tunneled to the access router where the binding is located. Upon arrival of a packet for a dormant mode host, the access router may use link layer paging to wake up the dormant mode host, or it may use simple IPv6 neighbor discovery, if paging is not supported.

For dormant mode location management, the cellular network is not a particularly good analogy for a Mobile IP network. Because cellular devices require a dedicated, active circuit, using the same location management protocol in active and dormant mode would require the dormant mode cellular device to drag around an active circuit switched connection, even though the connection is not used. In contrast, IP networks are connectionless by nature, so there is no cost to using the same location management protocol. By using Mobile IPv6 for dormant mode location management, the expense and management complexity of deploying different location management protocols for dormant and active mode mobile hosts is avoided.

5.6   AN INTERNET WITHOUT BOUNDARIES

The vision of the wireless Internet is to provide the user with IP service anytime, anywhere, and to allow services defined on top of IP to be accessed while the user is moving. From the user's viewpoint, the effect is of an Internet without boundaries. As a practical matter, however, boundaries do exist in the Internet, and they complicate implementing the vision. Technologies under development today help reduce packet loss during subnet transitions, move IP service characteristics from one subnet to another, facilitate handover between interfaces supporting different wireless media, and allow location tracking of hosts in dormant mode. As wireless IP network deployments become more common, additional boundary transition problems, such as moving from one ISP to another, are likely to pop up. Solving these problems in a way that provides the user seamless mobility is a major goal of the wireless Internet.

REFERENCES

Bluetooth Special Interest Group, "Specification of the Bluetooth System," Version 1.0B, V1 and 2, http://www.bluetooth.com, December, 1999.

Conta, A., "Extensions to IPv6 Neighbor Discovery for Inverse Discovery Specification," RFC 3122, June, 2001.

Dommety, G., et. al.., "Fast Handovers for Mobile IPv6," IETF draft, work in progress.

Droms, R., "Dynamic Host Configuration Protocol," RFC 2131, March, 1997.

Droms, R., et. al., "Dynamic Host Configuration Protocol for IPv6 (DHCPv6)," work in progress.

El Malki, K., et. al., "Low Latency Handoffs in Mobile IPv4," work in progress.

Finlayson, R., et. al., "Reverse Address Resolution Protocol," STD 38, June, 1984.

Funato, D., et. al., "Geographically Adjacent Access Router Discovery Protocol," Internet draft, work in progress.

Guttman, et. al., " Service Location Protocol, Version 2," RFC 2608, June, 1999.

Heine, G., GSM Networks: Protocols, Terminology, and Implementation, Artech House Publishers, MA, 1998.

IEEE, "Port Based Network Access Control," IEEE Std. 802.1x, 2001.

IEEE, "Wireless LAN Medium Access Control (MAC) and Physical Layer (PHY) specifications," IEEE Std. 802.11, 1999.

IEEE, "Wireless LAN Medium Access Control (MAC) and Physical Layer (PHY) specifications—Amendment 1: High-speed Physical Layer in the 5 GHz band," IEEE Std. 802.11a, 1999.

IEEE, "Wireless LAN Medium Access Control (MAC) and Physical Layer (PHY) specifications—Amendment 2: Higher-speed Physical Layer (PHY) extension in the 2.4 GHz band—Corrigendum1, IEEE Std. 802.11b, 2001.

Johnson, D., Perkins, C., and Arkko, J., "Mobility Support in IPv6," work in progress.

Kempf, J., "Dormant Mode Host Alerting ("IP Paging") Problem Statement," RFC 3132, June, 2001.

Kempf, J., "Problem Description: Reasons For Performing Context Transfers Between Nodes in an IP Access Network," RFC 3374, September, 2002.

Kempf, J., et. al., "Requirements and Functional Architecture for an IP Host Alerting Protocol," RFC 3154, August, 2001.

Kenward, G., "General Requirements for Context Transfer," work in progress.

Krishnamurthi, G., ed., "Requirements for CAR Discovery Protocols," Internet draft, work in progress.

Mamakos, L., et. al., "A Method for Transmitting PPP Over Ethernet (PPPoE)," RFC 2516, Feb., 1999.

Michael D. Gallagher and Randall A. Snyder, Mobile Telecommunications Networking with IS-41, McGraw-Hill, NY, 1997.

Narten, T., Nordmark, E., and Simpson, W., " Neighbor Discovery for IP Version 6 (IPv6)," RFC 2461, December, 1998.

Omae, K., et. al., " Performance Evaluation of Hierarchical Mobile IPv6 with Buffering and Fast Handover," submitted to Sigcomm 2003.

Perkins, C., "IP Mobility Support for IPv4," RFC 3220, January, 2002.

Perkins, C., and Wang, K., "Optimized Smooth Handoffs in Mobile IP," Proceedings of the IEEE Symposium on Systems and Communications, Red 340--346. IEEE Computer Society Press, June 1999.

Ramjee, R., et. al., "IP Paging Service for Mobile Hosts," Proceedings of ACM SIGMOBILE, July 2001.

Seshan, S., Balakrishnan, H., and Katz, R., "Handoffs in Cellular Wireless Networks: The Daedalus Implementation and Experience," Kluwer International Journal of Wireless Communication Systems, 1996.

Simpson, W., "The Point to Point Protocol (PPP)," STD 51, July, 1994.

Solomon, J., and Glass, S., " Mobile-IPv4 Configuration Option for PPP IPCP," RFC 2290, Feburary, 1998.

Thomson, S., and Narten, T., "IPv6 Stateless Address Autoconfiguration," RFC 2462, December, 1998.

Trossen, D., Krishnamurthi, D., Chaskar, H., and Kempf, J., "Issues in candidate access router discovery," Internet draft, work in progress.

Zhang, X., Castellanos, J., and Campbell, A., "P-MIP: Paging Extensions for Mobile IP," ACM Mobile Networks and Applications, 7(2), March, 2002.

Chapter 6

IP MOBILITY PROTOCOLS FOR WIRELESS INTERNET

Sajal K. Das, Nilanjan Banerjee and Wei Wu; Smitha Ganeshan and Jogen Pathak
The University of Texas at Arlington

Smitha Ganeshan and Jogen Pathak
Cyneta Networks, Inc

6.1   INTRODUCTION

Research in mobile communications has gained a lot of importance since the rapid growth of wireless networks and portable devices. The proliferation of Internet in every aspect of life has urged the service providers to provide seamless user mobility. While the Second-Generation (2G) wireless system has brought mobile telephony, the Third-Generation (3G) is expected to bring in high-speed data and multimedia communication to the mobile users. 3G not only promises much higher data rate than 2G technology, but also opens up the avenue of providing a rich set of applications like location-based services, multimedia messaging services and customized information. However, the evolution of 3G from 2G requires the change of core network from circuit to packet switching mode and building a core network that is independent of access technology. The Internet Protocol (IP) [1] has been the natural choice as a packet-switched protocol because of the increasing dominance. Another important reason for adoption of IP is because it is a network layer protocol that can be used with any kind of access network technology.

The primary step towards the evolution of 3G taken by Global System for Mobile Communications (GSM) Standards or Association [2] and operators is the choice of General Packet Radio Service (GPRS). As shown in Figure 6.1, GPRS [3, 4] introduces two IP components in the cellular networks, the Serving GPRS Supporting Node (SGSN) and the Gateway GPRS Supporting Node (GGSN). As the word Gateway in its name suggests, the GGSN acts as a gateway between the GPRS network and Public Data Networks such as IP and X.25. GGSNs also connect to other GPRS networks to facilitate GPRS roaming. The Serving GPRS Support Node (SGSN) provides packet routing to and from the SGSN service area for all users.

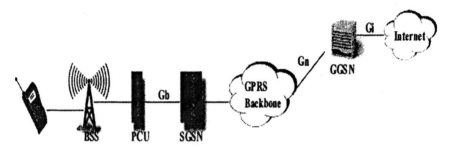

Fig. 6.1. GPRS Architecture

GPRS promises data rates from 56 up to 172 Kbps and continuous connection to the Internet for mobile phone and computer users. The higher data rates will allow users to take part in videoconferences and interact with multimedia Web sites and similar applications using mobile handheld devices as well as notebook computers. The next step has been to change the modulation scheme to increase the data rate. This system is called Enhanced Data Rate for GSM Evolution, or EDGE [5]. EDGE provides data rates in the range of 384 kbps or above. The advent of Wideband CDMA (WCDMA) [6] promises further increase in data rate. WCDMA is the technology used in Universal Mobile Telecommunication System (UMTS) [7], and with data rates upto 2Mbits and it has the capacity to handle bandwidth-intensive applications such as video, data and image transmission necessary for mobile Internet services. In fact, EDGE and WCDMA would exist as complementary technologies. In the USA, IS-136 [8] will evolve ultimately to EDGE. On the other hand cdmaOne would evolve to 3G through two phases, cdma2000 1xRTT (or IS-95B) and cdma2000 3xRTT [9]. cdma2000 1xRTT provides data at a speed of 153 kbps.

In addition to the above cellular technologies, there would be complementary technologies existing to provide higher data rates to the mobile users. The Wireless Local Area Network (WLAN) has become an attractive way for the public wireless broadband access since it provides superior bandwidth compared to the existing cellular access networks. For example, the IEEE 802.11b "Wi-Fi" WLAN standard [10] supports a maximum throughput of 11 Mbps, while the data rate of current GPRS networks is only up to 172 kbps. However, WLAN can only act as a complement to cellular networks because users need coverage while on the move, and WLAN cannot cover all the areas including rural areas. For example, the 802.11a [10] based WLAN product from Proxim has a range of only 100 meters. On the other hand, mobile operators can extend their services for WLAN access. From the point of view of the operators and service providers, the change to the current well-established cellular infrastructure should be kept as minimum and seamless as possible. Users, on the other hand should be able to operate seamlessly between the two complementary access technologies. Also, it should be easy for the existing customers to upgrade and subscribe to new services. Generally, all the current mobile wireless Internet services are applicable to WLAN. Some of the most important classes of existing and emerging end user services are:

- E-mail (including large size of attachment)
- Web browsing (ticket booking, news and weather report, train schedule, city maps)
- Messaging
- Multimedia streaming (online music and movie)
- Rich call (including pictorial communication)
- M-commerce transactions (online banking)

IP is the common network protocol used in both WLAN and GPRS. But IP was meant for fixed hosts. Attempts have been made to modify popular IP to support user mobility. The Internet Engineering Task Force (IETF) has proposed Mobile IP (MIP) [11] to accommodate this. Mobile IP may not be the best solution for real-time multimedia applications that require some definite Quality-of-Service (QoS) guarantees. IP Mobility can be broadly classified into inter-domain and intra-domain mobility. When the movement occurs between different IP domains, it is referred to as *inter-domain* or macro-mobility. When the Mobile Terminal (MT) moves within an access network typically under one IP domain it is termed as *intra-domain* or micro-mobility. To provide a complete mobility solution some kind of integration between these two types of protocols is necessary to establish synergy between the two even if they work in their own domain satisfying their objectives.

6.1.1   IP Mobility

Mobile telephony and the Internet are the fastest growing business in telecommunications market. The employment of the Internet technology, with its novel mobility and security extensions, can make several wireless networks co-exist. It will also be possible to deploy a common IP backbone to interconnect heterogeneous wireless IP networks and to rely on Mobile IP to manage user mobility among them. The development of Administration, Authentication and Authorization (AAA) [12] infrastructure over the Internet is required to accommodate user mobility and create further enhancements to Mobile IP to gain roaming capabilities between different administrative domains. Both cellular operators and Internet Service providers (ISPs), together with the relevant standardization bodies (like 3GPP and IETF) are eagerly looking at the possibility of employing IP and its mobility and security extensions even within a specific wireless network. Several architectural enhancements can be identified to end up with more IP oriented solutions.

IP-Mobility protocols have been proposed mainly for a MT, roaming from one wireless access network to another, so that they can make use of the standard legacy IP mechanism. The access networks are implicitly assumed to be under the same administrative domain. It is also assumed that each MT has a home network, from which it obtains its static IP address. MT visits the foreign network while roaming. Some of the most important issues related to the design of IP-Mobility protocols are i) Handover management, ii) Address management, iii) Packet forwarding and iv) Support for idle hosts, etc.

Mobile IP was proposed for nomadic users and the goal was to enable an MT retain its IP address while moving from one domain to another. Mobile IP provided application layer transparency to the user but was found to have some serious drawbacks when route optimality, service quality, assurance and infrastructure burden were considered. Due to the huge potential of Wireless Internet a lot of research is going on in the area of supporting inter-domain and intra-domain mobility across the access networks. Proposals such as Handoff Aware Wireless Access Infrastructure (HAWAII) [13,14], Cellular IP [15,16], Intra-Domain Mobility Management Protocol (IDMP) [17,18,19,20] and Terminal Independent Mobile IP (TIMIP) [21,22] are products of research in this direction. With the wide deployment and popularity of DHCP (Dynamic Host Control Protocol) [23], POP3 (Post Office Protocol) [24] or IMAP (Internet Message Exchange Protocol) [25] mail services, roaming user profile support for web browsers through LDAP (Lightweight Directory Access Protocol) [26], application layer transparency is also losing its significance. Micro-mobility protocols are required for emerging multimedia applications like IP Telephony, audio and videoconferences that have more stringent requirements for IP mobility support.

The remainder of the chapter is organized as follows. Section 2 describes the Mobile IP, a representative macro-mobility protocol. It also covers the extensions like Regional Registration, proposed to reduce the inherent latency and signaling overhead. In Section 3, we discuss the origin and classification of micro-mobility protocols. We also present some micro-mobility protocols representing this classification. Section 4 provides a qualitative comparison of the micro-mobility protocols, described in Section 3. The comparison is based on some performance parameters like handoff latency, path update and packet forwarding, address management, etc. Finally, Section 5 concludes the chapter.

6.2   MACRO-MOBILITY PROTOCOLS

Macro-mobility refers to user mobility that is not frequent and also spans over considerable space. Mobile IP has been the most obvious and popular choice for managing macro-mobility. Any user mobility can be of two types Discrete Mobility and Continuous Mobility. In discrete mobility, the user breaks the connection before making another or a new connection. In continuous mode, the user maintains the connection while on the move. Mobile IP suits most to support discrete mobility.

6.2.1   Mobile IP (MIP)

Mobile IP or MIP was proposed with the objective of supporting mobile user with application transparency and possibility of seamless roaming. This was done by modifying the legacy Internet Protocol itself. Figure 6.2 illustrates routing of datagrams to and from a mobile node, when it is visiting a foreign network. Typically, a mobile node is a host or router that changes its point of attachment from one network to another while keeping its IP address constant. The router at the home domain of a mobile node keeps all the information on the mobile node and forwards the datagram sent to it to any other visiting foreign network. The foreign agent that is the router at the visiting network cooperates with the home agent to forward datagrams to any visiting mobile node.

MIP consists of the following important elements i) mobile agent discovery, ii) registration with home agent, iii) delivery of datagrams using tunneling. The mobility agents make themselves known by sending advertisement messages. The mobile nodes on receiving the agent advertisements get to know whether it is in the home network or in a foreign network. When the mobile node moves to a foreign network, it obtains a *Care-of-Address* (CoA) on the foreign network either by listening to the advertisement from the foreign agent or by contacting the DHCP or Point-to-Point Protocol (PPP) [27]. The mobile node then registers its new CoA with the home agent. The datagrams sent by the correspondent host, destined for the mobile node are intercepted by the home agent and forwarded to the mobile node after encapsulation using a tunnel, with the endpoint either at the foreign agent or at the mobile node itself. The encapsulated datagrams are decapsulated and delivered to the mobile node.

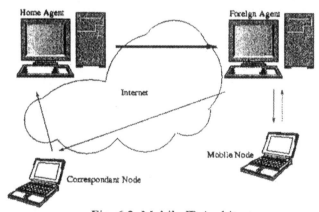

Fig. 6.2. Mobile IP Architecture

MIP requires no protocol changes to existing fixed hosts and routers. It is also not wise to assume that the massive installed base of IPv4 hosts and routers can be upgraded to support mobility functions. Thus, it is required that MIP implementation be limited only to mobile nodes themselves and the few nodes that provide special routing functions on their behalf. Mobile computing implies new security threats and hence the final requirement of MIP is to address these threats. MIP is specifically designed to prevent denial-of-service attacks, which would be possible if there were no authentication of messages by which a mobile node reports its current location.

There are various design goals yet to be met by MIP requiring the transmission of routing updates between the various nodes in the network.

- In order to make MIP suitable for use over a wide range of wireless links, one of the design goals was to make the size and frequency of these updates as small as possible.
- Another design goal is to make MIP as simple as possible to implement mobile node software. This increases the number of nodes that can potentially make use of it like memory and process-constrained devices.
- Yet, another goal is to avoid solutions which require mobile nodes to use multiple addresses, or which require large pool of addresses to be made available for use by mobile nodes.

For WLAN users, MIP being a standard solution is used as the default protocol to handle discrete mobility. For continuous mobility, MIP can be used depending on the distance of the home agent from the foreign agent and the handoff latency. In other words, if handoff latency is large, GPRS macro-mobility protocol can be used to reduce latency [28].

The scalability of the global Internet depends on network-specific routing as opposed to host-specific routing. This requires all the nodes to be connected to the same link to share a common network prefix portion of their IP address. If a node were to move from one link to another, then the network prefix of its IP address would no longer be equal to the network prefix assigned to its current link. Currently available solutions like host-specific routing and changing the node's IP address are not acceptable solutions for node mobility in the global Internet. However, Mobile IP is a solution that is scalable, robust, secure and allows nodes to maintain all ongoing communication while changing links unlike host-specific routing and changing the node's IP address. MIP provides a mechanism of routing IP packets to mobile nodes that may be connected to any link while using their permanent IP address.

Mobile IP's fundamental assumption is that unicast packets, those destined a particular recipient, are routed without regard to their IP Source Address. MIP assumes that the unicast packets are routed based only upon the IP destination address and typically only the network-prefix portion of the address. MIP does not consider the routing problems or the dynamic routing protocols that are being employed in the Internet.

MIP is designed for extensibility. All of the protocol messages defined by Mobile IP consist of small, fixed length portion followed by one or more extensions. The extensions allow virtually any useful information to be carried by the existing protocols. This provides a framework for people to innovate and provide new interesting applications for Mobile IP that its designers might not have envisioned.

Mobile IP works fairly well, except when the distance between the visited network and the home network of the mobile node is large, the signaling delay for these registrations   may be long. Regional Registration [29] has been proposed to solve this problem. The number of signaling messages to the home network, and the signaling delay, when a mobile node moves from one foreign agent to another, are reduced by regional registration, within the same visited domain. This Regional Registration based proposal employs a hierarchy of foreign agents to locally handle Mobile IP registration. The architecture is shown in Figure 6.3. When a mobile node first arrives at a foreign or visited domain, it performs a registration with its home agent. At this registration, the home network generates   a registration key for the mobile node. This   registration key is distributed to the mobile node and to the visited domain, and can be used for authentication of regional registrations.

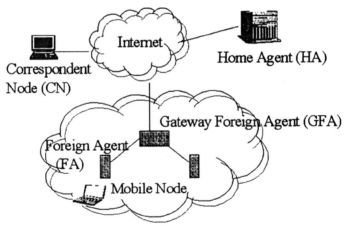

Fig. 6.3. Mobile IP Regional Registration

During a home registration, the home agent registers the care-of address of the mobile node. When the visited domain supports regional tunnel management, the CoA that is registered at the home agent is the publicly routable address of a Gateway Foreign Agent (GFA). This CoA does not change when the mobile node changes foreign agent under the same GFA. When changing GFA, a mobile node performs a home registration; when changing foreign agent under the same GFA, the mobile node may perform a regional registration within the visited domain. This helps in saving signaling cost and reducing handoff latency. Although, the proposed regional registration protocol supports one level of foreign agent hierarchy beneath the GFA, the protocol may be utilized to support several levels of hierarchy as well.

6.2.2   Mobile IPv6

To support mobile devices, IETF standardized a version of Mobile IP for IPv6 [30] known as Mobile IPv6 (MIPv6) [31]. It allows an IPv6 host to leave its home subnet while transparently maintaining all of its present connections and remaining reachable to the rest of the Internet. The mobility management in MIPv6 is very much similar to MIP in IPv4, and also the route optimisation is incorporated as an integral part. All new messages used in MIPv6 are defined as IPv6 Destination Options:

- *Binding Update*: Used by a mobile node to inform its Home Agent or any correspondent node about its current CoA.
- *Binding Acknowledgement*: This is used to acknowledge the receipt of a Binding Update message.
- *Binding Request*: Any node that requests a mobile node to send a Binding Update with the current CoA uses this option.
- *Home Address*: This option is used in a packet sent by a mobile node to inform the recipient of the mobile node's home address.

The data structures that need to be maintained for the operation of MIPv6 are

- *Binding Cache*: Every IPv6 has a binding cache that holds the binding for other nodes. When the packet is sent, the cache is searched for an entry corresponding to the receiver. If an entry is found, the packet is sent to the CoA directly.
- *Binding Update List*: Every mobile has a Binding Update List used to store information about each binding update sent by this mobile node for which the lifetime has not expired yet.
- *Home Agent List*: For each home link a node serves as Home Agent, it generates a list that contains information about all other

home agents on the link. This information is used by the Dynamic Home Agent discovery mechanism.

The packet forwarding in MIPv6 is very much similar to that in MIPv4. To avoid triangular routing, route optimization technique has been incorporated in MIPv6. This is done by sending Binding Updates to any corresponding node, which cache the current CoA and send packets directly to the mobile node instead of going through the Home Agent.

The Hierarchical Mobile IPv6 (HMIPv6) [32], an extension of MIPv6, uses the IPv6 address space and neighbour discovery mechanisms to support flexible, scalable and robust mobility management. A new entity called Mobility Anchor Point (MAP) is introduced to assist the MIP handoff and reduce the signaling overhead for mobile nodes in the local domain. A MAP can be located in any level in a hierarchical network of routers. However, unlike Foreign Agents as in IPv4, a MAP is not required in each subnet. The Access Routers (ARs) are the direct access points for the mobile nodes to the outside networks. A Regional Care-of-Address (RCoA) is an address obtained by the MN from the visited domain. It may be assigned to one of the MAP's interfaces   (in the Extended mode of operation) or auto-configured by the mobile node when receiving the MAP option (Basic mode of operation).

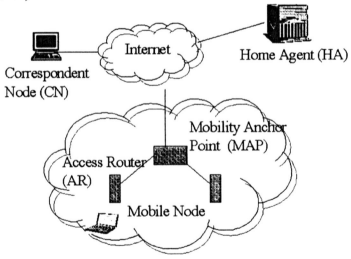

Fig 6.4. Hierarchical Mobile IPv6

6.3 MICRO-MOBILITY PROTOCOLS

As mentioned earlier, MIP was proposed only to support mobility in IP domain and provision of QoS was not considered. Mobile IP displays poor performance when the latency of handover during mobility, signaling cost and transmission latency are considered. It generates significant signaling traffic in the core network even for local mobility. For a mobile terminal, that frequently changes location, protocols such as MIP would generate considerable signaling traffic and would cause severe degradation in the QoS pledged to the user. Therefore, we need some kind of localized mobility protocol generally termed as *micro-mobility* protocols. These protocols operate in a restricted administrative domain and provide the mobile terminals within that administrative domain the connection to core network, while keeping signaling cost, packet loss, and handover latency as low as possible. There are several micro-mobility protocols proposed, such as HAWAII, Cellular IP, TIMIP and IDMP. These protocols are broadly classified into three classes based on their functionality and underlying architectures, as shown in Figure 6.5.

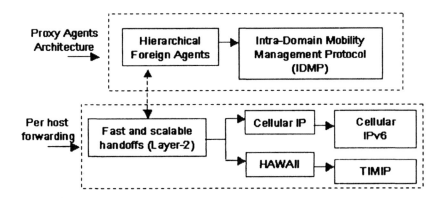

Fig 6.5. Classification of Micro-mobility Protocols

6.3.1 Handoff Aware Wireless Access Internet Infrastructure (HAWAII)

The Handoff-Aware Wireless Access Internet Infrastructure (HAWAII) was proposed to resolve the QoS and efficiency issues of Mobile IP. HAWAII operates above IP and hence the routers within a HAWAII domain act as a classical IP router with some additional HAWAII features. As shown in Figure 6.6, this protocol requires the intermediate routers in the access network to be organized in a tree-like structure with one node located at the root of the tree, performing the function of a gateway. The MT retains the co-located CoA assigned to it as long as it is within the domain. This makes the integration of HAWAII with Resource Reservation protocols, like RSVP [33], feasible. The routers in HAWAII network maintain mobility-specific location information on a per host basis in legacy routing tables, and use generic IP routing protocols to forward packets to the MT. HAWAII proposes four different path set up schemes that control handoff procedure between two Access Points (AP). The choice of path set-up scheme depends on the operator's priorities between handoff latency, packet loss and packet ordering.

HAWAII uses specialized path set-up schemes that install host based forwarding entries in specific routers to support intra-domain micro-mobility and defaults to using Mobile IP for inter-domain macro-mobility. These path set-up schemes deliver good performance by reducing mobility related disruption to user applications, and by operating locally, reduce the number of mobility related updates. Also, in HAWAII, mobile hosts retain their network address while moving within the domain, simplifying QoS support.

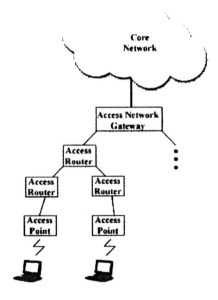

Fig 6.6. HAWAII Architecture

HAWAII operates entirely within the administrative domain of the wireless access network. In order to keep HAWAII transparent to mobile hosts, the mobile host runs the standard MIP protocol with Network Access Identifier (NAI), route optimization and challenge/response extensions. To reduce the frequency of updates to the HA and avoid high latency and disruption during handoff, in HAWAII, the processing and generation of MIP registration messages are split into two parts: between the mobile host and base station and between the base station and the HA. Note that this separation is required for any approach that desires to reduce updates to the HA. For example, similar separation at the foreign agent is proposed in the MIP Regionalized Tunnel Management approach [34].

The major design goals for HAWAII would entail achieving good performance by reducing update traffic to Home Agent and corresponding hosts, avoiding triangular routing where possible, and limiting disruption to user traffic. Another important feature of the design goal would be to provide intrinsic support for QoS in the mobility management solution, including per flow QoS and limiting the number of reservations that must be re-established when hosts move. HAWAII must be enhanced to provide reliability and good fault tolerance. These features must be included to enhance the robustness of mobility support. In HAWAII, the support for QoS is straightforward since a mobile host's address remains unchanged as long as the user remains within a domain.

HAWAII assigns a unique address for each mobile host that is retained as long as the mobile host remains within its current domain. In this context, maintaining end-to-end connectivity to the mobile host requires special techniques for managing user mobility. HAWAII uses path set-up messages to establish and update host-based routing entries for the mobile hosts in selective routers in the domain so that packets arriving at the domain root router can reach the mobile host. The choice of when, how and which routers are updated constitutes a particular path set-up scheme. In HAWAII, each host potentially needs two IP addresses: one to operate in its home domain, and a second when it moves outside its home domain. The address used by the mobile host in its home domain can be statically or dynamically assigned.

HAWAII handoff procedures are activated only when the mobile host's next hop IP node is changed during the handoff. In actual deployed networks, the mobile host's next hop IP node may or may not be a base station. The mobile host also sends periodic Mobile-IP renewal registrations to the base station. The base stations and routers in turn send HAWAII aggregate refresh messages periodically in a hop-by-hop manner to the routers upstream of the mobile hosts. HAWAII messages are sent to selected routers in the domain, resulting in very little overhead associated with maintaining "soft state". Soft state refers to a state established within routers that needs to be periodically refreshed; which otherwise is removed automatically when a present timer associated with that raises an alarm. The soft state condition within the routers increases the robustness of the protocol to router and link failures.

Reliability of Home Agents is a concern for any approach that is based on Mobile IP. In HAWAII as well as Mobile IP, this failure could be tackled through the configuration and advertisement of backup Home Agents. In HAWAII, in the common case of a mobile host not leaving its "home" domain, there is no HA involved. This greatly reduces HAWAII's vulnerability to HA failure as compared to Mobile IP schemes.

Security is also an important concern in HAWAII. There are two issues with security: user authentication by the DHCP server during address management that occurs during power up and inter-domain moves; and security authentication related to Mobile IP and HAWAII protocol messages. Authentication of HAWAII protocol messages is not a difficult issue since these messages are generated and processed only by nodes within a single administrative domain.

6.3.2   Cellular IP (CIP)

In a Cellular IP (CIP) network, the interconnected Cellular IP nodes route IP packets inside the Cellular IP network and communicate with the mobile hosts via wireless interface. CIP attempts to replace IP. Location management and handoff support are integrated in CIP. This protocol supports paging and a number of handoff techniques. CIP, as shown in Figure 6.7, also requires the intermediate routers, or nodes henceforth, to be organized in a tree-like structure with a Gateway (GW) node at the root of such a structure. Each of the nodes maintains a Routing Cache and some of the routes maintain a Paging Cache. The MT generated route updated packets update the routing cache, which maintains the dynamic path to active MTs. Paging cache stores the information for the idle hosts and thus helps in paging a host before transmitting data. Paging update packets refreshes the paging cache entries. While setting up a new connection with a MT, paging packets are sent to the APs in consultation with the paging cache. The corresponding MT upon receiving the paging packet sends the route update packet to the nodes in the access network to refresh its path from the gateway. The routing update packets move all the way up to the gateway to set up the path to the MT. Maintenance of two caches reduces the signaling overhead for idle hosts. There are two types of handoff defined for CIP, viz. hard handoff and soft handoff. Where hard handoff is the one having more packet loss than the soft handoff. The tradeoff is handoff latency vs. packet loss. CIP also makes use of Layer-2 information regarding access point signal strength in order to predict handoff to decrease the corresponding latency. One disadvantage of CIP is that all packets generated within a CIP domain are routed through the gateway even if the destination is located adjacent to the source.

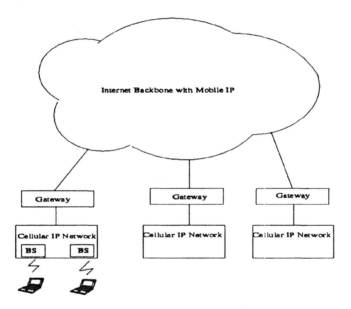

Fig 6.7. Cellular IP Architecture

CIP can interwork with Mobile IP to provide wide area mobility support, that is, mobility between Cellular IP networks. Cellular IP is a protocol that provides mobility and handoff support for frequently moving hosts. It can be used on a local level like in a campus or a metropolitan area network.

CIP assumes that a random Layer 2 protocol covers the air interface. It is applicable to networks ranging in size from Local Area Networks (LANs) to metropolitan area networks. A host connected to Cellular IP network must be able to send IP datagrams to hosts outside the Cellular IP networks.

These IP datagrams arriving at the Cellular IP network has to be delivered to the host with high probability regardless of its actual location. Mobile hosts migrating inside or between Cellular IP networks must be able to retain their own IP address regardless of the location. Hosts inside a CIP network are identified by IP addresses, but these addresses have no location significance. Hosts outside the CIP network must not need any updating or enhancements in order to communicate with hosts inside the Cellular IP network. Nodes sending or receiving datagrams to/from the mobile host must remain unaware of the host's location inside the CIP network.

CIP supports fast handoff, local mobility and paging techniques. Location management and handoff are integrated with routing in cellular IP access networks. It uses mobile originated data packets to reverse path routes. Nodes in cellular IP access network monitor mobile originated packets and maintain a distributed, hop-by-hop location database that is used to route packets to mobile hosts. Customized handoff procedures minimize the loss of downlink packets when a mobile host moves between access points.

6.3.3 Intra-Domain Mobility Management Protocol (IDMP)

Intra-Domain Mobility Management Protocol or IDMP is a stand-alone micro-mobility protocol. It supports intra-domain mobility by using multiple co-located addresses. The main objective of the protocol has been to reduce the handoff latency and signaling overhead for a highly mobile host. Similar to Regional Registration, the protocol localizes the mobility management in wireless access domains. IDMP supports additional mobility features, such as minimally interrupted handoff and paging [35], within the mobility domain for highly mobile users. This separation of intra-domain mobility from inter-domain mobility is intended to allow a common base protocol to coexist with multiple alternatives for global mobility management, including Mobile IP. An architecture called Dynamic Mobility Agent (DMA) [36] has also been recently proposed which uses IDMP as the base mobility management protocol to provide a scalable and robust mobility management framework.

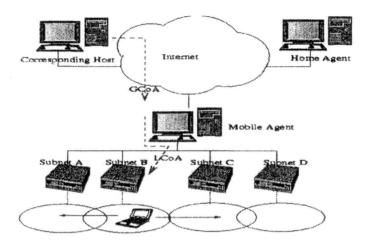

Fig 6.8. IDMP Architecture

IDMP uses a hierarchical structure as shown in Figure 6.8, whose two main components are the Mobility Agent (MA) at the top of the hierarchy and the Subnet Agent (SA) at the bottom. MA serves as the gateway to the Internet. In a foreign domain, an MT obtains two co-located address, the *Local Care-of-Address* (LCoA) and *Global Care-of-Address* (GCoA). LCoA identifies the Mobile Node's attachment to the subnet. Unlike MIPs CoA, the LCoA in IDMP only has local (domain-wide) scope. By updating its Mobility Agent of any changes in the LCoA, the mobile node ensures that packets are correctly forwarded within the domain. As long as the MT is within that domain it retains the Global Care-of-Address (GCoA). GCoA resolves the Mobile Nodes' (MN) current location only up to a domain-level granularity and hence remains unchanged as long as the MN is within a single domain. By issuing global binding updates that contain this GCoA, the MN ensures that packets are routed correctly to its present domain. The Home Agent is updated only when the MT changes administrative domains. This saves a lot of unnecessary location updates and hence reduces subsequent latency. Handoff procedure proposed under IDMP has also made provisions for minimizing the loss of in-flight packets during the transient handoff period. IDMP also provides the support for paging idle hosts using the concept of paging areas. Paging helps in power management of the MT by reducing unnecessary location updates.

Handoff procedure proposed under IDMP has also made provisions for minimizing the loss of in-flight packets during the transient handoff period. The MT sends a message to the MA before performing the handoff. After getting this notice, the MA will multicast the inbound packets destined to the MT to all the neighboring SAs as well as the current serving SA. Each of these candidate SAs buffers the packets during the transient period. After the MT has set up a new connection with one of these SAs, buffered packets will be forwarded to the MT by the new serving SA. The packets buffered in other SAs will be discarded finally. IDMP also provides the support for paging idle hosts using the concept of paging areas. The access domain is organized into several paging areas, each of which forms a multicast group. An idle MT only needs to update its location with MA when crossing the boundary of the paging area. To locate an idle MT, MA will multicast a solicitation message in the paging area, where each SA belonging to this paging group in turn broadcasts in its covered cell. Thus, paging helps in power management of the MT by reducing unnecessary location updates.

6.3.4   Terminal Independent Mobile IP (TIMIP)

All the above IP mobility protocols require the Mobile Terminal to have a mobility-aware protocol stack, since the MT notifies the corresponding AP about the handoff. With a large number of terminals already in the market with legacy protocol stack, it seems a daunting task to change them all. TIMIP was proposed in light of terminal independence. TIMIP achieved this terminal independence by coupling IP layer with Layer-2 handoff mechanisms at the APs with a suitable interface. As shown in Figure 6.4, a TIMIP domain is organized again in a tree-like structure like HAWAII, with the APs forming the terminal nodes of the tree, the Access Routers (ARs), constituting the internal nodes and the gateway called the Access Network Gateway (ANG) constituting the root. TIMIP requires the ANG to store all the information about the MTs visiting the corresponding domain. The authentication of a particular MT is done by an application layer authentication protocol that can again work over legacy protocols.

TIMIP is a mobility protocol that supports IP-mobility of mobility-unaware mobile nodes with legacy IP stacks, while fully interoperating with Mobile IP to provide macromobility across IP subnets. TIMIP does not assume that the mobile nodes have a mobility-aware IP stack.

TIMIP allows mobile nodes with legacy IP stacks to roam within an IP subnet. Macromobility, that is mobility across subnet boundaries, still relies on Mobile IP. The main features of TIMIP can be summarized as below:

- Routing of data packets within a TIMIP domain and routing re-configuration during handoff within a TIMIP domain does not need to reach the Access Network Gateway. This helps in using just the Access Routers located in the shortest path between the sender and the receiver.

- TIMIP does not require changes to the IP protocol stack of mobile nodes, either for micro-mobility or for macro-mobility.

- Refreshing of routing path is performed by data packets sent by the mobile terminals, with signaling being employed only when no traffic is detected at the routers for a certain time interval.

TIMIP is also applicable to larger corporate networks that form one subnet. To provide mobility between subnets (macro-mobility), Mobile IP should be supported in the Access Network Gateway of TIMIP domain. It is assumed that the data link layer of the AP can perform terminal power-up detection to notify the TIMIP layer.

TIMIP relies on MIP to support macromobility. The ANG is the Home Agent for Mobile Nodes whose home network is its TIMIP domain. When a foreign MN supports MIP, the ANG acts as a Foreign Agent for the local TIMIP domain. When the MT supports MIP but belongs to a different domain, the ANG plays the role of a foreign agent. The MT powers on in the same way as legacy MTs do in the TIMIP domain. A mobile terminal could also request for a Router Advertisement message and the AP forwards this to the ANG. After the MT receives the Router Advertisement message, it notifies its HA about the CoA through the ANG. The Home Agent then forwards the incoming packets to the CoA through an IP Tunnel. The ANG de-encapsulates the IP packets and forwards them normally to the MT according to the routing path established in the AR tree. Packets generated by the MT are also routed normally within the TIMIP domain.

6.4 COMPARISON OF MICRO-MOBILITY PROTOCOLS

Our qualitative comparison is based on some decisive parameters affecting the performance because all the micro-mobility protocols are designed with the same mobility scenario. These parameters are

- Path update and packet forwarding schemes
- Handover management
- Support for idle mobile hosts

- Address Management
- Routing Topology
- Security Issues
- Requirements for Mobile Nodes
- Handoff performance

We compared the micro-mobility protocols based on the evaluation framework presented in [37].

6.4.1   Path Update and Packet Forwarding

HAWAII: Packets are forwarded to the MT using dynamically established paths. In order to do this each domain in HAWAII is structured according to a hierarchy of routers and Access Points, forming a tree like structure, with the APs being at the terminal level of the tree. Each domain has a root gateway, called the domain root router. The packets are first forwarded to the domain root router and then forwarded to the MT using the dynamically established path. The mobile terminal establishes the path that triggers MIP registration messages whenever it moves between two APs. The mobile terminal retains the IP address, which it acquires when it enters a domain, as long as it is within the domain.

Cellular IP (CIP): In CIP, the handover between APs is done by using Layer-2 information regarding AP signal strength in order to predict handover. The terminals need to have CIP capability to accomplish this. A CIP domain is similarly composed of a number of CIP nodes in a tree topology, having a MIP gateway as the root node. Each CIP node maintains a   routing cache and a paging cache. The routing caches are used to track the mobile terminals and forward the IP packets destined for them. The IP packet transmitted refreshes the entries in the routing caches. The paging caches are maintained by paging update packets. The paging cache is used to locate idle terminals and hence saves unnecessary location update cost.

IDMP: Unlike host-specific routing used in CIP, HAWAII or TIMIP, IDMP uses the normal IP routing mechanism. The need to maintain the path using explicit route update messages is avoided. However, the mobile node needs to acquire a new LCoA address if it is attached to a new Subnet Agent, and update the Mobile Agent with the new LCoA. In other words, the change of LCoA is only known up to the MA and is transparent to HA, which is only aware of the GCoA. Incoming packets are forwarded to the MA, which then intercepts and tunnels them to the LCoA of the mobile node.

TIMIP: A TIMIP domain is also organized as a logical tree of access routers whose root is the ANG. Layer-2 association of the mobile terminal with the APs takes place here as well. For TIMIP, the APs need to be very sophisticated since they are responsible for sensing the mobile terminals at their wireless interface and sending route update message up in the hierarchy of ARs. The route information thus propagated is maintained in the route tables stored in the intermediate ARs. The routing path is based on soft-state information refreshed by either data packets destined to the MTs or by a special message sent to the ANG by MTs.

Evidently, all the above protocols except IDMP need the intermediate access routers and the access points to be organized in a tree like structure.

6.4.2 Handover Management

HAWAII: The handoff mechanism requires message exchange between the old and the new base stations. The total latency is thus twice the delay between the MT and its previous base station. The forwarding scheme causes the packets to be lost until the update message reaches the old base station. The non-forwarding scheme is faster since the packets are correctly forwarded as soon as the crossover router is aware of the handoff.

CIP: In CIP, the handoff mechanism triggers the MT to generate a packet that is forwarded hop-by-hop across the gateway, which must acknowledge it. Thus the latency involved is twice the average delay between the MT and the gateway. When a semi-soft handoff takes place there is no packet loss, but with hard handoff packets are lost from the moment the MT changes APs to the moment the route update message reaches the crossover base station.

IDMP: In IDMP, a multicasting scheme is used for fast handoff. The Mobility Agent multicasts all the incoming packets to the neighboring SAs for temporary buffering upon receipt of this message. SA with its new connection set up with the mobile node will forward the buffered packets to minimize the packet loss during handoff.

TIMIP: The Handoff mechanism is triggered in TIMIP as well. It also triggers route update message, that moves up the hierarchy of intermediate ARs up to the gateway, which then gets acknowledged down the hierarchy of the ARs. So, in this case the latency is twice the time, required by a message to reach the gateway from the MT. The packets are forwarded to the new location as soon as the crossover router gets to know about the route update.

6.4.3   Support for Idle Mobile Hosts

HAWAII: Soft-state information for the mobile hosts is maintained in the routers. The path updates are refreshed by special update messages.

CIP: CIP distinguishes between in-session and out-of-session mobility. CIP maintains Paging Cache and Routing Cache to handle the mobility. The Paging Cache is updated at a frequency equal to the migration frequency of a mobile host. Paging is done in consultation with the paging cache in charge of the concerned paging area.

IDMP: Paging in IDMP is an extension of multicasting scheme used in handoff. SAs are grouped into paging areas. The MA multicasts a PageSolicitation message to locate the idle mobile node and buffer the incoming packets. Once the mobile node re-registers with the Mobility Agent, it will forward the buffered packets.

TIMIP: The support for idle hosts in TIMIP is similar to that in HAWAII.

The main difference among these protocols is in the paging algorithm. In CIP, specific routers do paging and only these routers would maintain paging information. In HAWAII, the load of paging is dynamically balanced among the routers in the network. Based on the current load in the network, a particular router is designated to perform the paging. Any router in the network can do paging. In IDMP, the topology for multicasting needs restructuring.

6.4.4   Address Management

HAWAII: HAWAII requires an MT to acquire a publicly routable co-located CoAs in a foreign network. This feature requires a large pool of address and IPv4 may pose a threat to this.

CIP: There is no concept of location within a CIP domain allowing the use of any space of unique identifiers. Generally the host IP address is used as the host identifier for an MT. This helps in using the IP packets unchanged in the access network.

IDMP: In IDMP, each mobile node is assigned two distinct CoA. The GCoA is associated with a specialized node called the Mobility Agent, that provides a stable gateway to the Internet; Subnet Agents located at the wireless access edge provide the mobile node with an additional LCoA that could be the private address to handle intra-domain mobility.

TIMIP: TIMIP, like CIP, works with unique host identifiers. Here the MAC addresses of the MTs are used as the unique identifiers. The APs get the MAC addresses of the MTs when the Layer-2 association takes place.

In contrast, HAWAII requires an IP address in the foreign domain. CIP and TIMIP work with any unique host identifier and hence use the IP address pool more judiciously. IDMP requires two IP addresses - one is the public address and the other could either be a public or a private address.

6.4.5   Routing Topology

All the three protocols HAWAII, CIP, and TIMIP use a tree-like architecture within the domain. A dedicated node acts as the gateway to the outside network and as a root to the structure. The nodes closer to the gateway are heavily loaded in comparison to the nodes lower down the hierarchy. HAWAII keeps the paging information in a distributed fashion making it more robust to faults. CIP keeps the paging information in few intermediate routers making their failure vital to the system. Similarly, in TIMIP the route tables are maintained in the ARs and so their failure directly affects the robustness of the protocol. Since all these protocols depend heavily on the gateway node, its failure affects the system severely. On the other hand, IDMP is not limited to any particular topology because it uses normal IP routing.

6.4.6   Security Issues

HAWAII: There are three types of security specified in HAWAII between

- the mobile nodes and the base stations
- the base stations and the home agent
- the mobile nodes and the home agent

The AAA infrastructure is used to distribute three sets of authentication keys to the mobile nodes, base stations and home agents. The base station can generate registration replies to the mobile hosts on behalf of the home agent when the mobile nodes perform intra-domain handover. Also, the base station can send registration requests to the home agent on behalf of the mobile nodes when the mobile nodes perform the inter-domain movement. The Remote Authentication Dial-in User Server/Service (RADIUS) [38] protocol is for user authentication by the DHCP server during address assignment that occurs during power up and inter-domain moves.

CIP: There is a secret network key of arbitrary length known to all CIP nodes. The mobile nodes and other nodes outside the Cellular IP Network do not know this network key. A mobile node needs to register with gateway node before it powers up in the CIP network. After authenticating the entering mobile node, the gateway node concatenates the network key and the IP address of the mobile host and calculates the Personal Identification (PID) of the mobile node as an MD5 Hash, i.e., PID = MD5 (network key, IP address of MH). Then it acquires the public key of the mobile node from a trusted party, encrypts the PID and sends it to the mobile node. The PID is then the shared secret between the mobile node and the Cellular IP network, and used for authenticating packets transferred between the mobile nodes and CIP networks.

IDMP: IDMP uses same replay protection and security associations from Mobile IP for macro-mobility [39]. Security considerations for intra-domain mobility are not specified so far.

TIMIP: The mobile nodes need to register with ANG before powering up in the TIMIP domain. To prevent the entry of a mobile node with a fake MAC and IP address, the AP requests for the signature key from both the mobile node as well as the ANG. The reply contains a 128-bit MD5 message digest calculated with the authentication key of the mobile node, which is only known by the mobile node and ANG. If the mobile node is authenticated, the AP updates the route for the mobile node in the ANG.

6.4.7 Requirements for Mobile Nodes

HAWAII: This protocol requires the mobile nodes to be MIP capable, since it depends on the mobile terminal to trigger the path update by sending MIP registration messages, when it moves between two APs.

CIP: CIP also requires the MTs to implement specific CIP procedures like layer-2 handoff and periodic refreshing of routing and paging cache.

IDMP: IDMP defines its own message, originating from the MT, to convey the current location information among different entities in the IDMP-enabled domain.

TIMIP: TIMIP is independent of MT. In TIMIP, the AP attached to the MT does the path update on it. An authentication program running at the application layer authenticates the MT and hence the TCP/IP stacks of the MT do not require any change for TIMIP to operate.

HAWAII, CIP and IDMP require primitive support in the MTs. Considering a wide range of mobile devices that are currently available or will be present in the future; it is a daunting task to follow any particular standard. So it is preferable to design mobility protocols, which do not depend explicitly on the capability and functionality of the MTs. Hence, TIMIP is better than HAWAII, CIP, or IDMP in this respect.

6.4.8 Handoff Performance

There are two parameters that need to be considered to get an idea of the handoff performance. The first one is handoff latency, which is the interval between the time the MT starts moving and the time it receives the first packet in the new location. Packet loss is a measure of the packets lost during the movement since the MT remains disconnected on the move.

Both Cellular IP and IDMP can perform handoff without packet loss using smooth handoff schemes. However, system bandwidth will be wasted, since incoming packets may be delivered along multiple paths during the transient time. For the hard handoff schemes, the handoff latency is less than that of smooth handoff schemes for most protocols except IDMP. In IDMP, the MT has to acquire the new LCoA and update it with the gateway node for a new connection set-up.

Figure 6.9 represents the model for Handoff performance comparison and Table 6.1 shows the comparison of handoff performance of micro-mobility protocols in terms of handoff latency and packet loss. Typically, the relation, $T_{gw} > T_{old-bs} > T_{cbs} > T_{bs}$ is generally true for any topology.

6.4.9   Summary

For path update and packet forwarding, HAWAII, CIP and TIMIP replace the traditional IP routing protocol and implement their own host specific routing scheme, which incurs some additional overhead in maintaining current path information for each host. IDMP, on the other hand, uses normal IP routing scheme for forwarding packets. Since IDMP does not need an update in path information prior to handoff, the associated latency is

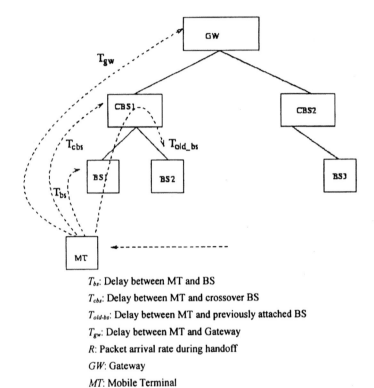

T_{bs}: Delay between MT and BS
T_{cbs}: Delay between MT and crossover BS
$T_{old\text{-}bs}$: Delay between MT and previously attached BS
T_{gw}: Delay between MT and Gateway
R: Packet arrival rate during handoff
GW: Gateway
MT: Mobile Terminal
BS: Base Station

Fig 6-9: Model of Handoff Performance Comparison

Protocol	Handoff Type	Handoff Latency	Packet Loss
HAWAII	Forwarding	$2T_{old\ bs}$	$R \times T_{old\ bs}$
	Non-forwarding	$2T_{cbs}$	$R \times T_{cbs}$
Cellular IP	Semi-soft	$2T_{cbs}$	0
	Hard	$2T_{cbs}$	$R \times T_{cbs}$

IDMP	Multicast	$2T_{gw}$	0
	Hard	$2(T_{gw}+T_{bs})$	$R \times (T_{cbs}+T_{gw})$
TIMIP	Hard	$2T_{cbs}$	$R \times T_{cbs}$

Table 6.1. Comparison of Handoff Performance

less than that of the other protocols. Moreover, IDMP is capable of minimizing the packet loss while doing a fast handoff. In CIP, when an MT wants to do a fast handoff, packet loss is not negligible. Except for TIMIP, all other protocols provide some kind of support for idle hosts in the form of a paging scheme. As far as address management is concerned, only CIP can use its home network address as its identifier in the foreign network and all other protocols have to get a new CoA in the foreign domain. In case of IDMP, the MT acquires two CoAs. Each of HAWAII, CIP and TIMIP needs to organize the access network in a tree like topology with a gateway node, at the root of the tree communicating with the Internet. For IDMP, there is no such requirement and a MT can communicate to the Internet through multiple gateway nodes. HAWAII uses some security extensions of MIP for authentication purposes. CIP has proposed its own security protocol for intra-domain communication. A network key specific to the CIP network is used for securing the control packets such as the route update packets. TIMIP proposes an application layer based authentication protocol that authenticates a MT using a key shared between the MT and the ANG. Security features for IDMP is not as developed as compared to other protocols. All other protocols except TIMIP require some change in the protocol stack of terminals.

6.5 CONCLUSION

We have presented a comprehensive survey of different mobility management protocols proposed for IP-based networks. IP mobility protocols are mainly classified into macro- and micro-mobility protocols. The micro-mobility protocols were developed to improve the performance of Mobile IP based macro-mobility protocols, which suffered from latency, routing and signaling overhead.

The salient features of the macro- and micro-mobility protocols as well as some of the other representative protocols are described in this chapter. A comparative study of different protocols with respect to different performance parameters, like path update and packet forwarding, handoff latency etc., for micro-mobility protocols have been presented. The comparison shows that each micro-mobility protocol has its own advantages and drawbacks.

REFERENCES

[1] ISI, University of Southern California, Internet Protocol, DARPA Internet Program Protocol Specification, RFC 791, Sept. 1981.

[2] *GSM Standards and Association*
http://www.gsmworld.com/about/index.shtml

[3] 3GPP 3G TS 23.060, "Technical Specification Group Services and System Aspects, General Packet Radio Service (GPRS), Service Description, Stage 2," Rel. 1999, v. 3.8.0, June 2001.

[4] R. J. Bates, GPRS: General Packet Radio Service, McGraw-Hill Professional Publishing, Nov. 2001.

[5] T. O. Halonen, J. Romero and J. Melero, GSM, GPRS and EDGE Performance: Evolution Toward 3G/UMTS, John Wiley & Sons, Jun. 2002.

[6] T. Ojanpera and R. Prasad, WCDMA: Towards IP Mobility and Mobile Internet, Artech House, Apr. 2002.

[7] H. Kaaranen, S. Naghian, L. Laitinen, A. Ahtiainen and V. Niemi, UMTS Networks: Architecture, Mobility and Services, John Wiley & Sons, Aug. 2001.

[8] L. J. Harte, A. D. Smith and C. A. Jacobs, IS-136 TDMA Technology, Economics and Services, Artech House, Apr. 2002.

[9] V. K. Garg, IS-95 CDMA and cdma 2000: Cellular/PCS Systems Implementation, Prentice Hall PTR, Dec. 1999.

[10] *IEEE P802.11 The Working Group for Wireless LANs*
http://grouper.ieee.org/groups/802/11/

[11] C. Perkins, IP Mobility Support for IPv4, RFC 3220, Jan. 2002.

[12] D. Mitton, M. St.Johns, S. Barkley, D. Nelson, B. Patil, M. Stevens and B. Wolff, Authentication, Authorization, and Accounting: Protocol Evaluation, RFC 3127, Jun. 2001.

[13] T. La. Porta, R. Ramjee, L. Lee, L. Salgerelli, and S. Thuel, "IP-based access network infrastructure for next-generation wireless data networks" *IEEE Personal Communications*, vol. 7, No. 4, August 2000.

[14] R. Ramjee, T. La Porta, S. Thuel and K. Varadhan, "IP Micro-Mobility Support Using HAWAII'', *draft-ramjee-micro-mobility-hawaii-01.txt*, July 2000, Work in Progress.

[15] C-Y. Wan, A. T. Campbell, A. G. Valko "Design, implementation and Evaluation of Cellular IP", *IEEE Personal Communications*, vol. 7, No. 4, pp. 42-49, August 2000.

[16] A. Campbell, J. Gomez, C-Y. Wan, S. Kim, Z. Turanyi and A. Valko, "Cellular IP'', *draft-ietf-mobileip-cellularip-00.txt*, January 2000, Work in Progress.

[17] A. Misra, S. Das, A.McAuley, A.Dutta, and S.K.Das, ``IDMP: An Intra-Domain Mobility Management Protocol using Mobility Agents", *draft-mobileip-misra-idmp-00.txt*, July 2000, Work in Progress.

[18] S. Das, A. Misra, P. Agrawal, and S. K. Das, "TeleMIP: telecommunications-enhanced mobile IP architecture for fast intradomain mobility", *IEEE Personal Communications*, vol. 7, No. 4, pp.50-58, Aug. 2000.

[19] S. Das, A. McCauley, A. Dutta, A. Misra, K. Chakraborty and S. K. Das, "IDMP: An Intra-Domain Mobility Management Protocol for Next-Generation Wireless Networks", *IEEE Wireless Communications*, vol. 9, No. 3, pp.38-45, June 2002.

[20] A. Misra, S. Das, A. McCauley, A. Dutta and S. K. Das, "Integrating QoS Support in TeleMIP's Mobility Architecture", *Proceedings of IEEE International Conference on Personal Wireless Communications (ICPWC)*, pp. 57-64, December 2000.

[21] P. Estrela, A. Grilo, T. Vazão, M. Nunes, "Terminal Independent Mobile IP (TIMIP)", draft-estrela-timip-00.txt, Mar. 2002, Work in Progress.

[22] A. Grilo, P. Estrela, and M. Nunes, "Terminal Independent Mobility for IP (TIMIP)", *IEEE Communication Magazine*, December 2001, pp. 34-41.

[23] R. Droms, Dynamic Host Configuration Protocol, RFC 2131, 1997.

[24] J. Myers, M. Rose, Post Office Protocol - Version 3, RFC 1939, May 1996.

[25] M. Crispin, Internet Message Access Protocol - Version 4rev1, RFC 2060, Dec. 1996.

[26] W. Yeong, T. Howes, S. Kille, Lightweight Directory Access Protocol, RFC 1777, Mar. 1995.

[27] W. Simpson, The Point-to-Point Protocol (PPP), RFC 1661, Jul. 1994.

[28] TR 22.934, Release 6 of 3GPPP specification for "WLAN/UMTS Interworking".

[29] E. Gustafsson, A. Jonsson, and C. Perkins, "Mobile IPv4 Regional Registration", *draft-ietf-mobileip-reg-tunnel-06.txt*, Mar. 2002, Work in progress.

[30] S. Deering and R. Hinden, Internet Protocol, Version 6 (IPv6) Specification, RFC 2460, 1998.

[31] David B. Johnson, Charles E. Perkins, Jari Arkko, "Mobility Support in IPv6", *draft-ietf-mobileip-ipv6-18.txt*, June 2002, Work in Progress.

[32] Hesham Soliman, Claude Castelluccia, Karim El-Malki, Ludovic Bellier, Hierarchical MIPv6 mobility management (HMIPv6), *draft-ietf-mobileip-hmipv6-06.txt*, Jul. 2002, Work in Progress.

[33] R. Braden, L. Zhang, S. Berson, S. Herzog, S. Jamin, Resource ReSerVation Protocol (RSVP) -- Version 1 Functional Specification, RFC 2205, Sep. 1997.

[34] Pat R. Calhoun, Gabriel Montenegro, Charles E. Perkins, Mobile IP Regionalized Tunnel Management, *draft-ietf-mobileip-reg-tunnel-00.txt*, Nov. 1998, Work in Progress.

[35] A. Misra, S. Das, A. McCauley, A. Dutta and S. K. Das, "IDMP-Based Fast Handoffs and Paging in IP-Based 4G Mobile Networks", *IEEE Communications*, vol. 40, pp. 138-145, March 2002.

[36] A. Misra, S. Das, A. McCauley and S. K. Das, "Auto configuration, Registration and Mobility Management for Pervasive Computing", *IEEE Personal Communications Magazine*, Vol. 8, No. 4, pp. 24-31, August 2001.

[37] J. Manner, et. al., "Evaluation of mobility and Quality of Service interaction", *Computer Networks*, vol. 38, pp. 137-163, 2002.

[38] C. Rigney, A. C. Rubens, W. A. Simpson and S. Willens, Remote Authentication Dial In User Service (RADIUS), RFC 2138, Apr. 1997.

[39] S. Glass, "Security Issues in Mobile IPv4", draft-glass-mobileip-security-issues-01.txt, March 2002, Work in Progress.

Chapter 7

AN INITIAL SECURITY ANALYSIS OF THE PERSONAL TRANSACTION PROTOCOL

Jari Veijalainen, Alexandr Seleznyov, Oleksiy Mazhelis
Computer Science and Information Systems Department, University of Jyväskylä

7.1   INTRODUCTION

Our society is becoming increasingly dependent on the rapid access and processing of information. The number of handheld mobile devices with access to the Internet and network-based software and services is exploding. Research indicates [1] that by the end of 2002 there will be over 1 billion mobile phone owners globally with Internet access, and that this number is going to grow exponentially in the nearest future. By 2006 the number of interconnected mobile device users is expected exceed the worldwide Internet subscriber population. It is estimated that in a few years there will be three times as many of these devices worldwide as personal computers.

Another research study [15] indicated that 90 percent of worldwide professionals and telecommuters will start to use wireless data solutions by 2005. The most probable reason for the dramatic increase in wireless usage is considered to be advances in wireless network speeds gained from 3G technologies. These speeds (up to 2 Mbps) mean that more and more resources will be available for a mobile device user. At the same time they will be more accessible, which means that more and more companies will provide products and services to take advantage of it.

Wireless mobility and associated functionality not only provide access to larger and more varied resources of data more quickly than ever before, it also provides an access path to the data from virtually anywhere in the World [17]. The research [15] showed that over the next three to five years, Internet users will be accessing the Internet through a variety of different wireless devices. For business Internet users, the top devices will be a wireless modem (56 percent), a cell phone (51 percent), a PDA (50 percent), or a laptop computer (43 percent). Furthermore, two-thirds of users are going to show a high level of interest in portable wireless Internet access.

These technological advances create also new security threats and possibilities for abuses. The wireless information systems have wireless network(s) and a number of different mobile devices as their components. Each of the system's components has its own security risks associated with it. The main risks are eavesdropping the wireless traffic and stolen mobile devices. To minimize the former risk, one must encrypt the traffic over the air interface. The stolen and abused terminals pose a new risk class. Statistic shows that handheld devices such as mobile phones, laptops and PDA's (Personal Digital Assistants) are now the most popular items at "lost and found" offices. For example, only in September 2001, 45000 handheld devices were lost or stolen in the UK [21]. According to other estimates [2], ten mobile phones are lost every minute in the world. Most of these devices were turned on when they were lost or stolen. Therefore, unauthorized calls can be made or sensitive personal data they contain may easily be abused. The risks are higher in case of PTDs that also contain personal certificates that can be used for various authentication and authorization purposes, such as granting access to corporate systems.

Thus, developers of mobile devices have to implement extra precautions at the development stage, which ensure sufficient security measures for mobile devices to prevent such potential abuse of information or unauthorized access to user resources. Also companies should develop a comprehensive security plan for mobile devices in general and for PTD in particular that includes protecting the physical device from theft, protecting the network from being accessed through a stolen device and protecting the information stored on the intranet and on the device. It is especially important that the stolen certificates can be revoked quickly.

It is not the purpose of this chapter to try to provide a full coverage of the wireless security issues no to speak about the entire network security field. The reader is urged to consult e.g. [1,5,7] for a more comprehensive overview and [23,18,11] for security flaws in some existing protocols. Instead, our goal is to perform an initial security analysis of the Personal Transaction Protocol currently being designed by Mobile electronic Transaction Forum/Open Mobile Alliance [16].

Personal Transaction Protocol (PTP) is intended to provide a practical mechanism over which the user can perform remotely security-related actions required by applications running at a Personal Computing Device (PCD). The central idea of the PTP protocol is that a personal telecom or another portable terminal has a so-called Security Element[1] capable of hosting PKI certificates. The current (draft) version of the PTP protocol specification assumes that the PTD hosts X.509 conformant PKI certificates, but it can be enhanced in the future to include support for methods with symmetric encryption keys. The practical significance of the protocol is that the user only needs to carry the PKI certificates within the PTD, wherefrom they can be used for various security- and privacy-related purposes at the PCD and other networked devices, with which the PCD interacts. At the same time the user can take advantage of the larger processor, memory, and display resources of the PCD while using the WWW-browser to access the contents and services in the web.

The certificates can be used over the protocol to identify a person carrying the terminal (in this case the PTD can reasonably have only one user to whom the certificate has been associated). Further, because a private key or keys are securely stored in the Security Element and can be used for encrypting data, PTD's Security Element can be used e.g. to:

- Register the user for a network service.
- Authenticate the user to a remote server or workstation.
- Perform digital signing of a document. This can be used e.g. for authorization of payment details in Web shopping, authorization of a Web banking transaction, signing an email,
- Encryption and decryption of the messages sent between the PTD and PCD, as well as other components in the network

In this work we make an attempt to analyze the PTP protocol in order to classify potential threats and protocol's security flows. We further make an attempt to analyze the PTP protocol from the perspective of the relative security, as compared to alternative ways of performing the same actions. At the current stage of the development we can mostly argue on the architectural level, because implementation details are not available. We still hope that we are able to identify security issues where additional attention should be paid during protocol's implementation. For potential users this will ensure that the PTP protocol does not provide unintended services giving opportunity to an attacker to abuse a user's sensitive information. Vendors, in turn, will minimize potential losses. They will not be forced to issue security patches afterwards, or correct some problems overhauling the basic structure of the protocol.

---

[1] In this work we attempt to use all definitions and terms in accordance with a terminology provided by the Mobile Electronic Transaction group [13].

7.2   PTP DESCRIPTION

In this section we are going to describe the main goals of the PTP protocol as well as overview techniques it uses to achieve them. The term *Personal Trusted Device* (PTD) refers to "a mobile communications device that can be personalized (through the registration of service certificates) and trusted by the user and service providers requiring application-level public key security" [13]. In practice, a PTD is a handheld terminal that hosts a *Security Element* (SE), supports several protocol stacks for data and voice communication, has some kind of a display, and a keyboard. SE is "a component of the PTD containing the user's key pairs and root certificates and responsible for performing encryption and authentication functions" [13]. The term *Personal Computing Device* (PCD) refers to "A personal communications and computing device used as a platform for applications requiring strong authentication" [13]. In practice, it can be a PC connected to the Internet via a modem or LAN, or a TV Set-Top-Box connected to Internet via CATV (down-link) and a suitable wired or wireless up-link.

7.2.1   Goals of the PTP protocol

The PTP protocol is intended to enable for PTDs the ability to handle different kinds of services and applications. It defines how these services can be built for PTDs for different environments providing a user to use different applications, such as banking, ticketing, email encryption, etc.

There are three potential environments defined where the PTP enables a PTD usage. These environments are:
- remote – connections made through the Mobile Internet;
- physical (or local) – connections made directly to a PCD of the service provider;
- personal (or home) – connections made from a user's PCD, where a PTD is used to authenticate/authorize the user.

As it is possible to see these environments represent possible scenarios of communications between a user's PTD, a PCD, as well as a service provider. In these scenarios the user may perform his/her transactions from the PTD using a PCD as a gateway, or from a PCD using the PTD as an authentication and authorization device. In addition, the PTD can be used as a pure authentication device in granting access to a PC or to another networked terminal.

Going into more detail, the intended usage of PTP includes the following:

- SSL/TLS authentication. A secure web server is requesting a client certificate. The PTD is used to get the client certificate and make the private key operation. PIN is entered to the PTD.
- S/MIME secure email. The PTD is used in private key operations: for signing sent messages and decrypting received messages
- IPSec VPN. The PTD is used for authentication.
- Desktop login. To make a PKI based login to a desktop, the PTD is used to get the client certificate and make the private key operation. PIN is entered to the PTD.
- Signing textual data (e.g. a payment contract).
- Storing content, like a certificate, in the PTD.

Figure 7.1 shows an overall structure of the PTP outlining the involvement of the OBEX protocol [8] and different bearers, such as infrared, cable, and Bluetooth. Each one of the used protocols includes its own sequence of the authentication and connection establishment. The PTP protocol is built upon the stack consisting of these protocols.

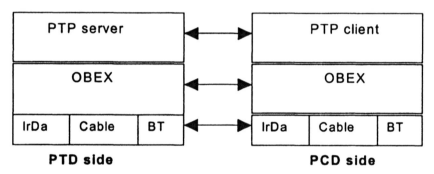

Figure 7.1. PTP overall architecture

7.2.2   PTP Functions

In the normal PTP usage all operations are initiated by the PCD. Before that, the PTD needs to be activated for PTP. A typical PTP session (used for authentication) contains following steps [14]:
1. Activation. The PTD is activated to use PTP with a certain PCD. This step involves creating the transport level connection and exchanging the device information. PCD is the initiator of the PTP connection.
2. Exchange of object information. The PCD requests information on objects contained in the PTD. The PTD returns the objects (or requested attributes).

3. Authentication. PCD sends a Sign Request. The PTD returns the signature.
4. PCD closes the connection (the transport level connection may remain active if it is needed for other purposes).

Figure 7.2 outlines a protocol stack structure when PTP and OBEX are run over the Bluethooth. It shows different layers of the stack on the server and client sides and interactions between them. The Baseband, LMP, and L2CAP are the OSI layer 1 and 2 Bluetooth protocols. RFCOMM is the Bluetooth adaptation of the GSM TS 07.10. Additionally, SDP is the Bluetooth Service Discovery Protocol.

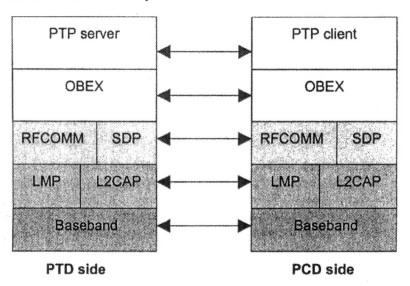

Figure 7.2. PTP run over the Bluetooth protocol stack [14, Fig. 2]

The PTP client shown in Figure 7.2 is a part involved in the mobile electronic transactions that provides the PTP functionality in the PCD. Similarly, the PTP server provides the PTP functionality at the PTD.

Apart from Bluetooth, different types of connections may be used between the PCD and PTD: infrared or cable (USB, RS232). The protocol stack differs in its lower part radically in these different cases. We are not going to provide the details of the low-level protocols and do not discuss different stack configurations when using different connections. This is because we are going to analyze the PTP on the higher level. For the purpose of the analysis, the assumption we make for the rest of the chapter is that the lower level protocols are designed and implemented in a correct manner, unless we explicitly deviate from this.

7.2.3   Summary of the PTP operations

In this section we provide an overview of the PTP protocol from a perspective of potentially involved parties. This would allow us to analyze the PTP design on a high level and discuss new potential threats that the PTP might create.

Generally speaking, the PTP is designed to provide means to use the PKI certificates carried within the PTD for various authentication and authorization purposes when the application runs at a PC or a set-top-box (cf. above).

There is a user on the one side of the channel. It is assumed that a PTD is used by a single user, i.e. PTD is indeed a personal device. Based on this linkage, the private-public key pair stored at the PTD really represents a particular person. If another person would use the device and the keys, i.e. the device-person liaison is broken, we say that the device is stolen and misused.

The PTD has the SE that contains information about its legal user, secret keys, and certificates required for the user authentication. The actual authentication and authorization are performed by the operations using the keys, but the operations require PINs to be entered. Through them, the attachment of the device to the right person is ascertained. Therefore, activated by its correct user that is able to enter the correct PIN, the security element indeed is able to perform the encryption/decryption operations for the correct person. If the device is stolen or lost the security element cannot be activated without the knowledge of the PINs. Further, the lost or stolen certificates can be revoked by the user so that the thief cannot use them, even if he or she somehow got to know the PINs.

Private key should never leave the SE, i.e. it should not be exposed under any circumstances. There is still a small but positive probability that someone manages to dig them out from the SE. We consider it very small, several magnitudes smaller than the risk of finding out some or all the PINs used to activate the SE (see [20] for further analysis of the PIN security in wireless contexts).

We perform the risk analysis in the following way. We take two other architectures, where PTD and/or PKI certificates are used for the same purposes as in the context of PTP, and look then at the authentication, authorization, and confidentiality risks in them and in the PTP architecture. We measure the risk levels with probabilities of security violations happening. We further deduce the differences in risk levels for the different cases and analyze them. Finally, we will look at the implementation-level issues and possible security threats at that level, although at this stage we cannot say very much about.

7.3   ASSETS, RISKS AND THREATS

7.3.1   The comparative environments

We envision three broader usage scenarios where PTDs or — instead of them — Personal Identification Cards are used. First, PTD can be directly used to access a server, such as an E-commerce server or a corporate server, over a wireless link. This is called *the direct PTD usage scenario.* An alternative way of accessing the servers in Internet is to use PCD (a PC or similar terminal directly). In this scenario the PTD is not involved, but rather a personal smart card (Personal Identification Card). It is inserted into a special reading device physically connected to the PCD. This scenario is called *the direct PCD usage scenario.* The third alternative is the case, where a PCD is used to browse the web, but the authentication/authorization is performed using PTD. This is called the *indirect PTD usage scenario.* The fourth alternative scenario is *password authentication/authorization scenario.* This is used to compare the current situation, where the passwords are used, with the situation, where the PTP protocol is used.

7.3.2   Assets, Risks, and Threats

In the above scenarios we can identify at least the following **assets:**
–   service provider servers (A-SPSRV),
–   PCDs (A-PCD),
–   PTDs (A-PTD),
–   Personal Identification Cards (A-PIC),
–   transport connections (A-TCONN),
–   private keys (A-PRKEY),
–   PINs (A-PIN),
–   message keys (A-MSGKEY),
–   passwords (A-PASSW),
–   transferred data (A-T_DATA),
–   stored corporate data at servers (A-S_DATA),
–   stored private data at the PTD (A-PTD_DATA),
–   certificate authorities (servers) (A-CASRV).

At least the following **risks** can be identified:
–   disclosure of corporate data at a server (R-DISC_SDATA),
–   modification of corporate data at a server (R-MOD_SDATA),
–   disclosure of private data on PTD (R-DISC_PTD_DATA),
–   signing of a forged message (R-SGN_FMSG),

- encryption of a forged message (data) (R-ENC_FMSG),
- forcing a PTD to decrypt a message for an adversary (R-DECR_MSG),
- translation of a message (described in the next section) (R-TRANS_MSG),
- false authentication misusing a PTD or PIC (R-AC_PTD, R-AC_PIC),
- false authorization misusing a PTD or PIC (R-AZ_PTD, R-AZ_PIC),
- false authentication using an exposed password (R-AC_PASSW),
- false authorization using an exposed password (R-AZ_PASSW).

At least the following **threats** can be identified:
- stealing a PIC (T-ST_PIC),
- stealing a PTD (T-ST_PTD),
- disclosure of a message key (T-MSGKEY),
- disclosure of an access PIN of the PTD (T-ACC_PIN_PTD),
- disclosure of an authentication PIN of the Security Element (T-AC_PIN_PTD),
- disclosure of an authorization PIN of the Security Element (T-AZ_PIN_PTD),
- disclosure of an authentication password of a PIC (T-AC_PASSW_PIC),
- disclosure of an authorization password of a PIC (T-AZ_PASSW_PIC),
- disclosure of an authentication password at PCD (T-AC_PASSW_PCD),
- installation of malicious software at PTD (T-MSW_PTD),
- installation of malicious software at PCD (T-MSW_PCD),
- malicious server (T-MSRV),
- man-in-the-middle attack on a connection:
    - PTD-PCD (T-MIM_PTD-PCD),
    - PCD-Internet servers (including trusted CAs) (T-MIM_PCD-SRV),
    - PTD-(Mobile) Internet servers (T-MIM_PTD-SRV).
- replay of a message on a connection (T-REP_PTD-PCD, T-REP_PCD-SRV),
- initiation of a connection impersonating PTD, PCD, or service provider (T-IPTD, T-IPCD, T-ISRVP),
- eavesdropping of a connection (T-E_PTD-PCD, T-E_PCD-SRV, T-E_PTD-SPSRV, T-E_PCD-CASRV).

As can be seen from above the lists of risks and threats are rather long. Still, we do not claim that they are complete. The further question is, what are the risks that diverse threats realize. This we will handle below. We proceed as follows. We look at the four scenarios introduced and relate the threats and risks while analyzing the scenario. We measure the risks and threats by probability with which they occur. E.g. P(T-AC_PASSW_PIC) denotes the probability with which the authentication password on a Personal Identification Card can be found out by a malicious person. As was discussed in [20], a PIN or password can be found out by directly observing a person while she is using it, by digging it out from the (stolen) device carrying it or its hash value, or simply by guessing. In the analysis below we assume that these individual threat probabilities do not change during the analysis. They might still be dependent on each other. One can ask: does the existence of the PTP protocol increase e.g. the threat of stealing a PTD running it, as compared to one that does not have it? One could argue that this is indeed true. This is because such a device is more usable for a thief that can be used also against a PCD to authenticate and authorize the user. Whether this argumentation is indeed true is for a further study.

The probability that a risk is realized, i.e. the risk probability, depends on the realized threats. The relationship between the threats and the risk can be complicated, because there might be several threats that realize the risk independently, whereas in other cases the risk is realized only if several threats actualize. A typical example, where several threats and risk can actualize is the stolen device. As such, the device in wrong hands does not yet cause much harm to other system components. The harms begin, however, if the thief can indeed use the device like the user. This requires often further PINs or passwords to be obtained by the thief.

7.3.3   Risks for the private data at PTD (R-DISC_PTD_DATA)

There are currently telecom terminals where the PDA and the data accessible through the user interface is unprotected. Thus, for such a terminal $P(T\text{-}ST\_PTD) <= P(R\text{-}DISC\_PTD\_DATA)$ in any case. For other terminals the same inequality holds, if they are stolen while "on", i.e. authentication performed by the real user for the PDA access prior to theft. If the terminal is stolen in a state where it is "off", and it requires user authentication, only one PIN or password is usually asked in order to grant access to the private data after booting or activating the device. In that case $P(R\text{-}DISC\_PTD\_DATA)$ is close to $P(T\text{-}ST\_PTD)*P(T\text{-}AC\_PIN\_PTD)$. We say close, because in reality private data can be disclosed e.g. by observation and other more sophisticated technical means.

Does PTP protocol affect this risk probability $P(R\text{-}DISC\_PTD\_DATA)$? One could argue that because it allows communication between the PCD and PTD, it might leak private data. Indeed, the protocol offers objectInformation Protocol Data Unit (PDU) using which the PCD can ask objects to be transferred from the PTD. The protocol also allows objects to be transferred from the PCD and stored at PTD (PTP-StoreObject primitives). These are meant to be used for fetching and storing cryptographic objects from/to the PTD. Thus, if the protocol is implemented correctly at the PTD, this should not allow private data to be fetched to the PCD. Should, however, malicious software run at the terminal, say, a false implementation of the PTP server itself, the situation is different. Thus, one can argue that existence of PTP raises the risk level from the above consideration, where stealing the device was a prerequisite. Intuitively, the increase in risk level P(R-DISC_PTD_DATA) due to PTP protocol is very small, but the more exact value is for further study.

7.3.4   False authentication risks R-AC_PTD, R-AC_PIC, R-AC_PASSW

Let us look at the authentication. We analyze cases, where a PTD is used over the PTP, PIC is used at PCD and a password is used at the PCD. Is it possible to infer about the risk level incurred in these various cases?

Firstly, consider a case of the stolen PTD. Assume that the device was "on" so that no access PIN is needed to be known. The thief has to find a PCD capable of running the PTP protocol This PCD should further be allowed to access corporate servers, should these be the actual target of the thief. PTP protocol can be run only over a short distance link, in practice only over a few meters distance, so that physical access restrictions to the enabled PCDs might already be an obstacle. We can assume that running the PTP protocol requires at least one authentication PIN to be known. This is used to allow the SE to encrypt the challenge, sent by the PCD, with the private key. While the thief is at work, the device owner could revoke the certificate so that access attempts would fail. Based on the above, it is possible to see that $P(R\text{-}AC\_PTD) < P(T\text{-}ST\_PTD)*P(T\text{-}AC\_PIN\_PTD)$ in this case. How much smaller the false authentication risk is? It is not possible to answer the question without more exact knowledge about the other factors, such as revocation speed, corporate policy, the configurations of the corporate firewalls, and etc. A further complication comes from the fact that a PTD could be used to verify identity of a wrong user. This is especially the problem, if there is malicious software at the PCD (T-MSW-PCD) or PTD (T-MSW-PTD). These and other types of attacks, like man-in-the-middle again raise the risk level $P(R\text{-}AC\_PTD)$.

Similar argumentation can be used in the case where the device is not "on" while stolen. The penetration probability includes in this case also the probability of finding the access PIN. Thus, $P(R\text{-}AC\_PTD) < P(T\text{-}ST\_PTD)* P(T\text{-}ACC\_PIN\_PTD)* P(T\text{-}AC\_PIN\_PTD)$ holds.

Similarly the analysis can be performed for a stolen PIC. We can assume that if a PCD supports PTP it also supports a smart card reader and PIC usage. Thus the other circumstances remain constant. Which risk is bigger,

P(R-AC_PTD) or P(R-AC_PIC), cannot be assessed without further empirical results.

Considering exposure of passwords, the malicious person should somehow obtain the password of the victim and using it authenticate him- or herself for the system. The pure guessing probability can be calculated based on the length of the password. This is, however, only the minimum risk level because people cannot easily remember long meaningless passwords they tend to pick up names, birth dates, license plates, etc. Thus, if the system e.g. requires six-character long password, the pure guessing probability is roughly 62^6 (Capital and small Latin letters plus numbers 0-9), but the real probability of guessing the password is considerably higher for the above reasons. Further, the system has to keep track of the chosen password(s) for each user. This data can be exposed to intruders. This further increases P(T-AC_PASSW_PCD) from the pure guessing. The real probability P(T-AC_PASSW_PCD) is thus difficult to assess without empirical data.

One can still say that the four-digit PIN often used in telecom terminals only offers 10^4 pure guessing probability (four random digits), but this does not say much of the overall risk level, because the intruder should get both the authentication PIN and the PTD in order to penetrate corporate systems. This results in the combined risk level P(T-ST_PTD) * P(T-AC_PIN_PTD). In contrast, in password-protected systems only the password and the user account name are needed to be known. Summing up, the relative risk analysis must be continued based on empirical data on the real values of P(T-AC_PASSW_PCD), P(T-ST_PTD), and P(T-AC_PIN_PTD).

7.3.5 False authorization risks R-AZ_PTD, R-AZ_PIC, R-AZ_PASSW

7.3.5.1 The indirect PTD usage scenario

Firstly, we will consider the R-AZ_PTD risks associated with the use of PTP in the *indirect PTD usage scenario*. The corresponding probability P(R-AZ_PTD) may be estimated as the sum of the following threat probabilities:

1) Probability of a PTD being stolen, P(T-ST_PTD), multiplied by the probability of an access PIN (if required) disclosure P(T-ACC_PIN_PTD), authentication PIN (if required) disclosure, P(T-AC_PIN_PTD), and the disclosure of authorization PIN P(T-AZ_PIN_PTD). Having stolen a PTD, an impostor is required to find out the access PIN (if the device is turned off), then to find out the authentication PIN (if certain time has elapsed since the last use of SE), and finally, knowing the authorization PIN is required

irrespectively of the necessity to obtain previous two PINs. Thus, in the worst (from the system security point of view) case the impostor will only need to obtain authorization PIN, and in the best case all three PINs should be obtained, i.e. the impostor should pass three independent hurdles.

2) Probability of malicious software being installed on the PTD P(T-MSW_PTD). Should it happen that a PTD is infected by malicious code, this code would be able to replace the real document to be signed by a faked one.

3) Probability of malicious software being installed on the PCD P(T-MSW_PCD). The risk could be realized if i) malicious code controls the PCD, and ii) the document to be signed is not present on the screen of the PTD (e.g. due to considerable size of the document as compared with the small screen size of the PTD).

4) Probability of the PCD impersonation P(T-IPCD) meaning that PTD is communicating with another PCD than what it assumes.

5) Probability of the man-in-the-middle attack P(T-MIM_PTD-PCD). In this scenario, the document to be signed is replaced by the faked one while the document is being transmitted from the PCD to the PTD.

6) Probability of the man-in-the-middle attack against the connection PCD–Internet Server P(T-MIM_PCD-SRV).

The above threats 1)-6) represent different kinds of attacks that could be performed independently. That is why we assume that the probabilities of their occurrence can be summed up. Given this, it is possible to give the overall expression for the false authorization risk:

(1) P(R-AZ_PTD) =
$$P(T\text{-}ST\_PTD)*P(T\text{-}ACC\_PIN\_PTD)*P(T\text{-}AC\_PIN\_PTD)*P(T\text{-}AZ\_PIN\_PTD)+$$
$$P(T\text{-}MSW\_PTD)+$$
$$P(T\text{-}MSW\_PCD)+$$
$$P(T\text{-}IPCD)+$$
$$P(T\text{-}MIM\_PTD\text{-}PCD)+$$
$$P(T\text{-}MIM\_PCD\text{-}SRV).$$

7.3.5.2   The Direct PTD usage scenario

For the direct PTD-Server communication (the *direct PTD usage scenario*), the probability of the R-AZ_PTD risk can be calculated in a similar manner:

(2)   P(R-AZ_PTD) =
$$P(T\text{-}ST\_PTD) * P(T\text{-}AC\_PIN\_PTD) * P(T\text{-}AZ\_PIN\_PTD) +$$
$$P(T\text{-}MSW\_PTD) +$$
$$P(T\text{-}MIM\_PTD\text{-}SRV).$$

The difference between the above two expressions (1) and (2) is due to the exclusion of the PCD from the analysis. Namely, the threats associated with the involvement of PCD (T-MSW_PCD, T-IPCD) are excluded, and the threats related to the communication links between PTD and PCD (T-MIM_PTD-PCD) and between PCD and Internet Server (T-MIM_PCD-SRV) are replaced by the threats related to the direct link between PTD and Internet Server.

7.3.5.3   The false authorization in the direct PCD usage scenario

Similarly as above, in order to calculate the risk of false authorization in the *direct PCD usage scenario,* R-AZ_PIC, we will take as a basis the risk in the indirect PTD usage scenario:

(3)   P(R-AZ_PIC) =
        P(T-ST_PIC) * P(T-AC_PASSW_PIC)* P(T-AZ_PASSW_PIC) +
        P(T-MSW_PCD) +
        P(T-MIM_PCD-SRV).

Here, the threats caused by the use of PTD (T-MSW_PTD, T-MIM_PTD-PCD) are excluded. This is because we assume that a PIC cannot in practice be altered software-wise, and because the card readers are always physically connected to the PCD thus making man-in-the-middle attack impossible.

The risk of using a stolen PIC, by analogy to the case of a stolen PTD, depends on the probability of i) PIC being stolen P(T-ST_PIC), ii) PIC access password being compromised P(T-AC_PASSW_PIC) and iii) PIC authorization password being compromised P(T-AZ_PASSW_PIC). In the current PICs the passwords are typically 4-8 digits long number sequences (i.e. they are PINs) that can be changed by the user. A PIC is typically locked if the user enters wrong password more than three times in a row.

7.3.5.4   False authorization in the password authorization scenario

Finally, in the alternative *password authorization scenario* the risk of the false authorization is caused by the PCD-related threats as well as by the threats related to the communication link between PCD and Internet Server. Thus, the risk probability P(R-AZ_PASSW) comprises the following threat probabilities:

1) Probability of the authentication password being exposed P(T-AC_PASSW_PCD) (if required), multiplied by the probability of the authorization password being exposed P(T-AZ_PASSW_PCD).

2) Probability of a malicious software being installed on the PCD P(T-MSW_PCD).

3) Probability of disclosure of a private key at PCD (T-PRKEY_PCD).

4) Probability of the man-in-the-middle attack against the connection PCD–Internet Server P(T-MIM_PCD-SRV).

Taking these together, the overall expression for the false authorization risk in this scenario is:

(4)   P(R-AZ_PASSW) =
       P(T-AC_PASSW_PCD) * P(T-AZ_PASSW_PCD) +
              P(T-MSW_PCD)+
              P(T-MIM_PCD-SRV).

7.3.5.5 Comparing the risks

Above, we introduced the expressions for estimation of the false authorization risk in four different scenarios. Now we estimate how risky is the indirect PTD usage scenario (which in essence is specified by PTP protocol), as compared to the others.

Firstly, we will compare *direct* and *indirect PTD usage scenarios* (1) and (2) By subtracting (2) from (1) we get

(5) = (1) – (2) =
              P(T-MSW_PCD) +
              P(T-IPCD) +
              P(T-MIM_PTD-PCD) +
              P(T-MIM_PCD-SRV)
              – P(T-MIM_PTD-SRV).

To obtain the value of this difference the values of threat probabilities need to be estimated. As well as with respect to the authentication risks analyzed in the previous section, this needs to be studied further. However, with a certain degree of confidence it is possible to assume, that

P(T-MIM_PTD-PCD) + P(T-MIM_PCD-SRV) >= P(T-MIM_PTD-SRV).

Based on this, the false authorization risk in the *indirect PTD usage scenario* is greater at least by P(T-MSW_PCD) + P(T-IPCD) as compared to the direct one.

This term corresponds to the threats posed by the use of PCD and PTP instead of using the PTD directly for authorization. In other words, it represents the increase in the risk level due to the use of the PTP. It is important to remember that a man-in-the middle attack between the PTD and a server might affect this conclusion, because we must assume that in the direct usage scenario the PTD is communicating over a wireless link with the server attached to a wired network. Wireless links tend to be more vulnerable to these kinds of attacks than wired ones. Again, further study is required to obtain the real values for the above risk levels.

Comparing *indirect PTD usage scenario* (1) with *direct PCD usage scenario* (3), one can again subtract (3) from (1).

(6) = (1)–(3) =
$\quad$ P(T-ST_PTD) * P(T-AC_PIN_PTD) * P(T-AZ_PIN_PTD) +
$\quad$ P(T-MSW_PTD) +
$\quad$ P(T-MSW_PCD) +
$\quad$ P(T-IPCD) +
$\quad$ P(T-MIM_PTD-PCD) +
$\quad$ P(T-MIM_PCD-SRV)
$\quad$ –(P(T-ST_PIC)*P(T-AC_PASSW_PIC)*P(T-AZ_PASSW_PIC)+
$\quad$ P(T-MSW_PCD) +
$\quad$ P(T-MIM_PCD-SRV)).

Assuming that it is roughly as probable to loose a PTD and a PIC to a thief and that finding out the two PINs are equally probable in both cases, the difference that remains is

(7)$\quad$ P(T-MSW_PTD) + P(T-IPCD) + P(T-MIM_PTD-PCD).

This expression represents the increase in the risk value due to the use of PTP protocol instead of Personal Identification Card in authorization against servers. We emphasize that this conclusion is based on assumptions about the relative risks of PTDs and PICs that seem plausible but have not been proven. Further, the absolute increase in risk level, i.e. the value of (7) must be evaluated, before the practical relevance of the increase can be estimated.

Finally, let us compare *indirect PTD usage scenario* with the *password authorization scenario*. Having excluded the common probabilities, the comparison should be made between the following risks:

(8) P(T-ST_PTD) * P(T-AC_PIN_PTD) * P(T-AZ_PIN_PTD) +
$\quad$ P(T-MSW_PTD) +
$\quad$ P(T-IPCD) +
$\quad$ P(T-MIM_PTD-PCD)
$\quad$ and

(9)$\quad$ P(T-AC_PASSW_PCD) * P(T-AZ_PASSW_PCD)

At first sight, (8) corresponding to the indirect PTD usage scenario is greater than (9), because there are more threats to sum up. That would indicate that the use of PTP for authorization is more risky than using conventional passwords. However, it may not be true, e.g. if the passwords are cashed on the PCD and can be used repeatedly. Should it be true, anyone who has managed to get access to the PCD has a possibility to find out the cashed passwords. Thus, we can assume that it is much easier to dig out the passwords at the PCD than the PINs at the PTD and therefore, the probability of a faked document being signed is greater. As a result, the overall risk level for the password authorization scenario using pure PCD is expected to be *rather greater* than smaller, as compared with the case when PTD is used.

It is necessary to note that the implementation of the authorization at the PCD can influence significantly the probability values of the relevant threats. For instance, if passwords are used only once and are not stored at the PCD prior to usage, the argumentation above is not valid. Thus, we emphasize that this is again the first conclusion that seems plausible. More real evidence in favor or against it must be delivered later on.

7.4   PTP TECHNICAL ANALYSIS

In this section we are going to discuss some technical aspects of the planned PTP. Here we are interested to identify potential unintended services that future PTP implementations may provide. Some of them are related to a usage of the untrusted software during electronic transactions.

At the beginning we discuss several general issues of authentication protocols related to the PTP design. After this we consider each PTP usage case separately identifying risks and suggesting potential solutions.

7.4.1   Protocol's environment

Before going into detail it is necessary to note that in the PTP specifications it is assumed that "the PCD is trusted to run a Web browser". In other words, it is necessary to trust a PCD to carry on some sensitive operations. Here we argue that it is a very strong assumption, which almost never holds in real life. For remote of physical environments it is not possible to check what kind of software PCD runs and can it be trustable. Moreover, a party that represents a service provider may maliciously tamper with software in order to collect some sensitive information about users, forge documents and receipts, etc.

Personal environment may have its own vulnerabilities. Since the PCD may be at home it usually means that it is not properly administered. Therefore, it is more vulnerable to viruses, break-ins, and Trojan horses. Not all people run firewall or traffic monitor software on their PCs.

The PTP is supposed to provide a secure channel for mobile electronic transactions by providing a reliable PKI based methods of authentication. It means that a secure channel has to be established between a PTD and a service provider. There might be some additional certificate authority servers involved in the establishment of the secure channel between a PTD and a service provider. Everything else in between is considered a hostile environment and the Dolev-Yao problem is thus relevant [3]. This problem refers to situation where a potential adversary may tamper with messages and use unintended services provided by a protocol without knowing cryptographic keys used to encrypt the messages. Because PCDs and the entire communication environment have to be considered as a part of the hostile environment the problem is relevant. The concrete risk levels are largely dependent on the logical and physical access possibilities of intruders to the PCDs and the network parts. These vary very much between, say, a corporate PC in Germany and a PC in an Internet café in China.

The PTP is developed for a next generation of mobile devices. Therefore, it is not possible to foresee or anticipate all future settings from where users will perform their transactions or login to a system. A channel established between a PTD and a service provider may go through purely wired or wireless network. The channel may be created combining wired and wireless networks in different combinations. Therefore, it is natural to expect that protocol messages will go through a hostile environment. We analyze below in the light of the current knowledge, where there might be security holes.

As we mentioned above a protocol may leak information or provide unintended services even if transport protocols provide ideal security and primitives used in the protocol are ideal. Thus, there are four kinds of potential unintended services that the PTD may offer to an attacker in the future:

- **Signature** A potential attacker may force a PTD's secure element or a server to sign an attacker's message.
- **Encryption** An attacker can use this service to force one of the communicating parties to encrypt some information submitted by the attacker.
- **Decryption** An attacker can use this service to force one of the communicating parties to decrypt some information submitted by the attacker.
- **Translation** Using this service an attacker submits some encrypted message and receives this message encrypted using a different key. An attack described by Lowe in [10] on the Needham and Schroeder protocol is an example of this.

We analyze closer the above issues below.

7.4.2   Freshness of Messages

The PTP is designed to use mobile devices as authentication devices. Very often such devices are used to authenticate transactions between a user and some service provider. It means that a secure element of a user's PTD and a service server do not share a common secret (long term key) that may be used to authenticate each of them. Therefore, at the beginning of a session it is necessary to authenticate them to each other and exchange session keys. One of the most "popular" attacks that are performed under such circumstances is replay attack [19]. This attack derives from the interleaving of the multiple protocol runs, where an attacker redirects messages among them.

The rely attack relies on the fact that sometimes there is no way provided to assure that received messages are fresh. Thus, an attacker can intercept messages sent during previously occurred sessions, take his/her time to break a session key, and use it later for initiation of an attack by impersonating one of the parties with whom a previous communication took place.

Syverson, in [19], provides a taxonomy of replay attacks. The article studies messages rerouted by an attacker and classifies replay attacks based on the forwarded messages' destinations (destination taxonomy). There are three possible cases of destinations:
– The message can be forwarded to the sender. This attack called a *reflection attack.*
– The message can be forwarded to a third party, which is neither intended recipient nor the sender. This is a *deflection attack.*
– An attacker may send the message to the intended recipient, but with delay or on different run of the protocol. This is known as a *straight attack.*

An authentication protocol itself and security primitives used in it ought to have some means to prove freshness of the transmitted messages. Otherwise, all communications between future parties involved into mobile electronic transaction will be vulnerable to different kinds replay attacks elaborated above.

It is necessary to make a distinction between freshness intervals. Sometimes it is required to assure that the received message was send during the current session (sometimes called epoch). In other situations it is required to check whether the message is fresh inside the current session.

Taking into account all said in this section it is possible to see that it is crucial for security primitives that are used in authentication protocols to contain some additional information proving the freshness of the received message. The freshness should be shown for the current session and/or inside it depending on the security goals that the protocol is specified to achieve. For the PTP protocol it is "a must" to be able to prove that the message was sent during the current session, whereas the necessity to have a proof of the inside-session freshness depends on a concrete future implementation of the PTP.

7.4.3   Fail-stop Criteria

Nowadays there are numerous kinds of attacks that may be used against network protocols, such as replay, reflect, DoS, etc. Almost all of cases where an authentication protocol fails to achieve its goals are due to active attacks in which a message or its part is somehow altered or substituted in one way or the other. For example, most protocols use random numbers as a nonce or/and a first message number. Thus, someone may start numerous connections in order to obtain this random number and study regularities among them to predict the next number, which may be used to forge or hijack connections [4].

In order not to be vulnerable to these kinds of attacks it is necessary to develop protocols in a fail-stop way establishing criteria as to when protocol has to be aborted if it is being attacked [6].

The main idea behind of the fail-stop design is organization of the messages of a protocol into an acyclic directed graph where each arc represents a message and each directed path represents a sequence of messages. If a message that was actually sent was altered or it is inconsistent with the protocol specifications, then all messages that come after the altered message on the current path of the graph will be discarded. In this way, the fail-stop design gives a protocol immunity against active attacks. Any active attack will cause the protocol to halt, therefore, no signature, encryption, decryption, and translation services can be possibly provided to an attacker.

7.4.4   PTP Use Case Analysis

7.4.4.1   Usage of PTD as an Authentication Device for SSL/TLS Client Authentication

This scenario is intended for authentication of the user to a remote server. The scenario assumes that the server has obtained the user certificate either through a registration process or otherwise. The server sends to the PTD (via PCD) a challenge, which is signed at the PTD with the secure authentication key (private key). To perform the operation, the user is assumed to enter the authentication PIN that unlocks the SE (if a certain time has elapsed since the SE was used). The signed challenge is sent to the remote server (again via PCD). Using the public key in the user certificate, the server can check the validity of the digital signature and ensure that the user who has signed the challenge is the one whose identity is carried by the certificate. Assuming further the rather high probability that the PIN above was indeed typed by the person on the certificate, the server can conclude that it is communicating exactly with that person.

As can be seen, during the communication the data is transferred via the PCD. Since the PCD is a non-trustworthy environment, it may contain some malicious software that runs while the above message exchange is performed. Such software could force user to sign a challenge from a wrong server. As a result, there is a risk of the sensitive information leaking to wrong people.

It is interesting to compare the PTP protocol with WAP protocol (versions older than 2.0). Similarly to WAP, PTP does not provide end-to-end security between PTD and the remote server. In WAP protocol 1.x., the gateway is placed between the PTD and a remote server. The fact that the information is decrypted and is in «clear text» at the gateway poses security risks to the data. Due to this, the protocol was strongly criticized (e.g. [9]). Similarly, considering PTP protocol, much of the sensitive data is in «clear text» at the PCD and is under the risk to be exposed and/or modified.

7.4.4.2   Desktop Login Using PTD

This scenario is a special case of the previous one. Instead of typing password on the PCD, a user is authenticated to the PCD by signing the challenge issued by it.

Again, since the PCD is considered to be a non-trustworthy environment, the risks associated with the SSL/TLS Client Authentication scenario are relevant there. The malicious software run on the PCD may force the PTD to sign the challenge that can be used so as to authenticate the user to an unintended remote server. Consequently, the sensitive information could be exposed.

7.4.4.3   Decryption

Here, the PTD is used so as to decrypt secure e-mail messages. The received message has been encrypted with a message key. This message key has been encrypted by the public key of the user and sent along the message or separately. The message key is decrypted at the SE of the PTD with the public key of the user. After that, the decrypted message key is sent back to the PCD and is used to decrypt the actual message.

A possible security problem here is the exposure of the secure message, after decryption, should the malicious software control the PCD. The risk can be especially considerable if a public PC is used. Another risk is that the PCD does not show what it decrypted but another document. To eliminate these risks the message should be decrypted and shown on the PTD only, without delivering the clear text to the PCD. This might, however, be impractical or even impossible if the message is large. Further, the PTD should be able to use message keys and the corresponding algorithms for decryption. This would require either SE to be enhanced or other software to be installed at the PTD.

7.4.4.4   Signing

In this use case, a user digitally signs the hash value of the document with the secure authorization key (private key) stored at the PTD. The document to be signed is shown on the screen of the PCD and PTD. The PTD calculates the hash value of the document and asks the user to enter the authorization PIN. Then, if the PIN is valid, the PTD signs the hash value and sends it back to the PCD. PCD then sends the document along the hash value to the receiver. The latter decrypts the hash value with the public key of the user and calculates the hash value from the document. If they match, the document is with high probability the same as signed by the user.

An alternative scenario is possible if the size of the document is rather big, and it is considered impractical to browse it on the PTD's small screen. In this case the hash value is calculated at the PCD and sent to the PTD. The user therefore is assumed to digitally sign the hash value of the document while the document itself is not present at the PTD.

Again, the problem can occur is the PCD is infected by malicious software. If the data to be signed is not presented to the user at the PTD, this data can be substituted e.g. while it is transferred to hash-calculation module of PCD. As a result, a faked document can be signed. Notice that even if the document was shown on the display of the PCD this does not help, because the malicious software at the PCD could display document A and let the user sign the hash value for document B. The receiver could not notice that the user thought to have signed A, because the encrypted hash value and the document B sent with it are indeed compatible.

7.4.5   Accountability of the PTP

A security-related protocol should be accountable in the sense that it is possible to later check what has happened. In the current PTP draft [14] this issue is not explicitly addressed. It is by no means a trivial question, where the logs tracking the interactions should be placed and what exactly should be logged. First, one can argue that the logs should be kept at the PTD. This is because the same PTD can be used with various PCDs for different purposes and it would be difficult to collect the partial traces from different PCDs or servers, if needed. Further, because a PTD is really a personal device, the PTP log should be easy to use for the individual controlling it. This is achieved if the log is carried within the device or is otherwise always accessible. There is, however, a critical argument against this based on the privacy. If namely the device is stolen and it contains all the audit logs, the thief might get valuable and private information based on them. What should be thus stored into the log, what left out? Should the log be encrypted so that only the legitimate user can open it? Should this be done e.g. using the public key stored at the SE?

Further, should the log be stored on the SE? This is probably not wise for several reasons. Security Elements should be enhanced with additional functionality and with a lot of memory, should this be implemented. Apart from this, it should be ensured that the description language of the audit log is universal (XML). If the audit log is stored outside the SE on PTD, it should be ensured that it could not be tampered with locally in order to be used as a proof (for example), in court. Further, because the PTD can be used both directly and indirectly to perform authorizations for m-commerce transactions and authentication operations, the PTP generated logs and other possible logs should form a logical sequence of events when the logs are combined. There are already functioning prototypes with XML-based logs [12] and also XML-based general log format has been suggested [22]. These issues should be drawn into consideration while implementing the software.

7.5   CONCLUSIONS AND RECOMMENDATIONS

PTP protocol is intended to make remote security operations possible on a PC or set-top-box in a situation where a PKI certificate is stored at a portable (telecom) terminal, called a Personal Trusted Device (PTD). The PTP protocol uses the possibility of a PTD to perform secure PKI-based cryptographic operations involving the secure keys and certificates stored in its Security Element. Similarly to the Personal Identification Card (PIC), it allows a user of a PCD to be authenticated using the PTD, to sign a document with the user's private key, or decrypt a message (email) encrypted with the user's public key. This functionality is expected to get a fast acceptance by corporate users; therefore, there is a great demand for this kind of a protocol to be designed and implemented.

We have performed an initial analysis of the protocol above from the security perspective. At this stage, when not even the protocol specification is final, it is impossible to do it in a detailed way. Therefore, we have concentrated mostly on the architecture level considerations and have tried to apply general theoretical results. We have attempted to list all conceivable risks and threats in the larger environment and then tried to answer the questions like: does PTP-type of protocol arguably increase some risks? Or decrease them? What are the weak points that should be paid attention to? Etc.

The analysis did not reveal any dramatic problems. Still, the current version of the protocol could pose some security risks to its potential users. These risks might increase as compared with the other PTD usage scenarios. The same holds for the situations, where the personal identification cards are used for similar purposes as PTDs over PTP. The rise of the risk level is caused by the inclusion of additional components onto the communication path. In comparison with the other PTD usage scenarios, the additional component involved is the PCD, while in comparison with the use of PIC, the PTD with the corresponding communication link is added. If stealing a PTD and a PIC is equally common, the above additional components would make the difference. If the PCD is not trustworthy, runs e.g. a malicious browser or PTP client then many further problems can arise. These pose additional risk for PIC and PTD usage, whereas they are not present in the direct usage scenarios of the PTD.

We also analyzed the problem of stolen devices. The initial results show that the PTP protocol does not open up new serious security holes in this respect. A more comprehensive answer still requires concrete analysis and evaluation of the absolute risk levels of certain threats. These are for a further study.

Different active attacks and possible unintended services of the PTP were not considered in the current draft[2] of the PTP. Some of these attacks (DoS attack, or infection of the PTD by malicious software) are unlikely to happen in the nearest future. Some (as e.g. replay attack) are relevant not only to the personal environment specified by PTP protocol, but also for the remote and local environments. The others are related to the PTP (e.g. signing a faked hash). All they have been discussed above with aim to draw the attention of potential protocol designers, implementers, and users to the associated security risks. We hope that the protocol designers and future implementers will circumvent the possible risks where possible. It concerns among others the provision of the messages' freshness and the provision of the auditing facilities ensuring that a created audit log of transaction cannot be tampered with locally and can be used as a proof (for example, in court). When the protection against certain threat is considered to be unachievable or the security measures are expected to bring a bad user experience, it is highly desirable to provide the user with information about potential losses or other negative consequences due to avoidance of a certain security step.

---

[2] May be it is intended to perform in the future. However, the PTP draft [14] does not clearly state this.

We found one issue that should perhaps be dealt with in the further development of the PTP protocol, namely accountability. In the current draft it is not addressed, but it seems necessary ingredient of the protocol in a form or another.

The further step of the research is to evaluate concretely the risk levels that now remained unknown. A more thorough analysis might also reveal new threats and risks that we overlooked at this stage. Finally, the real implementations of the PTP protocol will give more possibilities and insight for the detailed analysis of the risk levels.

We emphasize that this is really an initial analysis and we do not claim that it is complete. Neither do we claim that all conclusions we have made about the relative risk levels are final. For instance, we have assumed in a straightforward manner that all threats are independent while assessing the risks. This may not be fully true in all cases. These are for a further study.

In spite of the incompleteness and tentative nature of the analysis, we still hope that we have captured the key issues and that the analysis is helpful for the implementers and designers of the PTP protocol and other similar protocols.

REFERENCES

[1] "Access a Wireless Future", Trends Report, *http://www.trendsreport.net/wireless/2.html*, 2001.

[2] "Content Security at Hand", A White Paper on Handheld Device Security, White Paper, Available at *http://www.europe.f-secure.com/products/white-papers/hhsecurity020215.pdf*, February 2002.

[3] Dolev, D., Yao, A. On the Security of Public Key Protocols, IEEE Transactions on Information Theory, 29, pp. 198-208, 1983.

[4] Farrow, R. "TCP In The Crosshairs: Will a weakness in creating TCP connections open the way to new attacks?", Network Magazine, *http://www.networkmagazine.com/article/NMG20010620S0004*, 2001.

[5] Focardi, R., Gorrieri, R. (eds.) Foundations of security analysis and design, Lecture Notes in Computer Science 2171, Springer, Berlin, Germany, 2000.

[6] Gong, L., Syverson, P. Fail-stop Protocols: An Approach to designing secure protocols, 5th International Working Conference on Dependable Computing for Critical Applications, pp. 44-55, 1995.

[7] International Organization of Standardization: Information technology: Open Systems Interconnection: Security Frameworks for Open Systems, ISO/IEC DIS 10181, 1991.

[8] "IrDA Object Exchange Protocol – IrOBEX, Version 1.2", IrDA, March 1999. Available from *http://www.irda.org*.

[9] Juul, N.C., Jørgensen, N. Security Issues in Mobile Commerce using WAP. Proc. 15th Bled Electronic Commerce Conference, Slovenia, June 17-19, 2002.

[10] Lowe, G. An Attack on the Needham-Schroeder Public Key Authentication Protocol, Information Processing Letters, 56(3), pp. 131-136, 1995.

[11] Lowe, G. Some New Attacks Upon Security Protocols, 9th Computer Security Foundations Workshop, IEEE Computer Press, Kenmare, County Kerry, Ireland, pp. 139-146, 1996.

[12] Markkula, J., Katasonov, A., Garmash, A. Developing MLS Location-Based Service Pilot System. Proceedings of Smartnet'2002, IFIP 7th Conference on Intelligence on Networks, Kluwer, 2002.

[13] "Mobile electronic Transaction Forum. MeT terminology", *http://www.mobiletransaction.org/pdf/R11/MeT-Terminology-R11.pdf*, 2001.

[14] "Mobile electronic Transaction Forum. Personal Transaction Protocol (PTP) Specification Draft Version 0.14 (20-Aug-2002)".

[15] McNeill, K. "Wireless: You Will Be Assimilated", *http://www.crmdaily.com/perl/story/15238.html*, 2001.

[16] Open Mobile Alliance (OMA). See *http://www.openmobilealliance.org*.

[17] Power, R. "Current and Future Danger: A CSI Primer on Computer Crime and Information Warfare", Computer Security Institute, San Francisco, California, USA, 1998.

[18] Song, D., Wagner, D., Tian, X. Timing analysis of keystrokes and timing attacks on SSH, 10th USENIX Security Symposium, Washington, D.C., USA, 2001.

[19] Syverson, P. A Taxonomy of Replay Attacks, 7th Computer Security Foundations Workshop (CSFW), pp. 187-191, IEEE Press, 1994.

[20] Tang. J., Terziyan, V., Veijalainen, J. Distributed PIN Verification Scheme for Improving Security of Mobile Devices. ACM MONET, Special Issue on Security in Mobile Computing Environments. (Forthcoming).

[21] "Working in a Generation on the Move", *http://www.3elimited.com/OtherMenuPages/scoop.htm*, 2001.

[22] XML-Based Log Format (XLF). See *http://www.docuverse.com/xlf*.

[23] Ylonen, T, Kivinen, T., Saarinen, M, Rinne, T., Lehtinen, T. SSH Protocol Architecture, Available from *http://www.globecom.net/ietf/draft/draft-ietf-secsh-architecture-07.html*, 2002.

CHAPTER 8
NODE-CENTRIC HYBRID ROUTING FOR WIRELESS INTERNETWORKING

J.J. Garcia-Luna-Aceves and Soumya Roy
Computer Engineering Department
University of California, Santa Cruz *
jj@cse.ucsc.edu, soumya@cse.ucsc.edu

8.1.   Introduction

Multihop packet radio networks (or ad-hoc networks) consist of wireless routers that interconnect attached hosts without the need of any pre-established communication infrastructure. These networks play an important role in relief scenarios, battlefields and conference scenarios, where there is no base infrastructure.

Table-driven or proactive routing protocols can incur excessive signaling overhead in large ad-hoc networks (e.g., networks with hundreds of nodes or more), because each node in the network must maintain routing information for every other network node, even if the node never needs to handle traffic destined for some nodes and the path between any two nodes in a highly mobile environment changes frequently. Control overhead in proactive routing protocols increases with the size of the network and becomes redundant if the number of communicating peers is much less than the total number of nodes in the network.

To address the scaling problem of table-driven routing, on-demand routing protocols have been proposed for ad hoc networks. Nodes running such protocols set up and maintain routes to destinations only if they are active recipients of data packets and generate network-wide queries to establish routes to destinations. However, when only a few nodes of the ad hoc network must act as sources and sinks of data packets, and such choices are very stable, maintaining routing information to such destinations on demand and treating those nodes as any other

*This work was supported in part by the Defense Research Projects Agency (DARPA) under grant F30602-97-2-0338 and by the US Air Force/OSR under Grant No. F49620-00-1-0330

node may not be as attractive as a proactive approach. This motivates the interest in a hybrid approach to routing in ad hoc networks.

The Zone Routing Protocol (ZRP) [Haas and Pearlman, 1999] constitutes a framework for hybrid routing in ad hoc networks. ZRP adapts a hierarchical-routing approach based on clusters (called zones) and maintains routes proactively to destinations inside a zone, and on-demand routing is used to establish routing information spanning more than one zone. In this chapter, we advocate a different approach to hybrid routing that is node centric rather than based on zones or areas of the network.

The rationale for a node-centric approach to hybrid routing is that there are many cases in which certain nodes in an ad hoc network need to host special services to other nodes in the ad hoc network, specially when the ad hoc network is a wireless extension of the Internet (e.g., DNS services, Internet access, web proxies). We call those nodes that support special services for the rest of the nodes (and therefore that have a high likelihood of communicating with the rest of the ad hoc network) *netmarks*. This scenario is illustrated using Fig. 8.1, which shows an ad hoc network of mobile nodes $a, b, \ldots\ldots$, and t and a single netmark. The netmark can be fixed as well as mobile, depending on the application scenario. Under a node-centric hybrid routing approach, paths are constantly maintained between nodes $a, b, c, \ldots$, and t and the netmark. The forward and reverse paths between netmarks and nodes $a, b, c, \ldots$, and t (shown using solid lines) are maintained constantly. If node e needs to communicate with node f, while node j communicates with node p, the paths between those nodes are set up on demand and are shown in dashed lines in Fig. 8.1. Observe that node c does not need to know how to reach node r, so no route needs to be maintained at node c to reach node r.

The landmark hierarchy [Tsuchiya, 1988] is an earlier node-centric approach to hierarchical routing designed for proactive routing in large networks. The key difference between the node-centric routing described in this paper and the landmark hierarchy is that a landmark becomes the address of a common node, while a netmark is a destination that provides services. Hence, the node-centric routing is an orthogonal approach to the landmark hierarchy, which is aimed at aggregating routing-table entries.

Section 8.2 introduces two approaches to node-centric hybrid routing, in which a netmark is distinguished from normal nodes (which can be done through addressing or by having an explicit tagging mechanism). In one approach, a netmark forces the rest of the nodes to maintain their routes to it for long periods of time once they acquire such routes. This amounts to extending the caching of netmark routing information. In another approach, a netmark uses proactive routing updates to push its routing entry into the routing tables of the rest of the nodes in the ad hoc network. To describe our approaches, we assume that the nodes in the ad hoc network form a subnet and each node has a unique identifier, by which routing protocols and other applications can identify it. By looking at the address of the destination of any IP packet, any node can determine whether or not the packet is meant for a node outside the subnet. Links are assumed to be bidirectional, such that when a node hears from another node, it can assume that it can also forward packets to that neighbor. Nodes are assumed to operate correctly and information is assumed to be stored without errors.

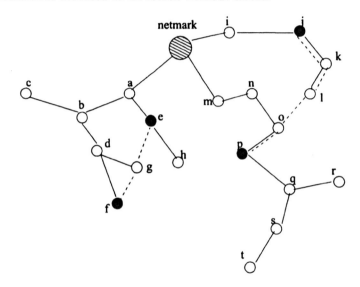

Figure 8.1.   Figure showing an ad hoc network with a single netmark

Section 8.3 shows how an on-demand routing protocol can be modified to adopt the node-centric approaches proposed in Section 8.2. Our choice of using on-demand routing based on partial link-state information for node-centric hybrid routing is based on the performance of on-demand link-state routing protocols and the ease with which node centric hybrid routing mechanisms can be incorporated in protocols based on link-state information. Section 8.4 addresses the issues that arise with having multiple netmarks in an ad hoc network. Section 8.5 presents the results of our performance comparison of node-centric hybrid routing with purely on-demand routing protocols. For our study, we use the ad hoc on-demand distance vector routing (AODV) protocol [Perkins et al., 2002], the dynamic source routing (DSR) protocol[Johnson and Maltz, 1994] and the source tree on-demand adaptive routing (SOAR) [Roy and Garcia-Luna-Aceves, 2001] protocol. Previous simulation studies [Roy and Garcia-Luna-Aceves, 2001], [Broch et al., 1998], [Das et al., 2000] have reported the performance of on demand routing protocols in mobile ad hoc networks with a very uniform traffic pattern, i.e., traffic flows are randomly distributed throughout the network, with no node being accessed more than others. However in practice, traffic patterns are not uniformly distributed and typically concentrate around nodes that offer special services to others, even though all nodes combine to forward data packets. This is true specifically when an ad hoc network is a wireless extension of the Internet and a few network access points are used to attach to the Internet as well as to provide such services as DNS and proxies. The traffic (mostly web traffic, which can be assumed to be heavy-tailed ON-OFF traffic) would exist mainly between the nodes of the ad hoc network and the network access points. This is also the case of battlefields or relief scenarios, in which a large amount of communication exists between a group leader and the rest of the group. The results of our simulation experiments illustrate the benefits of the node-centric hybrid routing approach.

8.2.   Node Centric Hybrid Routing

8.2.1   Hybrid Routing by Extended Caching of Netmarks

In pure on-demand routing protocols, routers set up paths to other nodes based on the existence of flows with them. Routes are cached once they are obtained using a route discovery mechanism and they are modified when they become invalid due to link failures. Among the on-demand routing protocols proposed in the literature, the basic difference is in how routes are cached and invalidated, and how route changes are reported to other nodes. Extending an on-demand routing protocol to support hybrid routing through extended caching of the routes to the netmarks entails the following two main changes to a pure on-demand protocol:

- A node sends a route request for a netmark whenever it looses all routes to it independently of its traffic to the netmark.

- Any node generates a route error whenever it detects the loss of a route to a netmark independently of the traffic for the netmarks.

We now summarize how extended caching can be applied to specific on-demand routing protocols, namely AODV, DSR and SOAR.

The ad-hoc on demand distance vector (AODV) protocol is based on distance vectors and uses sequence numbers to prevent temporary and permanent loops. AODV supports incremental routing because no routing table loops can be formed. Route Requests (RREQs) are generated by the sources of data packets and forwarded by intermediate nodes. When RREQs are forwarded, reverse routes for the source of the RREQ are installed. Route replies (RREPs) can be sent by the destination or an intermediate node with an unexpired route entry for the destination. The RREP message initiates the creation of a path for the destination in intermediate nodes that forward the RREP back to the sender of the RREQ. Each routing table entry has an expiration period (*active_route_timeout*) associated with it.

The specification of AODV [Perkins et al., 2002] states that AODV can use the periodic network-layer Hellos or link-layer notifications for determining connectivity with neighbors. When a node detects that its path to a destination has been broken due to a failed link, it sends a route error (RERR) packet to its active predecessors for that destination, i.e., those neighbors known to use the node to forward packets to the destination.

AODV can easily incorporate the idea of extended caching by increasing the expiration period of the route for the netmarks to a very high value or infinity. Every time the paths to a netmark fail due to link failures, the nodes can send RERRs to active predecessors, which travel upstream. When a node looses its route for a netmark, it starts the route discovery mechanism, irrespective of the presence of traffic for netmarks. Royer et al. [Royer et al., 2001] have suggested a similar foreign agent discovery mechanism using AODV where routes for foreign agents are discovered irrespective of the presence of traffic. However, routes for foreign agents are not maintained differently from the routes to common nodes.

The dynamic source routing (DSR) protocol uses source routes to forward data packets and exchanges routes in the form of paths. Routes are stored in a cache, until an indication that a link in the route is broken is obtained through route error (RERR) messages or link-layer notifications. A route discovery cycle is started by a source if it looses all routes to the required destination. DSR can determine on its own if a link is broken by doing multiple retransmissions, or can depend on the link layer for link-failure notifications during packet transmission failures. Route error containing information about failed links are sent towards the source of data packets. In DSR, if a node detects the route failure for a netmark while forwarding data packets for a common node, then it does not know how to propagate RERR information for the netmark.

Hence, generating route errors independently of traffic to netmarks is not trivial in DSR and flooding RERRs would be expensive. RREQs sent in response to route failures to netmarks cannot be used to replace the functionality of RERRs, because RREQs do not have the information about failed routes. However as in AODV, anytime the route to a netmark is broken, the basic route discovery mechanism of DSR can be easily modified to send RREQs for netmarks, irrespective of the presence of traffic for them.

The source-tree on demand adaptive routing (SOAR) [Roy and Garcia-Luna-Aceves, 2001] protocol is a link-state routing protocol in which routers exchange minimal source trees in their control packets. A minimal source tree consists of the state of the links along the paths used by the routers to reach active (important) destinations. Important destinations are active receivers, relays or possible relays. Each router uses the minimal source tree of its neighbors and its outgoing links to get the partial topology of the network. Routing table entries for known destinations are computed using a path selection algorithm on the partial topology and the data packets are forwarded hop-by-hop according to the routing entries. Links are validated using sequence numbers. SOAR uses queries and replies to create routes to unknown destinations. Update packets containing modified minimal source trees are generated when a node decides that its neighbors need to be updated to prevent erroneous forwarding or the formation of routing-table loops in the paths to important destinations.

The notion of *important node* in SOAR can be generalized to incorporate the concept of netmarks by always tagging netmarks as *important*, while the rest of the nodes become important on the basis of the traffic flowing to those nodes. Another variation of this approach is that netmarks would be considered important for longer periods of time than common nodes. Therefore, depending on the level of importance of a particular node, paths to nodes remain fresh for different time-spans. Hence, a simple modification to the basic mechanism of routing information exchange in SOAR enables incorporating the idea of extended caching of routes to netmarks. We call this modification netmark-aware on-demand link state routing (NOLR).

8.2.2 Hybrid Routing with Proactive Routes to Netmarks

The second approach to hybrid routing consists of maintaining proactive routes for the netmarks, while on-demand routes are used for other nodes. The modifications required for any on-demand routing protocol to adopt this approach are the following:

1 Adding a route for a netmark for the first time necessitates sending updates to neighbors, so that they can also set up new paths to the netmarks.

2 Depending on the protocol, route errors, route requests, or route updates are generated for netmarks, independently of the traffic to them.

3 A netmark can advertise its presence by sending Hellos to enable new neighbors to set up paths to it, and to find out whether the netmark is still reachable without having to depend on link-layer notifications. This enables paths to netmarks to be more proactive, rather than being data-packet driven.

Next we summarize how DSR, AODV or SOAR can be changed to incorporate the above concepts of hybrid routing. Implementing a network-layer Hello mechanism at the netmarks is easy in any of the three protocols.

As mentioned earlier, AODV uses destination sequence numbers to validate routes to destinations. Hellos sent by the netmarks in AODV must contain the highest sequence number for the netmarks, so that the receiving node can install new direct routes for the netmarks. Given that DSR sends RERRs only to the source of data packets when there is a failure of packet transmissions along a particular link, adding a Hello mechanism will not help DSR, because in such a case the protocol does not have a mechanism to decide how to propagate information about the loss of netmark-routes to other nodes in the network. However, if a node keeps track of the neighbors (i.e., predecessors) that use it for data delivery to netmarks for the last *pre-defined* amount of time, route failures to netmarks can be recursively reported to other nodes through the predecessors. Incorporating the Hello mechanism benefits both AODV and SOAR, because the paths to netmarks remain more up to date.

To propagate new route information for netmarks in SOAR requires only a change in the rules for sending an *update*. Rather than only sending an *update* for a destination when the path cost to it increases, updates are also sent when a route for a new netmark is discovered. Unlike SOAR, both AODV and DSR need the introduction of a new type of control packet that propagates route updates for netmarks to all the nodes in the network when a route to a netmark is first discovered.

8.3.   Hybrid Routing Using SOAR

We have chosen SOAR as the basic routing protocol to illustrate the benefits of node-centric hybrid routing over on-demand routing because of the following reasons because SOAR has been shown to be more efficient than DSR [Roy and Garcia-Luna-Aceves, 2001] and the results presented in Sec. 8.5 show that SOAR outperforms AODV. Furthermore, the modifications needed in SOAR to adopt hybrid routing are much simpler than the modifications required in DSR and AODV.

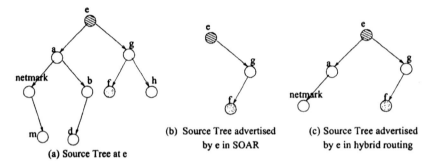

(a) Source Tree at e

(b) Source Tree advertised
by e in SOAR

(c) Source Tree advertised
by e in hybrid routing

Figure 8.2.   Figure showing difference in control information in SOAR and NEST/NOLR

Extended caching of netmarks can be adopted in SOAR by considering paths to different destinations as important for different periods of time. This modification of SOAR is called NOLR (netmark-aware on-demand link state routing).

The Netmark Enhanced Source Tree (NEST) routing protocol adopts the routing mechanisms of SOAR to maintain proactive routes for netmarks and on-demand routes for other nodes in the network. We provide the details of NEST in the rest of this section.

NEST and SOAR both exchange link-state information in the form of the minimal source trees, that contain the state of the links to reach important destinations in the network. However, there is a slight difference in the actual contents. In NEST, the advertised minimal source tree always contains a path to the netmark. Fig. 8.2 illustrates the difference in the control message advertised by node e in the network shown in Fig. 8.1, in which every node has a proactive path with the netmark and nodes e and i have on-demand routes set up for f and p, respectively. The source tree at node e (Fig. 8.2(a)) is the tree consisting of links that node e uses to reach the netmark and other nodes in the network. Node e advertises a portion of this complete source tree to its neighbors, which is called the *minimal* source tree. For SOAR, the *minimal* source tree would only consist of links needed to reach nodes with which it has active flows. In this example, node e has active flow with node f, the minimal source tree advertised by node e would be as shown in Fig. 8.2(b). In NEST, even if node e does not have active communication with the netmark, it advertises links in the path to it (as shown in Fig. 8.2(c)).

8.3.1   Netmark Discovery

In NEST, netmarks send Hello packets to inform their neighbors of their presence. This same effect is achieved by sending beacons at the MAC layer. At the routing layer, if the node does not receive the Hello packet for some predefined interval of time, then it can declare that the link to its neighbor is down. The link layer can also notify the routing layer about link failures when it cannot deliver data packets. In the presence of data traffic, this later mechanism can detect link failures faster.

When a node has a new entry for the netmark, it updates it neighbors about the new route using an update. Therefore, every node in the network eventually knows about at least one path to the netmark. For example, in Fig. 8.1, when a learns about the existence of the netmark in its neighborhood, it advertises its path to it, and its neighbors b and e re-advertise. Updates in NEST are broadcast packets and hence unreliable. If updates are lost, some nodes may not know about the route to the netmark. Furthermore, when a node comes up, it is not aware of netmarks. Under such circumstances, a node must initiate a *query*. However, because paths to netmarks are maintained proactively by all common nodes, there is a considerable reduction of queries required to establish routes to netmarks.

Another important issue regarding tagging some nodes as *netmarks* is how to disseminate the information about the nodes with netmark status. This can be achieved in one of the following ways:

- All common nodes can be pre-configured statically with the host addresses of the netmarks. This approach is beneficial if some nodes maintain statically their netmark status and all common nodes know about the netmarks. For example, netmarks hosting proxy services or DNS services will not change and the mobile routers can be statically configured with the address of the netmark.

- If the assignment of netmark status to nodes is not fixed, then the new netmarks can advertise their status by sending Hello Packets and all the nodes can mark the netmarks in the minimal source

trees while sending updates. Marking the netmarks means a special flag is attached in the minimal source trees to links with netmarks as tails. When a node ceases to be a netmark, it can notify
all nodes that it is no longer a netmark by flooding. This situation can happen in relief scenarios or battlefields, where the group
leaderships may change over time.

8.3.2   Maintaining Paths in NEST

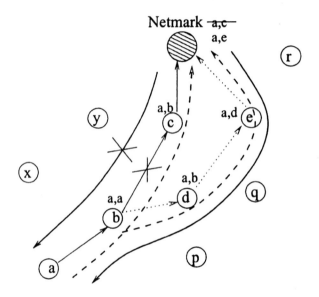

Figure 8.3.   Setting up of paths between netmarks and other nodes

Fig. 8.3 illustrates how the forward and reverse paths between a common node and netmark are set up. In Fig. 8.3, when node c learns of
the netmark, it advertises the netmark in its source tree and hence node
b knows about the netmark and neighbor c. When node b advertises its
own tree, it reports the information about nodes b, c, and the *netmark*,
i.e., the path to netmark consisting of the intermediate nodes b, c. Similarly, node b learns about an alternate path $[b, d, e, netmark]$ from node
d, but node b chooses the path through node c because it has a smaller
length. If link (b, c) fails, then node b can choose the alternate path
to the netmark through node d. The downstream nodes from a mobile
node towards the netmark may only know about the upstream predeces-
sor and not know about all upstream nodes (e.g., when node b advertises
its source tree, it may not advertise link to node a because node a is not

an important destination, in such a case, node c knows only about node b, but not about node a). Similarly, the netmark may know about node c, but not about nodes a and b. In order to send packets to node a, the netmark would have to send a *query* for node a in such a case. To prevent a *query* to be initiated from the netmark for every mobile node in the network, the following mechanism is adopted to set up the reverse paths from a netmark to any common node without introducing any extra control overhead.

When data packets start flowing from a node towards the netmark, the intermediate nodes along the path towards the netmark can set up paths towards the source of the data packets. For example, when the data packet from node a reaches node c and finds the destination is a netmark, then node c adds an entry in its routing table for node a as $[dest = a, nexthop = b]$. Similarly, the netmark keeps a routing-table entry $[dest = a, nexthop = c]$. These routing entries expire after a *Soft_State_Interval*. When link (b, c) breaks, data packets are forwarded along the path $[a, b, d, e, netmark]$ and the netmark replaces the entry $[a, c]$ with $[a, e]$ when the data packets arrive from node e. Similarly, when nodes d and e forward packets, they set up soft-state entries for destination a. Node c removes entry $[a, b]$ after the *Soft_State_Interval* due to the absence of any data packets from node a towards the netmark. Routes for different netmarks from the same common node can pass through same intermediate nodes. In that case, at any intermediate node, variations of the path traced by the data packets with the same source but with different destinations can lead to route flapping. Therefore, to prevent route flapping, the soft-state entry is not modified for *Soft_State_Interval* after the last change.

The reverse routes are set up towards the source based on the flow of data packets only if the destination of the data packets is a netmark. This is because each node maintains up-to-date paths only to netmarks. Hence, given that the paths from a node to any other node may not be current, the reverse path between any two common nodes would lead to data packet losses. In steady state, the reverse route from a netmark is essentially the same as the forward path. Therefore, if the forward routes to the netmarks are correct, the reverse routes are going to be also correct.

As discussed above, the reverse routes are set up using the path traversed by the data packets, without incurring extra control overhead. However, in the absence of a flow of data packets from a common node to the netmark, the netmark must resort to queries to find paths to the destinations. The number of these queries is reduced drastically if the flow between the netmark and the node is mainly bi-directional. This mechanism of reverse path set up and maintenance ensures symmetry of the paths taken by data packets in *tcp*-like bi-directional flows.

The path maintenance mechanism is similar in both NEST and SOAR; however, paths for netmarks in NEST are always maintained up-to-date, because the netmarks are always tagged as important. When a router finds that the distance to an *important* node increases, it reports modified minimal source trees in its updates to its neighbors in order to correct their link-state information. Link-state information is validated using sequence numbers. A link-state update is trusted if it has a higher sequence number, or the same sequence number but a smaller link cost.

When a node receives a packet from the application layer and it is meant for a node in the ad hoc network, it forwards it to the next hop as indicated in the routing table entry, provided that the node has a route to the destination. All nodes can determine when the packets are meant for a node outside the ad hoc network by looking at the IP address of the destination, and forward such packets towards the netmark.

8.4.   Multiple Netmark Scenarios

When multiple netmarks are present in the network, the way in which nodes affiliate themselves to netmarks has a direct impact on performance of routing protocols.

A node can be made to always communicate to a particular netmark irrespective of the location of the node with respect to the netmark in the network. This static affiliation can happen in a battlefield scenario or sensor net, where irrespective of the position of a soldier or sensor, data must be reported to a specific group leader or collection station. Though every node is only affiliated to a particular netmark, it has to know how to reach every netmark in the network, because every node may have to forward data packets for other nodes in the network and all nodes may not be affiliated to one particular netmark..

In a different scenario a node need not be affiliated with any particular netmark and it can be allowed to communicate with any netmark. This dynamic affiliation can happen when the ad-hoc network is an extension of the Internet, and there are multiple Internet access points. The common nodes in the network can communicate with any access point, because all access points are connected to the Internet. Given that packets are forwarded to any netmark, routing becomes efficient in terms of control overhead because (a) redundancy of routes to the Internet reduces the number of expensive route discovery cycles; (b) outgoing packets within the ad hoc network tend to be more optimal; and (c) an anycast route discovery mechanism can help to reduce control overhead, improve optimality of paths and reduce latency of path discovery. Queries are not required to be sent individually for each netmark. Instead *anycast* queries can be sent asking for a route to the anycast address of all the netmarks. In such a case, any router with a route to any one netmark can reply. The reply would contain the route to the nearest netmark if more than one routes were known to the responding router.

There can be scenarios in which netmark affiliations are both static and dynamic. Packets for some fixed destinations outside the ad hoc network are always forwarded to some particular netmarks, while packets for others can be forwarded to the nearest netmark.

Depending on how data packets enter the ad hoc network and how the outgoing packets are forwarded, paths followed by data packets can be symmetric or asymmetric. In the case of static affiliation, as shown in Fig. 8.4(a), if the incoming packets for a mobile router are always forwarded to the netmark with which the router is statically affiliated, then only the data-path can be symmetric. For the example shown in Fig. 8.4(b), due to the dynamic affiliation, though data packets arrive for router *i* from *netmark 1*, the outgoing packets are forwarded towards *netmark 2*, thereby making the data-path assymetric.

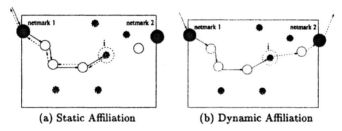

(a) Static Affiliation   (b) Dynamic Affiliation

Figure 8.4.   Figure depicting paths taken by data packets in an ad hoc network with multiple netmarks

When the netmark is an Internet access point, it advertises routes to subnets but not host routes to the Internet. Accordingly, in the example of Fig. 8.4 it may happen that *netmark 1* advertises path to a subnet that includes node *i*. However, due to network partitioning, *netmark 1* may not have any route for node *i*. Hence, packets for node *i* are dropped if they are forwarded to *netmark 1*. However, node *i* can still be reached through *netmark 2*. This can be remedied if the netmarks form a fully-connected overlay network using the wired Internet. In such a case, packets for node *i* can be forwarded using the overlay network from *netmark 1* to *netmark 2* and *netmark 2* can finally deliver the data packets to node *i*.

8.5.   Performance Evaluation

8.5.1   Simulation Model

We have compared the performance of node-centric routing approaches NEST and NOLR with the performance of pure on-demand routing protocols SOAR [Roy and Garcia-Luna-Aceves, 2001], DSR [Johnson and Maltz, 1994] and AODV [Perkins et al., 2002] using the ns2 network simulator. For DSR, we used the code available with the ns2 simulator [ISI-NS, 2000]. For AODV, we used the code available from the implementation of Marina[Marina, 2000] and the constants provided with the code. SOAR has been implemented according to the specifications provided by Roy and Garcia-Luna-Aceves [Roy and Garcia-Luna-Aceves, 2001]. NOLR is the modification of SOAR with extended caching of routing information for netmarks. NEST uses the same constants used in SOAR [Roy and Garcia-Luna-Aceves, 2001] along with three additional constants: Hello_Interval, Dead_Time_Interval and Soft_State_Interval. Hello_Interval (three secs) is the interval between sending of two consecutive Hello packets by the netmark. Dead_Time_Interval (nine secs) is the time interval for which if a node does not receive any packet from a netmark, the link to the netmark is considered to be down. When a broadcast control packet is sent, the netmark can defer the next transmission of Hello packet for a time equal to the Hello_Interval. Soft_State_Interval (one sec) is the maximum time a soft-state routing entry stays in the routing table, without being refreshed.

DSR, AODV, SOAR, NOLR and NEST do not depend on the link layer for neighbor discovery. All protocols use link-layer indications about link-failures when data packets cannot be delivered along particular links. Use of link-layer information for discovering neighbors can significantly improve the performance of routing-layer protocols. However, because our objective is to test the routing protocols as stand-alone protocols, we have not considered the effects of MAC layer interactions on the routing protocols' performance and promiscuous mode of operation has been disabled. The link layer protocol used is the IEEE802.11 distributed co-ordination function (DCF) for wireless LANs, which uses a RTS/CTS/DATA/ACK pattern for all unicast packets and DATA packets for all broadcast packets. The physical layer approximates a 2 Mbps DSSS radio interface. The radio range of the radio is 250m. We assume a netmark does not change its *netmark* status during the entire length of the simulation.

Nodal movement occurs according to the random waypoint model introduced in [Broch et al., 1998]. In this model, each node is at a random point at the start of the simulation and after a *pause time* seconds the node selects a random destination and moves to that destination at a constant speed. Upon reaching the destination, the node pauses again for *pause time* seconds, chooses another destination, and proceeds there. The speed of a mobile node during its movement is uniformly distributed between 0 and 20m/sec.

We have introduced two traffic models for performance evaluation, which we call (a) the INTNET model and (b) the RELIEF model. These traffic models are more realistic compared to the traffic models used in prior analyses [Broch et al., 1998], [Das et al., 2000], in which continuous CBR traffic flows exist between randomly chosen nodes making the traffic pattern more or less uniform throughout the network.

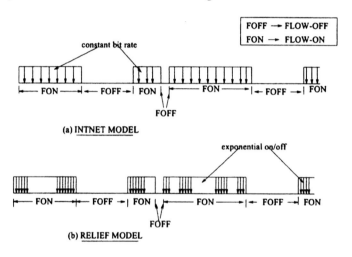

Figure 8.5.   Traffic Flow Models

The INTNET model is similar to the scenario of using ad hoc networks as wireless extensions of the Internet. The communication is mainly from each of the common nodes towards the netmark hosting commonly-accessed servers or acting as the access point to the Internet. The number of flows between mobile nodes only is much less compared to the number of flows between nodes and netmark. The traffic pattern is based on a FLOW_OFF/ON model, as shown in Fig. 8.5(a), with the parameters as given in Table 8.1.

During the FLOW_ON period, there exists *cbr* traffic and there is no packet flow during the FLOW_OFF period. The motivation behind simulating the FLOW_OFF/ON model, rather than a model in which the flows are on continuously, is that Web traffic consists of FLOW_ON/OFF periods, where the OFF periods correspond to the user's think time, and

FLOW_ON period	:	Uniform Dist (30,120) secs
FLOW_OFF period	:	Uniform Dist (50, 120) secs
Packet Size	:	66 bytes
Rate	:	3, 5 packets/sec per node

Table 8.1. Constants for Flows in INTNET model

the ON period represents download time. In our experiments with the INTNET model, there are four random flows between any two randomly selected common nodes at any time. The duration of these flows is always 200 secs and all the flows are bi-directional in nature.

The RELIEF traffic model is used to simulate traffic in relief or battlefield scenarios, where the group members report to the group leaders while the group members also exchange information. The group leader is the netmark contacted more frequently compared to other nodes. There are four random flows between common nodes and there are at most six random flows from a common node towards the netmark. We divided the set of common nodes into five groups and only one member in the group can talk at a time with the netmark. The traffic pattern per group is also like the FLOW_OFF/ON model. The packet arrivals during FLOW_ON period follow an interrupted deterministic process (IDP) as shown in Fig. 8.5. The IDP model is used to simulate the voice traffic. The ON/OFF periods during a FLOW_ON period correspond to talk-spurt/silence periods of the speaker. The parameters for the RELIEF model are as given in Table 8.2.

FLOW_ON period	:	uniform (30,150) secs
FLOW_OFF period	:	uniform (10, 20)
Packet Size	:	66 bytes
Rate	:	17 packets/sec (9kbps)
talkspurt	:	350ms
silence	:	650ms

Table 8.2. Constants for Flows in RELIEF model

We evaluate the routing protocols based on packet delivery ratio, control packet overhead, average hop count, end-to-end delay, and the number of queries and replies sent by each protocol.

8.5.2   Experimental Scenario 1

The first scenario consists of a network of 31 nodes moving over a rectangular area of 1000mx500m. There is a single netmark in the system, which is placed at coordinates (500, 250) and is fixed throughout the

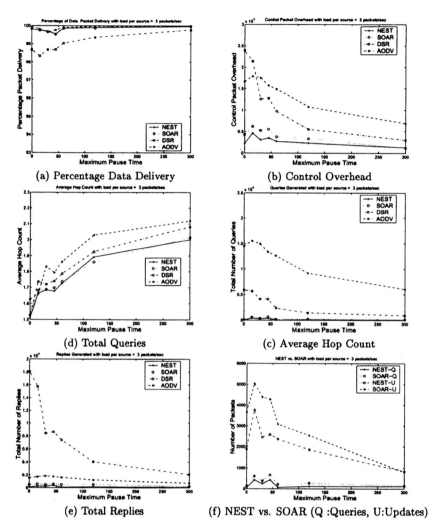

Figure 8.6. Performance of NEST, SOAR, DSR, AODV in a 31node Network at load per node of 3 packets/sec with fixed netmark

simulation time. The pause time of other nodes is uniformly distributed between zero and a maximum value, which can be one of 0, 15, 30, 45, 60, 120 and 300 seconds. The simulation length is 600 secs, while the results are presented on the basis of at least 3 simulation runs, where each run is with the same INTNET traffic model but with a different randomly generated mobility scenario (this is also true for subsequent

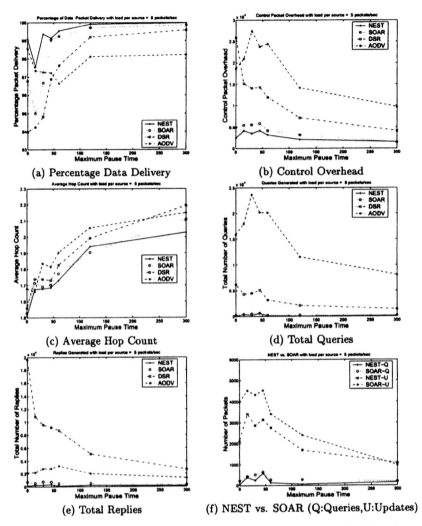

Figure 8.7.   Performance of NEST, SOAR, DSR, AODV in a 31node Network at load per node of 5 packets/sec with fixed netmark

experiments). Performance results are presented for two different load scenarios of three and five packets/sec per node during the FLOW_ON period.

Most of the results for AODV, SOAR and DSR conform to the results published previously for those protocols [Das et al., 2000], [Roy and Garcia-Luna-Aceves, 2001] and [Broch et al., 1998]. As shown in

Fig. 8.6(b) and Fig. 8.7(b), AODV's control overhead is found to be significantly higher than DSR's or SOAR's, except for the high mobility scenarios. AODV's control overhead consists primarily of queries (Fig. 8.6(d) and Fig. 8.7(d)), while the control overhead of DSR consists mainly of replies(Fig. 8.6(e) and Fig. 8.7(e)). This is because AODV resorts to the route discovery mechanism more often than DSR, while DSR sends multiple replies to queries. Contrary to the findings in [Das et al., 2000], [Broch et al., 1998] an interesting result for the INTNET model is that AODV's control overhead in highly mobile scenarios is lower than DSR's. Because each node in the INTNET model sends and forwards packets for a netmark, the number of cached entries for the netmark is comparatively higher in DSR than in scenarios where the traffic pattern is uniform. That effectively leads to significantly higher number of cached replies (many of which contain stale routes) which amount to higher control overhead in DSR than in AODV. In low mobility scenarios in which path information becomes stale less often, the effect of injecting old routes due to multiple replies is much smaller.

SOAR produces much fewer control packets compared to DSR or AODV under all mobility scenarios with varying loads. The reason behind this is that SOAR resorts to fewer route discovery queries than AODV or DSR, because of the redundancy in the exchanged routing information in the control packets specifying minimal source trees, and because of the use of mostly local updates to solve path breakage, rather than sending route error messages to the source of data packets. Because SOAR and DSR can use stale information, SOAR and DSR deliver slightly fewer data packets compared to AODV under heavy load scenarios and high mobility. The performance degradation is less in SOAR compared to DSR, because SOAR uses sequence numbers to validate link−state information and DSR uses explicit route error messages to invalidate link information.

The performance of NOLR was found to be almost identical to SOAR's. Accordingly, for clarity, Fig. 8.6 and Fig. 8.7 do not show the results obtained for NOLR. Unlike SOAR, NOLR maintains routing information for netmarks for longer periods of time compared to the time for maintaining information for other nodes. The reason why NOLR and SOAR exhibit the same performance in this scenario is that each node either sends or forwards packets for the netmark the vast majority of time; Therefore, any node in SOAR ends up treating the netmark as *important* throughout the simulation.

Under all scenarios, NEST performs much better compared to all purely on-demand routing protocols, both in terms of data delivery and control overhead. NEST (Fig. 8.6(a) and Fig. 8.7(a)) always delivers more packets compared to other protocols, with the effect being more prominent under heavy load. Overloading the network does not affect NEST, because each node always maintains correct paths to the netmark. Therefore, under heavy loads, NEST looses much fewer data packets than AODV, even though AODV attempts to avoid using stale routing information. SOAR maintains information for netmarks for significant periods of time; however, NEST paths are more accurate, because the netmark advertises itself periodically to force its routing information in other nodes and nodes using NEST update their neighbors when they first discover routes to netmarks. This conclusion is validated by the results of Fig. 8.6(f) and Fig. 8.7(f), in which we see that more updates are needed in SOAR compared to NEST to purge stale link-state information. On an average, NEST produces around 30% fewer updates than SOAR. We also find that NEST (Fig. 8.6(f) and Fig. 8.7(f)) produces fewer queries compared to SOAR, which leads to a reduction of replies in NEST (Fig. 8.6(e) and Fig. 8.7(e)). Queries are still sent by NEST for discovering routes on-demand with common nodes and for probing the netmark when the netmark becomes unreachable due to network partitions. We also see from Fig. 8.6(c) and Fig. 8.7(c) that the average hop count in NEST is the smallest, because NEST detects the presence of netmarks much faster. However, to conserve network bandwidth, given that the routers in NEST do not advertise route changes when distances to any node decrease, the path length in NEST can still be sub-optimal.

8.5.3   Experimental Scenario 2

The second scenario focuses on the effect of netmark mobility. It consists of a network of 30 nodes and one netmark with common nodes moving at a speed uniformly distributed between 5m/s and 20m/s. Pause times are uniformly distributed between 0secs and 30secs. Three different movement scenarios for the netmark are analyzed while keeping the mobility pattern for other nodes the same. The netmark in these scenarios is either static (model s), mobile (model m, the netmark moves over a rectangular area (250,250)) or very mobile (model vm, the netmark can move over the entire area (1000,500)). Because most of the traffic is towards the netmark, the routing protocols would be more stressed to maintain routes as the mobility of the netmark increases. The netmark moves with a speed similar to the speed of common nodes with pause time between 10 and 30 secs. We use the INTNET and RELIEF traffic models, which are indicated as REL and INT in Fig. 8.8. Accordingly, a static netmark model with INTNET traffic pattern is indicated as *sINT*, while a *vm* model with RELIEF traffic pattern is represented as *vmREL*, for example.

For this scenario, the results for AODV are significantly worse than for the rest of the protocols. Accordingly, the results for AODV are not shown in order to show in more details the performance differences among the other protocols. From Fig. 8.8 we see that there is no appreciable difference in the performance of the routing protocols, NEST, SOAR and DSR for the *m* and *s* model. This is because the radio range is 250 m and the netmark moves over an area of (250mx250m) for the *m* model, which does not contribute to too many additional path changes. The performance of all the routing protocols suffers when the netmark becomes highly mobile. From Fig. 8.8, we find that the INTNET model produces more stress on the routing protocols than the RELIEF model does, with DSR being affected the most. Though the traffic pattern is different in both cases, the total number of data packets sent throughout the simulation is the same and the network is not overloaded.

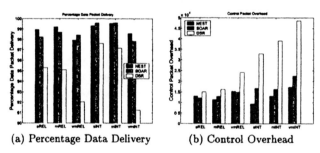

(a) Percentage Data Delivery   (b) Control Overhead

Figure 8.8.   Performance in a 31node Network with varying mobility models for netmark and two different traffic models

Table 8.3.   End to End Delay Distribution of the Voice Traffic For Mobile Relief Model (mREL)

Percentile	NEST (s)	SOAR (s)	DSR (s)
90	0.0091	0.0110	0.0087
95	0.1136	0.1503	0.0734
97	0.2682	0.3301	0.2226
range (max−min)	19.89	13.7335	19.329

From Fig. 8.8(a), we see that SOAR and NEST deliver on an average the same number of data packets in both traffic models, which we have also seen in our results of Sec. 8.5.3 for the low-load scenarios. DSR's percentage data delivery is 4%-7% smaller than the data delivery

Table 8.4. End to End Delay Distribution of the Voice Traffic For Mobile Relief Model (mREL)

Percentile	NEST (s)	SOAR (s)	DSR (s)
90	0.0101	0.0151	0.0121
95	0.1067	0.1768	0.0917
97	0.2150	0.3799	0.2087
range (max−min)	2.5041	14.4443	49.083

Table 8.5. End to End Delay Distribution of the Voice Traffic For Very Mobile Relief Model (vmREL)

Percentile	NEST (s)	SOAR (s)	DSR (s)
90	0.0611	0.0834	0.0986
95	0.3222	0.3930	0.3695
97	0.6832	0.8593	0.6371
range (max−min)	17.28	23.4054	13.119

achieved by SOAR and NEST, with the performance becoming worse with higher mobility of the netmarks. This is because of packet losses due to unavailability of routes to forward data packets at intermediate routers, which implies that DSR suffers due to stale path information.

DSR's control overhead is comparable to that of SOAR or NEST for the *sREL* or *mREL* models (Fig. 8.8(b)). DSR sends significantly more control packets for the INTNET model, where DSR utilizes redundancy in routing information less efficiently than SOAR or NEST.

SOAR and NEST have similar control overhead for the RELIEF model, though in the INTNET model NEST outperforms SOAR and DSR. This is because the RELIEF model has fewer flows (around six) towards the netmark compared to the INTNET model, in which theoretically any node can communicate any time with the netmark. When the number of flows is smaller, fewer links are used for active data delivery. Because detections of link failures are triggered only by the failure of transmission of data packets, with fewer flows more links remain stale in the topology table. Therefore, if the number of flows between common nodes is the same as the number of flows between netmarks and common nodes, the Hello mechanism does not improve the condition because SOAR and NEST require almost the same number of updates to purge wrong routing information. This indicates that the node-centric approach to proactive route maintenance can improve the performance of the network, however the degree of performance improvement depends on the amount of communication between common nodes and the netmark.

We also observe that netmark mobility does not impact the performance of NEST more than the performance of purely on-demand approaches, which could have been an argument for using an on-demand approach rather than a hybrid approach when netmarks are very mobile.

Because voice traffic is delay sensitive, we analyzed the delay performance for each of the routing protocols (Tables 8.3, 8.4, 8.5) for the RELIEF traffic model, where voice traffic is used. The results presented are for a randomly chosen run, so as not to average out the high frequency components of individual runs. This is important for voice traffic performance, because worst-case performance results are required for quality assurance. The following conclusions can be drawn from the data available from Tables 8.3, 8.4 and 8.5:

- Range (the difference between minimum and maximum delay) is significantly high under all cases. This is because the network becomes partitioned, and the node discovery mechanism is not very fast in on-demand routing protocols and can become really slow as the timeouts for resending queries increases non-linearly. The range for DSR for Mobile Relief Model (Tables 8.3, 8.4 and 8.5) is as high as 49 secs. Though NEST maintains proactive routes with the netmark, the range for NEST is also high because it uses on-demand routes to common nodes.

- As expected, there is an increase in delay when the netmark is more mobile, because nodes have more stale routes.

- NEST has better delay performance than SOAR in terms of percentile values because NEST has fewer queries and the paths tend to be more up to date, thereby spending less time queueing packets either at the routing layer or the link layer. This indicates that the hybrid routing approach helps to reduce the end-to-end delay of data packets.

8.5.4   Experimental Scenario 3

The third scenario is designed to address the impact of multiple netmarks on the performance of hybrid routing solutions. This scenario has two netmarks along with 30 mobile nodes. The netmarks are placed at coordinates (250, 250) and (750, 250), and are static throughout the simulation. The traffic pattern is according to the INTNET model. Packets from the Internet can enter the ad hoc network through any of the two netmarks. Because the netmarks advertise routes to the Internet for the

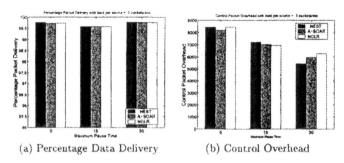

(a) Percentage Data Delivery   (b) Control Overhead

Figure 8.9. Performance of NEST, SOAR and NOLR for a network with 30 nodes and 2 netmarks

entire ad hoc network and not individual host routes, it can be safely assumed for the experiments that the incoming flows are randomly distributed between the two netmarks. Packets for the Internet are always forwarded to the nearest netmark. After a node decides which netmark to use, it encapsulates the IP packet with an IP header with the destination being the netmark's address. When routes to any netmark are not available, anycast queries (as discussed in Sec 8.4) are sent and routes are established based on the *anycast* replies. We compare the performance of NEST with SOAR (denoted as anycast-enabled SOAR or A-SOAR in Fig. 8.9) and NOLR. Like A-SOAR, if needed, NOLR and NEST also use *anycast* queries for netmarks.

From Fig. 8.9, we see that SOAR and NOLR both perform as good as NEST in terms of data delivery and control overhead under all three mobility scenarios. This is in contrast to the results presented in Sec. 8.5.2 and Sec. 8.5.3, where both SOAR and NOLR produce higher control overhead than NEST. This performance improvement in SOAR can be attributed to the following three reasons: (a) If a route to a given netmark is not available, packets can still be sent to other netmarks, which helps in reducing the number of queries; (b) anycast route replies help to prevent flooding of queries and faster route discovery; and (c) reducing the number of *reply-query* packets helps to prevent old link-state information to be injected into the network, which helps to reduce the number of updates. However, the proposed node-centric hybrid approach with proactive routes for netmarks is still attractive for this small network if the static or the hybrid method (Sec. 8.4) is used for forwarding outgoing packets, because in such situations every node might need a path to every netmark.

8.6.   Conclusions

We have presented node-centric approaches to hybrid routing for ad
hoc networks that distinguish between normal nodes and special nodes
called netmarks, which host popular network services or function as
points of attachment to the Internet. With node-centric hybrid routing,
netmarks force other common nodes to maintain routing information for
them by either advertising their routing information as in table-driven
routing protocols, or by requiring nodes to maintain routing entries towards them for extended periods of time. This reduces the network-wide
flooding and the corresponding delay for route set up every time a session needs to be established between a normal node and a netmark.
Routes between peer nodes are set up on-demand. We have evaluated
the changes needed to incorporate node-centric hybrid routing in the
basic mechanism of routing for some pure on-demand routing protocols, namely AODV, DSR and SOAR and compared the performance of
AODV, DSR and SOAR with the hybrid approaches, NEST and NOLR
(which have been adapted from SOAR) using ns2.

Our findings show that DSR's control overhead is higher than AODV's
for highly mobile and highly non-uniform traffic patterns. This is in
contrast to results of the previous studies [Das et al., 2000], [Broch
et al., 1998], where performance evaluations have been done with uniform
traffic pattern.

On the basis of ns2 simulations, we have found that, if a node in the
ad hoc network acts as a source or relay of data packets for significant
portion of its lifetime, the benefit of extending caching information in
a purely on-demand routing protocol is not noticeable. However, maintaining proactive routes as in NEST offers better performance than any
on-demand routing protocol, both in terms of data delivery and control
packet overhead when the traffic flow is mostly from common nodes towards the netmark. We have also found that the performance of NEST is
not affected by the mobility of netmarks. In a moderately-sized network
served by multiple netmarks, the performance of on-demand routing
protocols can be significantly improved by maintaining routes to any of
the netmarks and then sending anycast queries asking for a route to the
nearest netmark.

References

[Broch et al., 1998] Broch, J., Maltz, D. A., Johnson, D. B., Hu, Y. C., and Jetcheva,
    J. (1998). A Performance Comparison of Multi-Hop Wireless Ad Hoc Network
    Routing Protocols. ACM Mobicom.

[Das et al., 2000] Das, S. R., Perkins, C. E., and Royer, E. M. (2000). Performance Comparison of Two On-Demand Routing Protocols for Ad-Hoc Networks. IEEE Infocom.

[Haas and Pearlman, 1999] Haas, Z. and Pearlman, M. R. (June 1999). The Zone Routing Protocol (ZRP) for Ad Hoc Networks. http://www.ee.cornell.edu/ haas/Publications/draft-ietf-manet-zone-zrp-02.txt.

[ISI-NS, 2000] ISI-NS (25 May, 2000). The network simulator - ns-2.1b6, http://www.isi.edu/nsnam/ns/.

[Johnson and Maltz, 1994] Johnson, D. B. and Maltz, D. A. (1994). Dynamic Source Routing in Ad-Hoc Wireless Networks. Mobile Computing.

[Marina, 2000] Marina, M. (last updated on 12/07/2000). Aodv code for cmu wireless and mobility extensions to ns-2. http://www.ececs.uc.edu/ mmarina/aodv/.

[Perkins et al., 2002] Perkins, C. E., Royer, E. M., and Das, S. R. (March, 2002). Ad Hoc On-Demand Distance Vector (AODV) Routing. Mobile Ad Hoc Networking Working Group, draft-ietf-manet-aodv-10.txt.

[Roy and Garcia-Luna-Aceves, 2001] Roy, S. and Garcia-Luna-Aceves, J. J. (2001). Using Minimal Source Trees for On-Demand Routing in Ad Hoc Networks. IEEE Infocom.

[Royer et al., 2001] Royer, E. M., Sun, Y., and Perkins, C. (2001). Global Connectivity for IPv4 Mobile Ad hoc Networks. draft-ietf-manet-globalv4-00.txt.

[Tsuchiya, 1988] Tsuchiya, P. F. (1988). The Landmark Hierarchy: a New Hierarchy for Routing in very Large Networks. ACM Sigcomm.

Chapter 9

MOBILE MULTICAST

Elias C. Efstathiou and George C. Polyzos
Athens University of Economics and Business

9.1 INTRODUCTION

During the late-1990s, the popularity of Internet applications like B2C (business-to-consumer) e-commerce and web-browsing led traditional network operators to seek ways which allowed their customers to access these new services. Today (2002), IP-based packet switching technology has finally matured enough to replace the circuit-switched backbones of telecommunication carriers. Convergence on the IP protocol is now happening and it will continue to happen throughout this decade. It is obvious that this one common protocol will ease the management of all wired and wireless data networks. IP will also fuel the deployment of novel applications. An important feature these applications will expect from an all-IP internetwork is full support for multipoint communications. Multipoint, or group, communications are best described by the term *multicast*, a term associated with network support for efficient data delivery to more than one interested recipients. Multicast's objective is to place the least amount of burden on network and end-host resources. Applications that could exploit this feature include conferencing, on-line games, software distribution, and others. Although multicast can be emulated by letting the data sources themselves send packet copies to all intended destinations, this "multi-unicast" solution offers no scalability as resources on both the source host and its local network would eventually be depleted.

Today, the many players involved in what is a rapid expansion of the Internet still place "all-IP" convergence several years away. All of these players are, in one way or another, focusing on Internet protocols, with 3G licensees bearing much of the burden of bringing millions of new users and terminals to the Internet community. This worldwide deployment of the new 3G IP-based networks represents the Internet's biggest expansion yet. Meanwhile, digital broadcasters are also joining the IP bandwagon, with *DVB* (Digital Video Broadcasting – the digital TV standard) providing IP support and over 30 Mbps of shared bandwidth per DVB macro-cell. In addition, we must not forget all the older technologies that support IP over various link layers and which include Ethernet networks as well as all PSTN, ISDN, DSL and cable connections. Finally, all the wireless *IEEE 802.11* networks, also known as Wireless Local Area Networks or WLANs represent extremely important bridges between the older and newer technologies. The ease with which WLANs are deployed outweighs most of their limitations and, for many, WLANs represent the future in ubiquitous broadband wireless Internet access.

We can see, therefore, that although the "all-IP" goal has not yet been reached, it's clear that we are approaching it. All these converging, technologies were traditionally associated with particular services, e.g. bidirectional one-to-one audio (voice conversations) for cellular telephony, unidirectional one-to-many video (TV broadcasting) for DVB. The ideal all-IP internetwork must support each one of these services, integrate them, and "enhance" them with features such as higher interactivity and, most importantly, mobility and *tether-less* access. The requirement to support IP-based group communication is also dictated.

The IP suite was originally designed with one-to-one communications in mind. However, the advantages of one-to-many and many-to-many communications are numerous [10, 28] and we will not dwell on them further. Network technologies that provide "native broadcast" are better suited to support multipoint communications. Examples, such as the traditional Ethernet, IEEE 802.11 in all its flavors, and satellite, terrestrial or cable broadcast networks, allow for transmitted link-layer frames to be received efficiently by all hosts in a local subnet. Other network technologies, such as GSM and GPRS offer point-to-point links in the *user plane* [19], and only support broadcast in the *control plane*. With these cellular networks in particular, limited information can be delivered by *cell broadcasting*, but generally, users in a cell cannot engage in efficient group communication, although this is theoretically possible. Of course, there is absolutely no support in traditional GSM for inter-cell group communication. Finally, technologies like the PSTN, ISDN, and DSL networks are designed from the ground up to provide native support for point-to-point communications only.

IP currently (in both versions 4 and 6) supports efficient, local, network layer, many-to-many communications (i.e. multicast), wherever link level broadcast is provided. However, extending this support beyond a local subnet is non-trivial. Also, to support many-to-many communications in an internetwork topology that changes dynamically while IP hosts (and even entire subnets) are moving is even more challenging. Here we will focus on what we describe as *IP-based quasi-reliable mobile multicast*. IP multicast [10], which is the basis of our analysis, is a best effort multipoint communication protocol. Reliable, sequenced delivery extensions do exist [13]. However, applications don't always require this reliability, since they can adapt in their own more advanced ways. This, in particular, is the case with *streaming media* services (especially real-time media).

In an *"all-IP wireless Internet,"* IP hosts should be able to change their network point of attachment with minimal disruption of ongoing communications. We may assume that one or more of the following may occur while an IP host is moving:

– the host may temporarily disconnect from the network;
– the host's IP address, the one used for packet routing, may change;
– in a wireless scenario, a horizontal or vertical (cross-technology) handoff may occur;
– handoffs may occur between cells belonging to different administrations.

Here we will offer our perspective on the issues involved in combining multicast capability with host mobility in an all-IP wireless environment. We will do this while assuming the existence of a fixed, wired routing infrastructure. Also, whenever we discuss IP-related protocols, such as IP multicast and Mobile IP, our focus is on version 4 of the IP protocol. We are aware that the move to IP version 6 is happening, albeit slower than expected, so IPv6 is taken into account when it helps with our analysis.

The remainder of this chapter is organized as follows. Section 2 shows how early Internet assumptions influenced present-day protocols and made it harder to solve our current mobility problems. Section 3 outlines the IP multicast protocol. Section 4 presents two IP mobility protocols, Mobile IP, the standard IETF macro-mobility protocol and Cellular IP, a less well-known IETF micro-mobility protocol. Section 5 will mention the usefulness of *transcoding* filters to the mobile multicast problem. Section 6 deals with the coexistence of mobility and IP multicast. In Section 7, we present our own perspective on the problem; we mention requirements for future mobile multicast protocols; we outline our own, Cellular-IP based, mobile multicast solution; and we describe in more detail the all-IP wireless Internet environment we envisage. Section 8 presents some concluding remarks.

9.2   BASIC ASSUMPTIONS INFLUENCING TCP/IP

The Internet was originally built to support packet based data communications among pairs of stationary IP hosts. This assumption influenced the design of most protocols in the IP suite [26]. Although the overall design of the Internet protocols is "layered," in practice, it is often found that interlayer dependencies exist and become apparent when attempting to port existing services to newer network technologies.

We will use TCP as an example of the dependencies, design limitations and assumptions that cause service disruption when a TCP application like Telnet or FTP is used in a mobile environment:

Interlayer dependency - Each TCP connection is identified by a unique pair of *sockets*, with each socket being a *<Host_IP_address, TCP_port>* pair. The TCP connection ID is, therefore, the 4-tuple *<Client_IP_addr, Client_TCP_Port, Server_IP_addr, Server_TCP_port>*. It is obvious that lower layer identifiers (IP addresses) are used as part of higher layer ones and that TCP's internal session state relies on these identifiers. If either end-host changes IP address, the TCP connection will break because, not only does TCP rely on lower layer identifiers, but also, it assumes that they will not change during the lifetime of a TCP connection [26]. So, in practice, a user (who for example has initiated a lengthy FTP file download) has no way of changing his or hers network point of attachment without needing to restart the FTP session. Users with ongoing Telnet sessions and WLAN/GPRS enabled laptops cannot handoff these TCP/IP connections to the GPRS network as they exit from a WLAN hotspot, since, most probably, the WLAN and GPRS networks will have different IP network prefixes and the mobile host will be assigned completely different IP addresses in each one.

Design limitations - "Classic" TCP (we refer here to IETF's RFC 793, including the 1988 Jacobson refinements) is designed to interpret packet losses as a sign of congestion along the router path between the two end-hosts. TCP normally adapts by lowering its rate and allowing router queues to drain [17]. However, if the packet loss was caused by the increased BER (Bit Error Rate) of a wireless link, lowering the rate only decreases TCP performance. Studies and simulations have shown that the connection's effective bandwidth is greatly reduced even with moderate increases in BER, mainly because of TCP's "slow-start" algorithm. The Internet community's underlying assumption for TCP was, of course, that the two TCP endpoints would be stationary and that they would normally communicate over wired links with extremely low BER [30]. With wireless error rates in the 10^{-5} and

10^{-4} range, even approaching 10^{-2} and 10^{-1} in extreme cases, this assumption needs reexamination. There are many TCP variations, which address this issue, either by relying on associated protocols to explicitly deliver notifications about either congestion or loss events in addition to the simple TCP acknowledgment mechanism. One such notification mechanism is ECN, or *Explicit Congestion Notification*. Others try to recalibrate TCP timers and reconfigure the slow-start algorithm. Of course, there are also many solutions that try to increase reliability at the link-layer or attempt to add *Forward Error Correction* (FEC) information, usually combined with packet fragmentation and fragment reordering so as to minimize the effects of burst errors.

9.3   IP MULTICAST

In this section we present an outline of the basic IETF multipoint protocol, IP multicast. IP multicast is a many-to-many communication protocol. The *host group* service model defines its requirements: let H be the set of all IP hosts. Let E_G be a subset of H. Set E_G forms a *multicast group with group address G*, if and only if:

- members of H may join and leave E_G at any time;
- members of H can communicate unidirectionally with all members of E_G, using only identifier G. This identifier is also known as then *host group address*. This second requirement suggests that a host need not be a member of a multicast group in order to sent data to that particular group.

The definition above does not specify how to satisfy these requirements. They represent an idealized version of what group communication should be like and in practice the various implementations interpret these requirements liberally.

As IP hosts can belong to more than one group, mechanisms are needed in order to:

- associate hosts with groups;
- track group membership;
- route data to all group members [28].

Also ideally, IP multicast packet delivery should emulate the exactly-once semantics of packet delivery in a traditional (non-switched) Ethernet [2]. Since in an Ethernet all transmitted frames are received by all attached interfaces, exactly-once delivery comes for free. In an Ethernet, the only additional mechanism needed is the mapping of multicast group identifiers to MAC addresses. In this way, Ethernet hosts know which frames to process and which to discard. (Switched Ethernets must also provide support for multicast frames.)

The IETF set aside all legal IPv4 class D addresses to be used as multicast group identifiers, underlining the fact that IP multicast is a network layer protocol. The special IPv4 class D address *224.0.0.1* always identifies, according to the specification, the link-local *all-hosts* multicast group, which all multicast IP hosts are required to be a part of [9]. The mechanisms needed to implement multicast packet delivery can be divided into *global* and *local* [28].

9.3.1   Global Mechanisms

In the multicast model, the burden of delivering multiple copies of packets falls on the network. The sender need only transmit one copy of each packet addressed to a multicast group identifier (a host group address). Multicast enabled routers will forward the packet as needed, replicating it onto more than one of their outgoing interfaces only when paths towards the destination group members diverge. The global multicast routing protocols deliver a group's packets to multicast routers that have expressed interest in receiving packets for the particular group. This interest, in turn, is triggered by hosts in the router's local IP subnets that declare their wish to join the specific group.

Proposed multicast routing protocols include the *Distance Vector Multicast Routing Protocol* (DVMRP), *Multicast Open Shortest Path First* (MOSPF), *Core Based Trees* (CBT), *Protocol Independent Multicast-Sparse Mode* (PIM-SM), and *Protocol Independent Multicast-Dense Mode* (PIM-DM). All these protocols attempt to build a multicast delivery tree of routers for each multicast group. DVMRP and MOSPF are less scalable (they construct one tree per sender per group) than CBT or PIM (they construct one tree per group, which senders share). CBT and PIM are also independent of the underlying unicast routing protocols used. PIM-SM is optimized for *sparse* multicast groups and PIM-DM for *dense* multicast groups. PIM works by choosing a *Rendezvous Point* (RP) when it constructs the multicast delivery tree for a group, where multicast senders can "meet" multicast receivers. With all protocols, there is an associated *graft* delay when a multicast router joins an existing tree because the tree has to be adjusted. All these protocols implicitly assume stationary hosts.

9.3.2 Local Mechanisms

The global protocols we described above are concerned with multicast senders, multicast routers and multicast groups, but not with the individual multicast listeners. These listener hosts are "hidden" behind their local multicast router. The router can be thought of as the interface between the local and the global mechanism [28]. Based on a local group management protocol, this router builds a list with all the different multicast groups its hosts have joined (on each one of the IP subnets it may serve). Only this aggregate list is exposed to the global mechanism.

The local group protocols are the *Internet Group Management Protocol* (IGMP) [12] for IPv4 and the *Multicast Listener Discovery* protocol (MLD – derived from IGMP version 2) [11] for IPv6. IGMP assumes the existence of link-level native broadcast (e.g. Ethernet) and is designed around the soft-state principle, which traditionally leads to robust Internet protocols. IGMP works as follows: every *querying_period* the multicast router responsible for a local subnet sends out queries to the all-hosts group. Its objective is to refresh the group list it exposes to the global mechanism. All local listeners receive the query and respond (after a small *random_report_delay*) with a *group_report*, one for each of the groups they participate in. Each report is actually sent to the multicast address for the group reported so that all interested local listeners may learn of this fact.

The soft state principle in IGMP dictates that queries are router-initiated and that no explicit *leave_group* message is needed when a listener leaves a group: the router will discover it in the next query cycle. Extensions for unsolicited listener messages exist [12]. IGMP version 2 adds a *leave_group_message* and IGMP version 3 adds the ability to selectively join and leave multicast groups [1]. A good usage example of these extensions comes from the need to lower the *join delay* (the local equivalent of the global graft delay). A host joining a group not already present on the local link may send an unsolicited *group_report*. If not, it could very well wait up to *querying_period* + *random_report_delay* + *graft_delay* before multicast packets start arriving.

9.4   IP MOBILITY PROTOCOLS

9.4.1   Mobile IP

The goal of *Mobile IP* (M-IP) [24] is to allow internetwork host mobility in a manner transparent to the transport layer. The objective is to leverage the investment on existing TCP/IP applications and avoid the need to redesign new, mobility-aware ones. This mobility transparency can be achieved by assigning two different IP addresses to every *mobile host* (MH), a permanent one used by applications for identification, and another one, which may change, used for routing. The permanent IP address is called the MH's *home address*. If the MH changes its point of attachment and moves to a *foreign* IP subnet, a *Correspondent Host* (CH) may still deliver packets to the MH using the MH's home address as the destination. This is possible because a new entity at the MH's home subnet - a router known as the *Home Agent* (HA) - intercepts these packets and *tunnels* them (reroutes them using *IP-in-IP encapsulation* [23]) to the MH's current network address. To achieve this, HAs advertise reachability to the home network using standard IP routing protocols. This network could very well be a virtual one, i.e. have no physical instantiation: it could be represented by just a HA which keeps track of many MHs roaming around the Internet. In case the network is not a virtual one, the HA may use *proxy ARP* (ARP is the Address Resolution Protocol, the protocol for resolving IP addresses to link-layer MAC addresses in Ethernet and Ethernet-based LANs) in order to fool local network nodes into sending packets addressed to the MH to the HA instead.

The MH's current network address is known to the HA as a result of a registration procedure: when the MH first moves to the foreign subnet, it communicates back to its assigned HA its new *Care-of Address* (CoA). This new address is either a *co-located* CoA or a *Foreign Agent* (FA) CoA. In the co-located case, this new address is usually obtained through *DHCP* (Dynamic Host Configuration Protocol) on the foreign subnet. In the FA case, this address corresponds to a special router - the FA - that resides on the visited subnet. In both cases, this address is communicated back to the HA (either by the MH or the FA), where a new address *binding* with an associated lifetime is created, which binds the MH's home address with its current CoA.

In the co-located case, *reverse tunneling* (decapsulation of packets) is carried out by the MH, whereas in the FA case, reverse tunneling is the responsibility of the FA which will then forward each packet to the MH (FAs and MHs are normally assumed to have link-layer connectivity). In both cases, IP-in-IP from the HA will deliver a packet with an outer destination address matching the CoA and an inner destination address matching the MH's permanent home address.

The aforementioned mechanisms describe how CHs can always send packets to MHs. For the reverse path, the standard IP mechanism can be used: the MH will send packets addressed to the CH's address and place its home address in the source address field. This "trick" has no adverse effect since standard unicast IP routing normally depends only on the destination address (if no *ingress security filtering* is used).

The M-IP communication mechanism results in an inefficiency called *triangle routing* [28]. Extensions [25] to the original M-IP RFC allow CHs to make address bindings of their own which subsequently gives them the ability to discover the MH's new address and route packets directly to the MH bypassing its HA and home subnet. This feature can only be used if the CHs are also M-IP-aware. This particular feature has been slow in its adoption and will not really be used before IPv6 is deployed sufficiently, since it raises security concerns not easily solved within the confines of the M-IPv4 specification.

M-IP is built around the soft-state principle which dictates that all registrations with the HA (as well as all local registrations of MHs with FAs) should have an associated lifetime after which they expire. MHs need to refresh the bindings periodically. There is no explicit deregistration procedure. When an MH moves to yet another subnet or when it disconnects completely, the old FA (if one existed) and the HA in charge will soon learn of this fact. FAs maintain a list of all the visiting MHs they serve and HAs maintain a list containing the bindings of all the MHs for which they are responsible.

9.4.2 Cellular IP

M-IP was designed with "slow" macro-mobility in mind. The protocol incurs several delays, the most important one being the time MHs take to discover an FA in the foreign subnet, register with it, acquire a CoA, and register with their HA. This *agent discovery time* can be significant. If a moving MH traverses many *pico-cells*, each one controlled by a different FA (co-located with the cell's base station), the signaling required during every handoff may inhibit the normal reception of user data packets. In the case of a fast moving host, the so-called *mobility assumption* [3] may be violated: the total registration delay may be more than the time an MH spends inside a cell. As a result, no user packets will find the time to get rerouted to the MH.

Cellular IP (C-IP) [6] is a best effort *micro-mobility* protocol designed with campus-wide, partially-overlapping micro-cells and pico-cells in mind. Its objective is to provide seamless local mobility. C-IP can rely on M-IP for *macro-mobility* support (for example, an MH arriving at a campus internetwork will still register this fact with its M-IP HA). For intra-campus mobility and local handoff control, C-IP takes over and uses its own internal network of packet forwarders (which may be co-located with base stations). As far as the MH's HA is concerned the whole campus internetwork is controlled by only one FA (M-IP has to be used in FA mode for this to work).

A C-IP *gateway* can be co-located with M-IP's FA. This gateway is the interface between the M-IP routing infrastructure and the C-IP access network (see Figure 9.1). Base stations within a C-IP network are fixed, interconnected IP forwarders with at least one wireless interface. They operate at layer 3 and can be connected via any number of layer 2 bridging nodes. To forward packets to MHs, they employ non-standard IP routing techniques: MHs are only identified by their M-IP home addresses and, unlike M-IP, no tunneling is employed [6]. The C-IP route learning mechanism greatly resembles the MAC bridge auto-learning mechanism and it follows the soft state principle: routes have to be refreshed periodically. This happens whenever MHs send regular user data packets towards the gateway. Therefore, there is no need for explicit signaling to notify about an MH's current location. In addition, there is no need for yet another type of end-host identifier since the simple and flat forwarding architecture, which resembles layer 2 bridging, uses the host's permanent home IP address as an identifier which is mapped to a particular port on each forwarder. These bindings are "auto-learned" whenever an MH sends regular data packets towards the gateway, by "snooping" and looking at the source IP address field in the IP packet. Although this IP-to-port mapping is not a very scalable solution, the size of real C-IP access networks is not expected to be unmanageable, since they will be limited to a certain geographical area and will serve a relatively small number of local, visiting or home, IP hosts.

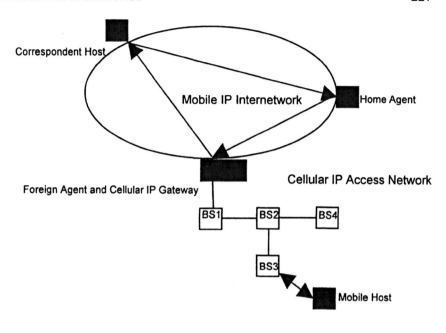

Figure 9.1. Cellular IP Access Network

Certain key notions at the heart of C-IP are based on existing mechanisms used in cellular telephony networks. These include the notions of *active* and *idle* nodes and the notion of *paging* [6]. To minimize signaling overhead, MHs only update their location information when they are active, i.e. engaged in communication. Since C-IP is as connectionless as IP is, MHs can only be deemed idle after a certain period of inactivity. When they are idle, the fixed base station infrastructure has no knowledge of their whereabouts if the soft state routes have also expired. When new incoming packets have to be delivered to an MH, paging is used to discover its exact cell: the C-IP forwarders flood all their outgoing interfaces (unless a cached route entry exists), except the one they received the packet from. C-IP supports smoother handoffs by allowing MHs to send *route_update* packets as soon as they detect that they have moved to a different cell.

9.5   TRANSCODING FILTERS

There exist software components, collectively called *media filters* or *transcoders*, which, given an input media stream, can produce a different, but similar, output media stream. Their purpose is to help maintain a level of *User-Perceived Quality of Service* [14] as the media stream traverses many internetwork links of varying bandwidth. These filters can be divided into *smart* and *simple* [14]. Smart filters do not have to decode the media stream (e.g. an MPEG-2 video stream) to a raw format and then re-encode it. Rather, they operate on the encoded stream directly. Smart filters differ from simple ones in that they require more processing power, but, generally, because they take advantage of a previous encoding stage they are also faster and produce better perceived output than their simpler counterparts.

Filters can be used to reduce the bit rate of a media stream before it is injected into a bandwidth limited link, such as a wireless link [22]. Filters can support multicast for wireless communications (many believe they are absolutely necessary) by allowing wireless receivers that cannot receive the source stream at its original bit rate to receive a "cut-down" version of it. Usually, this means lower resolution, fewer frames per second and smaller frame-size. It can also mean lower color-depth, less smooth movement and the tweaking of a variety of parameters that depend on the media encoding format. This sort of *flexibility* represents a subset of the requirements for fully programmable and *reconfigurable networks* at all layers. Technologists are planning ahead for these "software" networks, whose appearance will change many of today's accepted networking principles. The vision also involves dynamic reconfiguration of radio modulation parameters, dynamic spectrum allocation and dynamic protocol downloading and configuration.

On the same front, dynamic encodings like *layered coding* and *multiresolution layered coding* [15] can be used to separate a media stream into more than one streams, each carrying, progressively, more information. Listeners can receive as many of the streams their incoming bandwidth allows. This can be used in conjunction with filtering. A proposal also suggests using different multicast groups to transmit each one of these substreams [20]. Listeners may then "tune into" as many groups as possible.

Filters become very useful to multimedia communications in wireless environments. A natural location to place such filters is at the boundary between the wired and the wireless part of the internetwork. There, they can modify the passing stream based on actual measurements of the conditions on the wireless link. In a multicast scenario, different receivers may require different filters. In the case where many receivers at the same multicast sub-tree request the same transcoding function, the relevant filters can propagate upstream (towards the sub-tree root - their nearest common ancestor - closer to the stream source), where they can combine into one and serve all of the sub-tree's requesting nodes [21]. In the general case, an *active* or *configurable* IP router (one that allows third-party code uploading and execution) is a natural location for these mobile filters to reside in.

9.6   COMBINING IP MULTICAST AND MOBILITY

The IP multicast protocol was designed to bring multipoint communication capability to the Internet. At the same time, Mobile IP and Cellular IP were designed to allow transport-layer-transparent mobility. Trying to integrate the protocols into a coherent IP framework for multicast mobility exposes several fundamental issues. Existing approaches [16, 18, 24, 27] offer some functionality, but further simulation and deployment is needed before the scalability of each proposed solution can be judged.

9.6.1   "Fixed Host" Assumption – IP Addressing Issues

We already mentioned that the "fixed host" assumption influenced the design of several protocols in the IP suite. When considering mobility, we need to take into account that mobile devices can be very different from fixed hosts connected to an Ethernet network. Some basic differences are listed below:

- limited battery life dictates that unnecessary operations should be avoided, so, protocols that rely on constant monitoring of network traffic are impractical;
- protocols that assume high bandwidth, low latency connectivity may become inoperable;
- protocols that assume low BER and no disconnections are faced with a hostile wireless environment;
- the notion of handing-off the connection to a different cell, neighboring or overlaid, is non-existent in the wired world;
- protocols that rely on link-local broadcast capability are not always easy to port to a wireless environment because the techniques used in several wireless networks cannot be easily modified to support link-level multicast. GSM's time division multiple access, CDMA's power control, and the fact that, in most current cellular systems, the base station is not an IP-layer router but a lower layer entity, make such attempts difficult.

IPv4 address shortage is also a problem now. For example, GPRS operators usually have to rely on *Network Address Translation* (NAT) and are forced to assign private IP addresses (usually over PPP) to the mobile devices that register with their networks. All the usual NAT limitations affect IP multicasting although work-arounds do exist [31]. Also, some cellular operators have their own interpretation of 2.5G and 3G multicasting (simple cell broadcast usually) which is more limited than the IP-based multicasting envisaged for an all-IP wireless Internet.

9.6.2   IETF Mobile IP Multicast Support

The current IETF proposed standard for Mobile IPv4, RFC 3220 [24], devotes no more than a single page to multicast packet routing in conjunction with Mobile IP. Two methods are mentioned, which are referred to by [8] as *Remote Subscription* (MIP-RS) and *Bi-directional Tunneled Multicast* (MIP-BT). Both methods are only relevant when MHs are visiting a foreign subnet. We only describe them with M-IP operating in FA mode, since this is the preferred mode of operation both for wireless communications and for interoperability with C-IP.

Remote subscription - This may be used only when a multicast router is present in the visited subnet (this router may be co-located with the FA). The MH can use IGMP to subscribe to any number of groups using this router.

Bi-directional tunneled multicast - Another option is for the MH to setup a bi-directional tunnel to its HA (this HA must also be a multicast router). The HA will join groups on the MH's behalf. The tunnel is used to send IGMP messages and receive multicast packets. In this case, double encapsulation is required: first, the HA has to encapsulate the multicast packet inside another packet addressed to the MH's home address, and then, encapsulate once more as specified by Mobile IP. This means that the MH must be able to decapsulate the multicast packet even if it uses an FA for the standard Mobile IP decapsulation procedure.

Both approaches have their disadvantages. MIP-RS requires that the MH re-subscribes to a potentially large number of multicast groups after every subnet move. This delay will cause packet losses for the MH and will require rearrangements of the multicast tree with each move. MIP-RS assumes that a multicast router will exist in the visited subnet. Also, MIP-RS generally causes *get-ahead* and *lag-behind* effects (terms borrowed from [6]). These may happen when MHs register with a new multicast router and re-subscribe to multicast groups. Because of the complex nature of a multicast router tree, some edge routers may lag behind others in their reception of multicast packets. One solution to the get-ahead problem is for the host to accept the loss of some packets and rely on higher layer protocols to recover from this loss. More sophisticated solutions involve buffering at the multicast routers. The lag-behind problem may be solved locally at the host, where higher layer protocols can discard already received packets, or at the router if somehow the router is signaled to drop packets before forwarding them to its internal subnet (which is important if the subnet is a wireless, bandwidth limited one).

With MIP-BT, the multicast tree does not have to be rearranged, since the MH's HA will be a stationary multicast receiver. Also, if the HA buffers multicast packets, no get-ahead or lag-behind problems will exist for the MH. MIP-BT, however, is associated with the three following unwanted phenomena:

- If an HA supports many MHs, all visiting the same foreign subnet and all subscribing to the same groups, then multiple point-to-point channels will have to be setup between the same HA and the foreign subnet, each one carrying the same information. The duplicate packets will strain the (potentially wireless and bandwidth limited) visited subnet.
- If, somehow, the HA detects this and sends only one packet copy per foreign subnet, then a problem known as *tunnel convergence* (see figure 9.2) may still occur if *many* HAs have MHs, all visiting the same foreign subnet and all subscribing to the same group. This will also lead to unnecessary packet duplication.
- Another way for unwanted duplicate packets to appear on the foreign subnet is if a host, local to that subnet, already subscribes to one or more of the groups some MHs wish to subscribe to. Since MHs will receive multicast packets encapsulated in unicast packets addressed to their home address (see MIP-BT description), it would not be easy for the FA or some multicast router in the foreign subnet to detect this duplication and stop the unnecessary transmission.

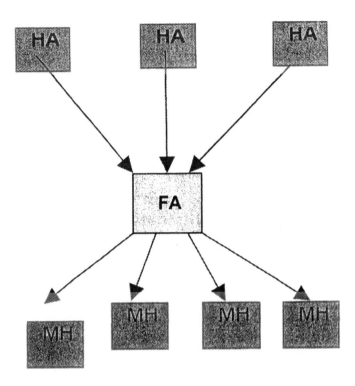

Figure 9.2. The Tunnel Convergence Problem

9.6.3   Extensions to the IETF Approach

The following examples are alternatives or extensions to the basic Mobile
IP and IP multicast interoperability approaches proposed by the IETF. Each
one of these examples improves on the basic mechanisms but we believe that
further refinement is needed before these solutions can be widely deployed.

Mobile Multicast (MoM) *Protocol*

MoM [8, 16] is based on MIP-BT and its key extension is the use of a
Designated Multicast Service Provider (DMSP). DMSPs attempt to solve
the tunnel convergence problem (see section 6.2). A DMSP for a given
multicast group is an HA chosen by the visited subnet's FA out of the many
HAs that forward packets for the specific group to the visited subnet. MoM
supports choosing more than one DMSP for redundancy and it also supports

DMSP-handoff, which is necessary when a DMSP has no more MHs of its own in the visited subnet that require packet tunneling.

MoM-specific algorithms are executed every time:

- MHs arrive at a foreign subnet;
- MHs depart from a foreign subnet;
- MH registrations with the FA time out;
- unicast or multicast packets from the HAs arrive at the FA.

The subnet's FA keeps track of HAs, MHs, and multicast groups, so it always has enough information in order to choose a DMSP and to perform DMSP-handoffs whenever required. MoM is the most cited alternative to MIP-BT and MIP-RS.

Mobile Multicast with Routing Optimization (MMROP)

MMROP [18] is based on MIP-RS and its key extension is the introduction of the *Mobility Agent* (MA) entity, which attempts to solve the get-ahead problem due to handoffs (see section 6.2). This is done to "ensure routing efficiency and no packet losses from roaming" [18]. MAs are FAs that route missing packets (via tunneling) to neighboring subnets. MMROP works as follows: let FA1 and FA2 be an MH's old and new foreign agents respectively. Let's assume the MH was subscribed to group G through FA1. The MH will attempt to resubscribe to group G through FA2, at which point FA2 will start buffering packets. Upon joining the new subnet, the MH will look at the sequence numbers of the packets for G and decide whether or not it should ask for cached packets from FA2. (MMROP assumes packets are somehow numbered.) If FA2 cannot supply these packets, it will request FA1 to continue transmitting packets to the MH through a tunnel between FA1 and FA2, until FA2 and the MH are synchronized, at which point FA2 will start delivering packets to the MH through its own multicast subscription.

Constraint Tree Migration Scheme (CTMS)

CTMS [7] is an attempt to design a new global multicast routing protocol that would improve on CBT [5] when it comes to highly dynamic multicast configurations, such as those found when multicast listeners are mobile. CTMS "automatically [migrates multicast trees] to better ones, while maintaining the QoS guarantees specified by mobile users" [7]. CTMS uses fewer resources per multicast tree and, as a result, packet losses due to reconfigurations and join delays are reduced. CTMS is a good alternative to existing mobile routing protocols but its adoption will be difficult, considering most multicast routers still run DVMRP.

Multicast Scheme for Wireless Networks (MobiCast)

MobiCast [27] is based on MIP-RS and its key extension is the introduction of the *Domain Foreign Agent* (DFA) which serves many small adjacent wireless cells. A hierarchy is introduced, with small cells being organized into one *Dynamic Virtual Macrocell* (DVM). Micromobility is thus hidden from the global multicast mechanism, which does not require re-configuring when handoffs occur within the same DVM.

9.6.4   IGMP Mobility Support and IGMP Assumptions

IGMP was designed with Ethernet networks in mind. Its basic functionality assumes link-level native multicast. Also, its soft-state timers require that multicast listeners repeatedly announce all the multicast groups to which they are subscribed. This happens every time the subnet's designated multicast router issues a query. We present IGMP's two main problems with respect to mobility support:

IGMP is not suitable for point-to-point links – If the local multicast router is not only connected to an Ethernet subnet, but also has interfaces that connect to *point-to-point* links, then, IGMP queries have to be issued to each one of these interfaces [28]. The IGMP replies will not be heard by all other participants unless the router specifically multi-unicasts them to every one of the point-to-point links that it supports. This increases delay, data traffic and state information needed at the router. Instead of just the group list which the global multicast mechanism requires (see section 3), the multicast router must record per-host information [28]. Indeed, as we mentioned, many cellular networks currently offer only point-to-point links for user data. The PPP protocol that is usually used on these networks (as well as on most Internet home connection technologies) to support IP packet transfer does not have the same semantics nor does it use the same techniques as the shared Ethernet protocol.

IGMP is not suitable for mobile hosts – We already mentioned (section 6.1) that mobile hosts don't have the luxury to monitor network traffic constantly. That would place a burden on their battery and processing power. Also, mobile hosts should not be forced to keep resending unnecessary information if this can be avoided. The IGMP soft state timers, although they are simple to implement and they contribute to IGMP's robustness, force the hosts to keep repeating the same data for as long as each one of their group subscriptions is active. A solution proposed in [29] suggests using explicit *join_group* and *leave_group* messages which would require from hosts (which are only multicast listeners) to send significantly fewer packets. This scheme may interoperate with "traditional" IGMP.

9.7 NEW PERSPECTIVES ON MOBILE MULTICAST

9.7.1 Multicast Semantics and Mobility

An extension to the issues raised in section 6.1 is that multicast semantics need to be reexamined in the presence of mobility. We present a simple example based on ideas raised on [8] that exposes this problem. Let's assume X and Y are two Ethernet-based IP subnets. Let's also assume that MHs with home addresses in subnet X are designated Xi and that MHs with home addresses in subnet Y are designated Yj. Some Xi MHs are visiting subnet Y and some Yj MHs are visiting subnet X. If an IP multicast packet addressed to the link-local all-hosts group 224.0.0.1 is directed towards subnet X, then there are three possibilities, according to [8], about what should happen:

- the packet should only be delivered to MHs in subnet X, regardless of whether they belong to the Xi or Yj set;
- the packet should only be delivered to all MHs in subnet X that belong to the Xi set;
- the packet should only be delivered to all MHs belonging to the Xi set, irrespective of their current location.

Obviously, there exist techniques for each one of the three possibilities. But which one is semantically correct? There is no right or wrong answer. This depends on the service protocol that originated the link-local all-hosts multicast. We refer to three service examples, each one assuming a different interpretation:

- an advertisement for public network printers that are present on a specific floor;
- an advertisement for available high-quality color photo printers, to be used as part of subnet X's core business;
- an administrator's advertisement, describing the new authentication procedure for subnet X's SMTP server.

Currently, there is no clear IPv4 mechanism that would help mark the advertisement packet and allow the mobility protocol to make an informed decision. IPv6 can, however, differentiate between *link local*, *site local* and *organizational local* multicast scopes.

9.7.2   Mobile Multicast Requirements

Some general issues that should affect all mobile multicast solutions are the following:

Significant vs non-significant moves - Let MH be a host subscribed to a set of multicast groups. If the move of MH to a new subnet causes the subnet's multicast router to subscribe to new multicast groups, then the move is said to be *significant*. Otherwise, if due to existing subscriptions packets addressed to the MH's set of groups are already being transmitted on that subnet, then the move is *non-significant*. Mobile multicast protocols will have to differentiate quickly between these two types of moves. Ideally, non-significant moves must have no effect at all on the global mechanisms. All the necessary information to identify non-significant moves can be found within the local subnet and at the subnet's multicast router. In addition, significant moves should appear identical to non-significant moves from a user's perspective.

Multicast packet buffering – Although, in theory, IP multicast is a best effort protocol, in practice, if mobile multicast schemes are to work efficiently, packet buffering should happen at the IP multicast layer. For example, with MIP-BT, the HA may buffer packets before tunneling them to the MH. This is necessary in order to achieve smooth handover when the MH moves to a new subnet and reestablishes the bi-directional tunnel. With MIP-RS the situation is more complicated. Depending on how MIP-RS is used, both the FA and a local multicast router on the visited subnet are candidates for buffering packets. The main problem with buffering is the following: buffer packets until when? In a wireless environment, with significant and non-significant moves, disconnections due to handoffs, disconnections due to physical layer problems and disconnections due to user intent, it will not be easy to judge how long buffering should go on. If multiple entities buffer simultaneously complexity increases. Soft-state timers must adapt based on a number of parameters and on input from both lower and higher protocol layers.

Mobile subnets - Ships, planes, trains, and even cars can be thought of as mobile subnets, each one with several local mobile hosts. Dealing with these as one logical entity will greatly assist routing protocols, ease tunnel convergence problems and minimize state information kept throughout the internetwork.

Roaming - The problem of global roaming between different administrations is very difficult to solve, even for point-to-point communications (the same applies to vertical handoffs). Even if mobile multicast routing protocols are simulated and tested, true mobile multipoint communications will ultimately rely on sophisticated authentication mechanisms and pricing schemes.

9.7.3   Cellular IP and Mobile Multicast

In this section we present a model that integrates local multicast mechanisms to Cellular IP (C-IP). By using C-IP in conjunction with Mobile IP (M-IP) in a hierarchical manner similar to the MobiCast scheme (see section 6.3) we outline a mobile multicast scheme based on C-IP and M-IP interoperability ideas developed within the IETF. In our scheme, a MobiCast DFA is a C-IP gateway and a MobiCast DVM is a C-IP subnet. Introducing multicast support to C-IP is relatively straightforward considering that the basic C-IP forwarding mechanism is simple. However, a real deployment would be necessary in order to test the scalability of our proposed architecture.

We chose C-IP and a method based on MIP-RS because we took into account not only the current evolution of the Mobile IP and Cellular IP specifications but also the real network configurations that people deploy. These include campus-wide 802.11 internet-works, UMTS cells and the future DVB-T macro-cells. It is our position that the MIP-BT based tunneling schemes (although friendlier to current multicast routing protocols) are simply not scalable enough. The current *Content Delivery Network* (CDN) trends strengthen our belief that content and services need to be pushed as much as possible to the edges of the Internet and that hosts should first try to exploit whatever resources they have available in their immediate environment (i.e. follow the *locality* principle) and only when this fails should hosts try to access more distant resources.

As we have already mentioned, the way Cellular IP forwarders view the MH address space is "flat". Inside a Cellular IP access network, the MHs use their home address as identifiers. Since the relatively simple mapping inside a forwarder maps these IP identifiers to forwarder ports, we can safely say that multicast (Class D) IP addresses would not appear inherently different from unicast addresses, at least as far as a potential mapping is concerned.

In addition, C-IP has keep-alive mechanisms that look similar to IGMP's keep-alive soft-state mechanisms. These mechanisms serve similar purposes: C-IP maintains the mappings that concern active hosts as long as these hosts send data packets or the special C-IP route_update packets. On the other hand, IGMP responses to router queries serve to maintain the subscription to a specific multicast group. It is theoretically possible to adapt the C-IP route update mechanism to supplant IGMP's operational semantics. In practice, to achieve this, MHs that are subscribed or wish to subscribe to a particular multicast group can send C-IP route_update packets, but instead of using their own IP address in the source IP field, they can use the multicast group address instead. These packets will not only update the forwarders' mappings, but also, when they reach the C-IP gateway (which we assume to be a M-IP FA and a multicast router as well) they can cause the gateway to subscribe to the particular group using the global multicast mechanisms. Of course, if the gateway is receiving the group already, there is no need to graft to the particular multicast tree again. In this way, we replace IGMP by native C-IP mechanisms.

One may say that we view multicast groups as virtual C-IP hosts. In order to complete the picture, C-IP forwarders should be able to handle *IP address-to-multiple port mappings*. Also, MHs should be able to use their IP stack in a rather unconventional way (placing the multicast address in the source field can be considered "non-standard"). A result of all the above is that packet duplication inside the C-IP access network due to the multicast transmission will only happen when paths towards receivers diverge, because of the way the C-IP forwarders update their mappings. This is a basic requirement for efficient multicast protocols. The following figure (figure 9.3) depicts a multicast transmission source (sending data to multicast "channel" 224.1.2.3) somewhere on the Internet and 3 MHs inside a C-IP access network that are subscribed to the 224.1.2.3 group. Until they reach base station/C-IP forwarder BS2, there is no need for the multicast data packets to be duplicated.

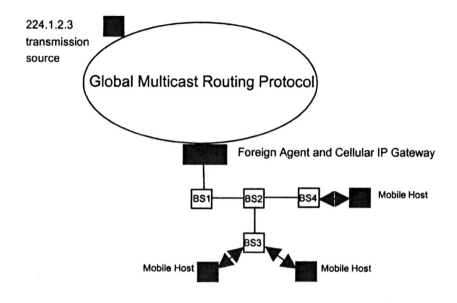

Figure 9.3. Combining Cellular IP and Mobile Multicast

9.7.4 The All-IP Wireless Internet

Having completed our discussion on the basic mobile and multicast protocols and their interoperability issues, we describe the all-IP wireless internetwork that we envisage will support improved versions of all the aforementioned protocols. It will be made up of many, superimposed, cellular technologies. If we ignore the satellite component, which is usually the least cost-effective way to create wireless cells, we still have many promising technologies with which to build this hierarchical cell structure. DVB-T may be used for metropolitan size (1–100 Km) macro-cells, 3G systems, such as UMTS, may be used for neighborhood size cells, and the 802.11 variations, wherever available, for local micro-cells and pico-cells. Although DVB-T is unidirectional, extensions do exist that provide bidirectional functionality. Even without these extensions, 3G networks are perfectly suited to provide the necessary return channel [4].

DVB-T, with 5-30 Mbps of shared bandwidth and excellent support for mobility within a DVB-T macro-cell, could become the technology of choice for delivering IP multicast traffic, bypassing most of the problems we described in previous sections. Going down the cellular hierarchy, we would have overlapping 3G cells (running versions of Mobile IP) and, then, localized 802.11 networks (running versions of Cellular IP). This structure needs to be augmented with umbrella cells that can handle traffic for fast moving receivers in trains or in cars. Mobile subnets could also be supported by installing Cellular IP gateways on ships and on trains, which would support micro-mobility within the moving subnets.

Devices with multiple interfaces are starting to appear. Common offerings include support for 802.11, combined with either DVB or GPRS. As long as IP and Mobile IP are accepted standards we can expect that all future devices and networks will support them. For unicast applications, alternatives to TCP will be used most of the time and improved variations of TCP will provide backwards compatibility. Transcoding functions and filter mobility protocols will be standardized and they will be put to use by all wireless network providers.

9.8   CONCLUSIONS

We examined the multipoint communications problem assuming an all-IP wireless Internet with mobile hosts, a fixed network infrastructure, and a best effort network layer. We focused on reusing IETF protocols where possible. IP multicast and Mobile IP were obvious choices. We showed how existing transcoding techniques may be used. However, we saw that the proposed IETF IP multicast and Mobile IP interoperability solutions are not perfect and that they require extensions. We presented a number of additions and alternatives to the basic scheme. We mentioned an approach involving Cellular IP and we offered our perspective on additional multicast mobility issues. Finally, we described the all-IP wireless Internet we envisage.

Still a lot of functionality needs to be added to the basic IP multicast and Mobile IP offerings before infrastructure-less networks, strong reliability and security are also supported. Ultimately, global roaming and pricing agreements will be necessary to complete this ideal vision of mobile multicast in an all-IP wireless Internet.

REFERENCES

[1] D. Agrawal, C. Cordeiro, H. Gossain, "Multicast: Wired to Wireless," *IEEE Communications Magazine*, 40(6):116-123, June 2002.

[2] A. Acharya, A. Bakre, and B.R. Badrinath, "IP Multicast Extensions for Mobile Internetworking," *Proceedings of 1996 IEEE INFOCOM*, pp. 67-74, San Francisco, CA, March 1996.

[3] A. Acharya and B.R. Badrinath, "A framework for delivering multicast messages in networks with mobile hosts," *ACM/Baltzer Journal of Mobile Networks and Applications*, 1(2):199-219, October 1996.

[4] Ad hoc Group DVB-UMTS, "The Convergence of Broadcast & Telecomms Platforms," *Public Document, www.dvb.org*, March 2001.

[5] A. Ballardie, J. Crowcroft, and P. Francis, "Core based Trees (CBT) – An architecture for scalable inter-domain multicast routing," *Computer Communications Review*, 23(4):85-95, October 1993. (Proceedings of the ACM SIGCOMM'93.)

[6] A.T. Campbell, J. Gomez, and A.G. Valko, "An Overview of Cellular IP," *IEEE WCNC*, New Orleans, September 1999.

[7] K. Chen, N. Huang, and B. Li, "CTMS: A novel constrained tree migration scheme for multicast services in generic wireless systems," *IEEE JSAC.*, 19:1998-2014, October 2001.

[8] V. Chikarmane, C.L. Williamson, R.B. Bunt, W. Mackrell, "Multicast Support for Mobile Hosts Using Mobile IP: Design Issues and Proposed Architecture," *ACM/Baltzer Mobile Networks and Applications*, 3(4):365-379, Jan. 1999.

[9] S. Deering, "Host Extensions for IP Multicasting," *RFC 1112*, August 1989.

[10] S. Deering, "Multicast Routing in a Datagram Internetwork," *Ph.D. Thesis, Department of Computer Science, Stanford University*, 1991.

[11] S. Deering, W. Fenner, and B. Haberman, "Multicast Listener Discovery for IPv6," *RFC 2710*, October 1999.

[12] W. Fenner, "Internet Group Management Protocol, Version 2," *RFC 2236*, November 1997.

[13] S. Floyd, V. Jacobson, and S. McCanne, "A Reliable Multicast Framework for Lightweight Sessions and Application Level Framing," *IEEE/ACM Transactions on Networking*, 1995.

[14] G. Gardikis, E. Pallis, A. Kourtis, "Beyond 3G: A Multi-Service Broadband Wireless Network with Bandwidth Optimisation," *IST Project "Multi-Services Management Wireless Network with Bandwidth Optimisation," Public Document*, www.openmux.com-/mambo/Public/Beyond 3G.pdf, 2000.

[15] J.K. Han and G.C. Polyzos, "Multi-Resolution Layered Coding for Real-Time Image Transmission: Architectural and Error Control Considerations," *Real-Time Imaging*, 4(4):275-298, Academic Press, August 1998.

[16] T. Harrison, C.L. Williamson, W.L. Mackrell, and R.B. Bunt, "Mobile Multicast (MoM) Protocol: Multicast Support for Mobile Hosts," *Proceedings ACM MOBICOM*, Budapest, Hungary, September 1997.

[17] V. Jacobson, "Congestion Avoidance and Control," *Proceedings ACM SIGCOMM*, CA, USA, August 1988.

[18] J. Lai, W. Liao, M. Jiang, and C. Ke, "Mobile Multicast with Routing Optimization for Recipient Mobility," *Proceedings IEEE ICC2001*, pp. 1340 - 1344, June 2001.

[19] Y. Lin, "A Multicast Mechanism for Mobile Networks," *IEEE Communications Letters*, 5(11), November 2001.

[20] S. McCanne, V. Jacobson, and M. Vetterli, "Receiver-driven Layered Multicast," *Proceedings ACM SIGCOMM*, Stanford, CA, August 1996.

[21] J.C. Pasquale, G.C. Polyzos, E.W. Anderson, and V.P. Kompella, "Filter Propagation in Dissemination Trees: Trading off Bandwidth with Processing in Continuous Media Networks," *NOSSDAV, D. Lecture Notes in Computer Science 846*, Shepherd et al., eds., Springer-Verlag, Berlin, Germany, pp. 259-268, 1994.

[22] J.C. Pasquale, G.C. Polyzos, E.W. Anderson, and V.P. Kompella, "The Multimedia Multicast Channel," *Internetworking: Research and Experience*, 4:151-162, 1994.

[23] C. Perkins, "IP Encapsulation within IP," *RFC 2003*, October 1996.

[24] C. Perkins, editor, "IP Mobility Support for IPv4," *RFC 3220*, January 2002.

[25] C. Perkins and D.B. Johnson, "Route Optimization in Mobile IP," *Internet Draft draft-ietf-mobileip-optim-11*, September 2001.

[26] A.C. Snoeren, H. Balakrishnan, and M.F. Kaashoek, "Reconsidering Internet Mobility," *Proceedings HotOS-VIII*, May 2001.

[27] C.L. Tan and S. Pink, "MobiCast: A Multicast Scheme for Wireless Networks," *ACM MONET*, 5(4):259-271, 2000.

[28] G. Xylomenos and G.C. Polyzos, "IP Multicast for Mobile Hosts," *IEEE Communications*, 35(1):54-58, January 1997.

[29] G. Xylomenos and G.C. Polyzos, "IP Multicast Group Management for Point-to-Point Local Distribution," *Computer Communications,* 21(18):1645-1654, Elsevier Science, December 1998.

[30] G. Xylomenos, G.C. Polyzos, P. Mahonen, and M. Saaranen, "TCP Performance Issues over Wireless Links," *IEEE Communications*, 39(4):52-59, April 2001.

[31] "How Does Multicast NAT Work on Cisco Routers?" *Technical Note, www.cisco.com/warp/public/105/multicast_nat.html*, 2001.

Chapter 10

MULTIPATH ROUTING IN AD HOC NETWORKS

An-I Andy Wang*   Geoffrey H. Kuenning✧   Peter Reiher*
*University of California, Los Angeles

✧Harvey Mudd College

10.1   INTRODUCTION

In an ideal network, a source always knows how to reach the destination, and the network connection is always reliable.   In an ad hoc wireless network where any node can be mobile, a source needs to update the location of the mobile destination and intermediate nodes constantly, and network connections may break frequently due to the changing network topologies and unreliable wireless connectivity.

Routing is a major challenge in this wireless mobile environment. Mobility renders standard Internet routing methods (e.g., single-path shortest route) inappropriate. Typically, ad hoc networks operate on wireless links with limited bandwidth and transmission range, and the nodes constituting the network often operate off batteries, placing a further premium on efficient operations.

One implication of mobility is that frequent location updates make the conventional approach of maintaining source-destination distance tables impractical, since refreshing table entries requires a high rate of updates. In addition, mobility causes many table entries to become out of date before ever being used. Newer on-demand approaches maintain table entries only when a communication session is initiated, thereby reducing the table maintenance overheads.   However, such approaches are still inefficient, since mobility can frequently break those on-demand routes and trigger expensive repair mechanisms to reestablish table entries.

Clearly, handling mobility demands protocols that have higher resiliency in the face of rapidly changing network topologies. Such resiliency can be achieved by using multipath routing solutions, which create several redundant routes for a source-destination pair. If one route fails, a backup route will still be available. Combined with on-demand approaches, multipath routing can handle mobility efficiently by tracking intermediate nodes and destinations only when necessary. The combined approach offers a greatly reduced route recovery time when a main route fails.

Multipath routing also offers other advantages. In a conventional network infrastructure, classical multipath routing allows load balancing among multiple routes, reducing network traffic congestion and improving the overall quality of service (*QoS*). Transmitting data through multiple paths in parallel also permits aggregation of network bandwidth. Higher resiliency can be achieved by transmitting data either redundantly or with error-correcting information through separate routes simultaneously.

In the context of ad hoc networking, all the classical applications of multipath routing still apply, but ad hoc multpath routing provides additional benefits. First, in a mobile environment, a pre-established route is likely to break often, and reducing the failure recovery time by having standby alternative routes can significantly affect the QoS perceived by end users. Alternating paths to transmit information can also spread the energy use among network nodes and prolong the battery life for the ad hoc network as a whole. In addition, transmitting encrypted data across multiple routes can significantly reduce the likelihood of man-in-the-middle, replay, and eavesdropping attacks. This property is especially important in mobile environments, since wireless communication is inherently more vulnerable to security failures.

This chapter describes the problems involved in constructing multiple paths in wireless mobile networks and surveys the approaches being taken to overcome those problems.

10.2   DESIGN SPACE FOR AD HOC MULTIPATH ROUTING PROTOCOLS

The typical problems encountered in designing single-path protocols for a mobile wireless environment also arise in designing multipath protocols. Mobility makes it difficult or impossible to maintain a global view of the network. Mobility also has implications in terms of caching policies. If cached information at intermediate nodes is frequently out of date, caching can degrade routing performance because detecting inaccurate routing information is not instantaneous. In addition to the dynamic topology, unreliable and range-limited wireless transmission makes resiliency a requirement rather than an enhancement in a routing solution. Since mobile transmitters are likely to be battery powered, routing protocols need to minimize the communication for coordinating network nodes.

At the protocol level, the design of multipath routing needs to consider failure models, characteristics of redundant routes, coordinating nodes to construct routes, mechanisms for locating mobile destination and intermediate forwarding nodes, and failure recovery.

10.2.1   Failure Models

Most multipath routing protocols are designed for **independent, isolated failures** in terms of network components. More precisely, each node and link in a route has a probability of failure p_f during some small interval T. The probability of a route failure is defined as the probability of at least one failed component in a route during the interval T. The isolated failure model is quite realistic, especially for hardware components. Redundant routes can handle this failure model gracefully.

Another form of failure is the **geographically localized and correlated failure**, which results in all nodes failing within a circle of a radius R. The choice of R is somewhat arbitrary and depends on the possible causes of failure. This model of failure reflects various environmental factors, such as poor weather conditions or natural disasters. Multipath routing protocols that form routes spanning a large geographical region are more likely to survive this type of failure. Protocols that construct paths that are adjacent to one another usually resort to reconstruction of multiple paths when faced with this type of failure.

10.2.2   Characteristics of Redundant Routes

The degree to which a multipath routing algorithm succeeds in building useful multiple paths depends not only on the design of the algorithm, but on why multiple paths are wanted. To illustrate various designs, *Figure 10.1* shows a sample ad hoc network. The dashed lines represent the available wireless links between nodes. The figure represents relative geographical positions, in addition to connectivity. The connectivity varies depending on physical distance, radio characteristics, and environmental conditions. In this figure, the source node *S* wants to send traffic over the network to *D* through multiple paths.

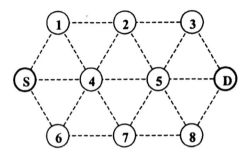

Figure 10.1. An example of an ad hoc wireless network

To achieve reliability, one possibility is to construct ***node-disjoint routes***, where each route constitutes a different set of intermediate forwarding nodes (*Figure 10.2*). With *k* node-disjoint routes, a multipath scheme can tolerate at least *k − 1* intermediate network component failures without disconnecting a source-destination pair. Node-disjoint routes can be geographically adjacent to the shortest path, if a goal is to minimize end-to-end delays and maximize network bandwidth aggregation at the same time. *Figure 10.2* shows two routes, S-1-2-3-D and S-4-5-D. In this case, multiple routes constructed include the shortest path (S-4-5-D).

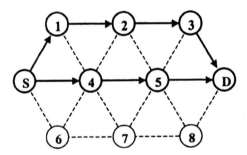

Figure 10.2. Two node-disjoint routes, one of which is the shortest path

Routes can be also widely separated geographically, if the predominant goal is to reduce the probability of multiple routes being disrupted by a single regional, correlated failure. *Figure 10*.3 shows a different pair of routes, S-1-2-3-D and S-6-7-8-D, which will require a failure that affects a larger geographical region before both paths can be broken.

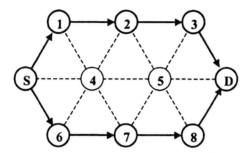

Figure 10.3. Two node-disjoint routes that are separated geographically

Another multipath approach is to form ***link-disjoint routes***, in which links are not shared, but intermediate nodes can be shared when constructing multiple paths for a source-destination pair. (Note that node-disjoint routes are automatically link-disjoint.) Link-disjoint routes are not as resilient to geographically localized and correlated failures. However, detecting and repairing a single point of failure can be a more localized operation, because node-disjoint routes need to propagate network failures back to the source before an alternate can be deployed for recovery. In contract, an alternative route in link-disjoint schemes can be set up by an adjacent node in the upstream direction. In an energy-constrained environment, constructing and maintaining link-disjoint routes can be more energy efficient than maintaining node-disjoint routes, because link-disjoint routes tend to be adjacent to the primary route, which is often the shortest route. On the other hand, node-disjoint routes tend to cover a wider geographical region and spread battery consumption more evenly throughout the network.

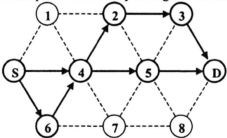

Figure 10.4. Two link-disjoint routes (S-4-5-D and S-6-4-2-3), with node 4 shared

Figure 10.4 shows two link-disjoint routes, S-4-5-D and S-6-4-2-3-D. If the main route S-4-5-D fails due to node 5, the route recovery process can be handled at node 4 to re-establish the route S-4-2-3-D. If an additional link failure occurs at the link between node *S* and node 4, node *S* will re-establish the route S-6-4-2-3-D. On the other hand, if node 4 fails first, both routes can be destroyed in a single failure. In general, *k* link-disjoint routes do not provide as much resiliency as *k* node-disjoint routes. However, since the total number of link-disjoint routes in a given network almost always well exceeds the number of node-disjoint routes, a sufficient number of link-disjoint paths can achieve the resiliency of node-disjoint paths, assuming independent failures of network components.

A more relaxed form of the multipath routing scheme is to construct *partially disjoint routes*. The formation of partially disjoint routes is mostly for reliability purposes. The primary route is used for data transmission, while other partially disjoint routes are standby routes for failure recovery.

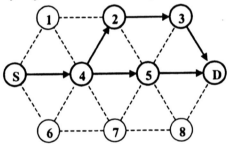

Figure 10.5. Partially disjoint routes (S-4-5-D and S-4-2-3-D), with link S-4 shared

Figure 10.5 shows a primary route (S-4-5-D) with one partially disjoint route (S-4-2-3-D). If node 5 fails, the primary route is recovered by switching to the partially disjoint alternative.

10.2.3   Construction of Multiple Paths

Multiple paths can be constructed by relying on global knowledge, incomplete global knowledge, or only local knowledge.

Routing protocols that rely on global knowledge allow a route-forming node to access the current status of all network nodes and links. At the time of formulating multiple paths, the node can produce multiple routes in a centralized manner. Obtaining such information requires an abundance of local node resources (unlikely for wireless mobile nodes), or the existence of a somewhat centralized database (which is difficult to scale). In any large-scale network, obtaining global knowledge about the entire network is very difficult and costly, even disregarding the additional constraints of wireless mobile environments. For that reason, wireless multipath routing methods avoid any reliance on global knowledge.

At the other extreme, relying on local knowledge means that a source constructs multiple paths in a distributed manner that is largely based on access to the current status of neighboring nodes and links. Relying solely on neighboring information may seem insufficient for coordinating nodes to form multiple paths. However, since wireless communication is broadcast-based, a node can potentially overhear information from neighboring nodes. In addition, certain communication can be implicit. Although routing protocols that rely on local knowledge can scale well due to their distributed nature, constructing multiple routes based on localized coordination is generally difficult because nodes need to determine locally whether global invariants are met (e.g., disjointness of routes, loop-free routes).

Thus, the majority of multipath schemes rely on incomplete global knowledge because of the ease of constructing and verifying centralized solutions. Incomplete global knowledge can be easily obtained by constrained flooding of the network. However, a few multipath schemes are distributed and rely on local knowledge gained through periodic exchange of information with neighboring nodes or by overhearing information through the promiscuous mode.

Without complete global information, one greedy approach to constructing multiple paths is to apply variants of the Dijkstra or Bellman-Ford pair-wise shortest-path algorithms iteratively. The partial network topology is represented as an undirected graph $G = (V, E)$ with wireless mobile nodes as vertices V, and network links as edges E. Each iteration on G will yield a pair-wise shortest path, which is added to the list of multiple routes and removed from the original G before the next iteration. After k iterations, this approach will yield a greedy solution of the k shortest node-disjoint routes, which is useful when multiple routes are needed to maximize network bandwidth. If, after each iteration, the nodes in the shortest path are not removed from G, the same method will yield the k shortest link-disjoint routes. Multiple routes with different characteristics (e.g., minimal congestion) can be obtained by adjusting the lengths of the edges according to various constraints. For example, to avoid congestion, the length of edges can be increased for congested links.

Most well-known distributed or localized multipath routing protocols are inspired by biological and physical models, such as ants [Braginsky and Estrin 2001], water flow [Park and Corson 1997], and diffusion models [Estrin et al. 1999; Intanagonwiwat et al. 2000]. In the water flow method, for example, a source can send network traffic through multiple paths to a destination by properly defining heights of *terrains* at each intermediate forwarding node. Whenever the traffic is trapped within a section of terrain (an intermediate node), the terrain is modified so traffic can flow outwards again.

Distributed multipath schemes can be elegant and scale well, and each node can perform routing by keeping only local state. However, these schemes need to overcome the challenge of oscillations. For example, multiple nodes can independently trigger mechanisms for network detection, recovery, erasing routes, and creating routes, resulting in unstable network routes that change frequently due to both mobility and node behaviour. Although centralized multipath approaches are more intuitive to construct and verify, each route typically has a predefined set of forwarding nodes. In ad hoc networks, such predefined routes can be easily broken due to high mobility.

10.2.4   Location Discovery

Before forming multiple routes in a mobile environment, a source node needs to approximate the current state of the network so it can locate the destination and intermediate forwarding nodes. Location discovery methods for ad hoc multipath protocols are largely based on single-path protocols for such networks.

Proactive approaches actively maintain variants of routing tables for each source-destination pair. Commonly, these routing tables are organized in a hierarchy for scaling. The advantage of proactive approaches is that a source can immediately use local or nearby tables to construct multiple routes. However, mobility renders proactive approaches impractical, since keeping these distributed tables up to date requires high messaging overhead. Constructing multiple routes on top of these distributed tables also means that a source-destination pair might have multiple table entries and thus higher storage overhead. Multiple entries are more likely to become inconsistent and lead to broken paths and routing loops. In addition, mobility can cause many table entries to become out of date before ever being used, and those unused entries waste precious storage resources and update efforts on resource-poor mobile nodes.

Reactive, or *on-demand approaches* flood the network right before a source initiates a communication session with a destination, so that only states of active routes are being maintained in the network. The flooding process also allows a source to update its view of the network to construct multiple routes. Since flooding mechanisms become prohibitively expensive as they scale up, most on-demand schemes impose constraints on flooding.

For example, under certain conditions a node can decide to drop redundant route requests as opposed to forwarding them [Lee and Gerla 2001]. A source can also enlist the help of a geographic location service such as the global positioning system (*GPS*). With knowledge of the destination's prior location and mobility characteristics, a source can limit the flood to an area where the destination is likely to be located [Pei et al. 2000]. However, these constraints also imply that the source may obtain only a partial view of the network state, resulting in the building of multiple routes that are not as effective in achieving a particular set of goals.

A hybrid of the proactive and reactive approaches is also possible. For example, in a large ad hoc network, small groups of nodes can use proactive routing for neighbouring nodes. Each group can elect a head node to represent the group, so that reactive routing is performed among head nodes. This approach is based on two observations. First, the information from distant nodes is less likely to be correct. Therefore, an on-demand approach to obtain remote information is more appropriate. Also, nodes within the same vicinity are more likely to communicate among themselves, so maintaining the complete state at that scale may not differ from the cost of flooding initiated at the beginning of each communication session.

A reverse composition is also possible. Since head nodes experience relatively less mobility due to the distance between them, a proactive approach can work well. Lower-level nodes can use on-demand routing to accommodate frequent and local topological changes.

10.2.5   Route Recovery

Since one of the major motivations for having multiple routes is to reduce route recovery overhead, the route recovery process for multipath routing protocols is relatively simple.

The failure recovery approach largely depends on how redundant routes are used. If redundant routes are used as backup routes, failure recovery simply means that one of the backup routes will perform normal data delivery. On the other hand, if multiple routes are being used for achieving a certain QoS (such as bandwidth), the failure of one route will prompt the construction of a new redundant route, while the remaining routes can provide a graceful degradation of QoS. If multiple routes are used for load balancing and congestion control, the lack of response from one route may mean that a network is overly congested at the moment. Reestablishing a new redundant route may not be a desirable choice. Fortunately, as long as a source-destination pair has a well-defined priority for a given connection, the decision on the recovery action should be straightforward.

10.3   EXAMPLES OF AD HOC MULTIPATH ROUTING APPROACHES

Ad hoc multipath routing has generated considerable research interest over the past decade, and a number of algorithms have been developed to address the problem. We will summarize major classes of multipath approaches and consider a few examples of each. Since most ad hoc multipath routing protocols are extensions of single-path approaches, we will start with a brief review of single-path routing protocols.

10.3.1   Review of Ad Hoc Single-Path Routing Protocols

Many ad hoc multipath routing protocols are direct descendants of two popular single-path approaches: dynamic source routing (*DSR*) [Johnson and Maltz 1996] and ad hoc on-demand distance vector (*AODV*) [Perkins and Royer 1999]. We will briefly review each scheme.

Both DSR and AODV are on-demand approaches and establish routes as needed. Under DSR, a source locates a destination via flooding. Duplicate route-request messages are discarded at intermediate forwarding nodes. Once the destination is located, the destination will respond to the first request message and use the path recorded in the request packet to acknowledge back to the source. DSR uses a *source routing protocol*, in which the source precomputes the entire communication route, and the routing information is encoded in each packet header being transmitted.

One major advantage to using source routing is that intermediate nodes can perform *stateless forwarding*, in the sense that each intermediate forwarding node maintains no state regarding the routes being forwarded. Therefore, the overhead of forwarding is not as sensitive to the size of the network. However, DSR does use caching to speed up the process of locating a destination. As mobility increases, caching contributes negatively because cache entries are often invalid. Stale routes, if used, may start polluting other caches [Li et al. 2000]. Also, in the case of network failures, new routing requests and associated flooding are required to recover the route.

AODV uses a table-driven approach instead. AODV uses the same on-demand flooding and route recovery mechanisms as DSR. However, each node maintains a routing table that lists the next hop for each reachable destination for each active route. A sequence number is associated with each entry to prevent routing loops. Periodic beaconing is required to keep those tables up to date.

10.3.2   Extensions of Ad Hoc Single-Path Routing Protocols

DSR and AODV have been modified in various ways to provide multiple routes. The key observation is that during the on-demand flooding phase, enough information can be gathered to form redundant routes without additional overhead. DSR and AODV can be extended to provide multiple routes by relaxing route-request broadcasting constraints during the flooding phase and aggregating network states at the source, the destination, or intermediate nodes. With additional knowledge of the network states, a node can make a more informed decision regarding disjoint route construction.

The *diversity injection* approach [Pearlman and Haas 1998] modifies DSR to compute multiple routes. The key observation is that the flooding process is typically constrained; therefore, replies to multiple route requests at the destination tend to produce routes that share many links. One potential fix is to relax the flooding constraint of the request messages to discover more routes, but the traffic produced by flooding is prohibitive. Since each route request contains a potentially different route back to a source, caching recent requests can build a library of routes back to different sources. When a destination replies toward the source, an intermediate node can inject diversity by probabilistically selecting a route from the library to the source.

Nasipuri and Das [1999] propose an on-demand multipath scheme that modifies the destination under DSR—causing it to reply to multiple route requests selectively. The goal of this approach is to construct partial disjoint routes, where alternate routes are connected from various nodes on the primary route to the destination. *Figure 10.6* shows an example. The primary route is S-4-5-D, and partially disjoint routes are S-4-2-3-D and S-6-7-8-D. If the primary route is broken at node 5, once node 4 detects the failure, it will alter packet headers to replace the primary route with the alternate route S-4-2-3-D. This process continues until all routes break; then a fresh route discovery is initiated. Although the intent of the design is to construct many alternate routes along the primary route, the quality of routes constructed is largely a function of the thoroughness of flooding. Also, a failure point near the source will render many downstream alternate routes unavailable.

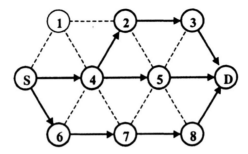

Figure 10.6. Multipath protocol proposed by [Nasipuri and Das 1999]

Split multipath routing (SMR) [Lee and Gerla 2001] is another on-demand, ad hoc multipath routing protocol using source routing. SMR is similar to DSR except that SMR tries to increase the probability of finding more disjoint routes during the route discovery phase by avoiding the use of cached routes and relaxing the constraint on forwarding duplicate route-request messages during the flooding phase. The choice of routes is based on the minimal overlapping of nodes and links among routes.

SMR-GPS [Prier et al. 2002] enlists the aid of GPS information to improve the disjointness of backup routes in SMR. SMR-GPS maximizes the minimal pair-wise distance between nodes within a route. SMR-GPS can outperform SMR in terms of surviving geographically localized and correlated failures.

Ad hoc on-demand multipath distance vector (*AOMDV*) [Marina and Das 2001] extends AODV by adding mechanisms to detect link-disjoint routes. A source can initially send different versions of a route request to each of its neighbor nodes. Based on the version stamp of the route request messages, a destination or an intermediate forwarding node can deduce the number of potential disjoint routes to the source. Based on the hop count of the route request, an intermediate node can decide whether to rebroadcast a certain version of the route request. Multiple routes are built incrementally during the forwarding process.

One challenge to modifying the existing AODV algorithm to support multipath routing is avoiding routing loops. The difficulty lies in the distributed storage of network states; maintaining a loop-free invariant is difficult because tables may become inconsistent. In addition, since each source-destination pair may contain multiple routes, a node may reach a destination through different hop counts, which further complicates verifying the correctness of the algorithm. Marina and Das [2001] deal with this problem by ensuring that nodes receiving a route request only forward it if its hop count is lower than the hop count of any route request they have already received. This requirement ensures that a route request looping back on a node will not be forwarded again, though it may also prevent forwarding of non-looping routes. Other approaches rely on multiple entries in the routing table and version numbers on the route requests. These approaches must contend with difficult bookkeeping issues.

10.3.2   Other Ad Hoc Multipath Routing Protocols

The temporally ordered routing algorithm (*TORA*) [Park and Corson 1997] maintains a destination-oriented directed acyclic graph (DAG) to construct multiple routes. The use of a DAG assures that the algorithm is loop-free. TORA uses a height-based algorithm, with traffic flowing like water from the source to the destination through multiple paths. When traffic is trapped within the terrain, the terrain is modified so traffic can flow again. The use of this gravitational model enables TORA to compute multiple routes in a distributed fashion. Its localized computation allows TORA to scale and be responsive to changes in dynamic topologies. However, TORA may potentially encounter oscillations of multiple routes, especially when multiple sets of coordinating nodes are concurrently detecting partitions, erasing routes, and building new routes based on each other [Royer and Toh 1999]. Also, TORA needs to flood the network to erase invalid routes due to proliferation of states. In addition, the assumption of reliable, in-order delivery of routing control messages imposes high overheads [Broch et al. 1998]

Directed diffusion [Estrin et al. 1999; Intanagonwiwat et al. 2000] is designed in the context of sensor networks, where minimizing energy consumption is a top priority. Unlike conventional routing approaches, the diffusion model is a data-centric and application-specific approach to directing data from sources to destinations, or sinks. A sink may disseminate its interest in data with certain attributes. Nodes that have the data of interest or information on how to obtain the data will backtrack the trail of interest to the sinks. A group of sensor nodes can cluster, and nodes rotate roles to allow batteries to recharge. The motivation for this diffusion model is the use extensive caching to avoid end-to-end communication, thus prolonging the battery life of individual nodes and the life of the sensor network as a whole. However, the energy consumption is traded off against the storage needed to cache data, and the effectiveness of data duplication and caching is highly dependent on the mobility of sensor nodes.

10.4 RECENT ADVANCES IN AD HOC MULTIPATH ROUTING PROTOCOLS

Existing multipath approaches often involve storing additional state. Keeping distributed state consistent is usually complex. Therefore, recent advances try to move toward distributed schemes where decisions are made with local knowledge. Ideally, no coordination is required to build multiple disjoint paths in parallel, and each forwarding node performs stateless routing for better scaling. In addition, recent advances are also more energy-aware.

10.4.1   Braided Multipath Routing

Braided multipath routing [Ganesan et al. 2002] was developed in the context of sensor networks, stressing energy conservation. A shortest alternative path is created for each node in the primary path, resulting in braided paths (*Figure 10.7.*). Location discovery is through low-rate dissemination of source and destination information throughout the network.

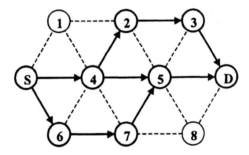

Figure 10.7. Braided multipath routing

Each node can use the promiscuous mode to overhear nearby routing information and form local detours around the nodes in the primary route. The total number of routes between the source and destination is proportional to the nth Fibonacci number, where n is the number of nodes in the primary path. With a large number of alternative routes, the primary path under this approach can potentially sustain many independent failures.

Figure 10.7. shows an example of braided routing with one primary route (S-4-5-D) and two partially disjoint routes (S-4-2-3-D and S-6-7-5-D). If node 5 fails, the primary route will fall back to the alternate route S-4-2-3-D; if node 4 fails, the primary route will fall back to the alternate route S-6-7-5-D. However, if any of two neighbouring nodes fail simultaneously, braided routes can no longer rely on alternate routes for recovery. Therefore, braided multipath routing is not resilient to geographically localized and correlated failures.

10.4.2 Magnetic-Field-Based Multipath Routing

Magnetic-field-based multipath routing (*MFR*) [Nguyen et al. 2002] is an on-demand protocol that exploits the shape properties of magnetic force lines to build node-disjoint paths (*Figure* 10.8). For each communication pair, a source represents the positive pole, and a destination represents the negative pole. Multiple paths are formed by following or approximating a designated set of magnetic force lines going from the positive pole to the negative pole. By choosing field lines with different initial angles at the source node, MFR can control the distance between disjoint paths. An angle of 0 degrees represents the straightest and (most likely) the shortest path from the source to the destination. Destination discovery is based on flooding, and MFR assumes the assistance of GPS to identify the source and destination locations.

MFR is quite different from the foregoing approaches. Since knowing the position information of the source, the destination, and the node itself is sufficient to compute the direction of a magnetic force line, no explicit control messages are needed to coordinate the formation of multiple routes. Although each node makes local decisions to forward traffic, constructed paths are likely to be node-disjoint. In addition, each node can perform stateless forwarding without maintaining information for each route.

Figure 10.8 shows a pair of communicating nodes using MFR. The three routes shown are based on magnetic field lines with initial angles -60, 0, and 60 degrees. For independent failures, the multiple paths can serve as alternate routes. The node-disjoint routes formed under MFR can also tolerate geographically localized and correlated failures. However, since the disjoint routes can be significantly longer than the shortest route, energy consumption may be suboptimal. In terms of mobility, each route in MFR has no fixed set of nodes; therefore, the node membership for each route can change dynamically without breaking the multiple routes. For example, nodes 5 and 8 can exchange positions to form new routes S-4-8-D and S-6-7-5-D without affecting the multiple routes.

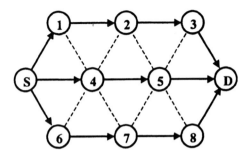

Figure 10.8. Magnetic-field-based multipath routing

10.4.3   Future Challenges

As we can see, no single routing approach can currently address all of the various requirements of ad hoc networks—resiliency, energy constraints, stability, scaling, and extreme density and mobility of nodes. As a result, there is not yet a consensus on the proper way to perform multipath routing in ad hoc mobile wireless networks. Substantial research remains on either finding a better alternative protocol than those already devised or making key improvements to an existing protocol.

Even within the range of existing protocols, some important issues are still inadequately addressed. For example, many ad hoc routing protocols face particular challenges when the radio transmission characteristics of the environment are difficult. Urban areas with many nodes located indoors or distributed in tunnelled areas are some examples. In these cases, the routes actually available may be rather serpentine, and not all protocols are capable of finding even one of them, let alone several.

Few of the existing approaches have considered security, since getting basic services deployed for the mobile wireless environment is already challenging, and security mechanisms often impose high overhead. However, the ad hoc wireless environment has security problems not present in fixed wired networks or even more conventional uses of wireless, and many of the applications for ad hoc wireless networks have strong security requirements. These factors make security correspondingly more important for ad hoc wireless networks. Attacks on wireless networks are becoming increasingly popular, based on the insecurity of commonly used protocols. The protocols proposed for ad hoc networks are not substantially more secure than wired protocols, particularly when compromised network nodes are participating. Whether there are special security issues related to building multiple paths, rather than a single path, remains to be seen.

Certainly the design of the protocols should guarantee that in spite of the actions of malicious participants, good paths will still be successfully constructed.

Advances in ad hoc multipath routing are moving at a rapid pace, in anticipation of the increasing need for such networks. A consensus regarding protocol design requirements may be reached in the near future. Much more research remains to be done in this area to ensure that the eventual choice of multipath routing algorithms for mobile ad hoc wireless networks is a wise one.

REFERENCES

[Braginsky and Estrin 2001] Braginsky D, Estring D. Rumor Routing Algorithm for Sensor Networks. *International Conference on Distributed Computing Systems (ICDCS-22)*, November 2001.

[Broch et al. 1998] Broch J, Maltz D, Johnson D, Hu YC, Jetcheva J. A Performance Comparison of Multi-Hop Wireless Ad Hoc Network routing Protocols. *Proceedings of the IEEE/ACM MOBICOM*, 1998.

[Estrin et al. 1999] Estrin D, Govindan R, Heidemann J, Kumar S. Next Century Challenges: Scalable Coordination in Sensor Networks. *Proceedings of the Fifth Annual ACM/IEEE International Conference on Mobile Computing and Networking*, 1999.

[Ganesan et al. 2002] Ganesan D, Govindan R, Shenker S, Estrin D. Highly Resilient, energy Efficient, Multipath Routing in Wireless Sensor Networks. *Mobile Computing and Communications Review (MC2R)* 1(2), 2002.

[Intanagonwiwat et al. 2000] Intanagonwiwat C, Govindan R, Estrin D. Directed Diffusion: A Scalable and Robust Communication Paradigm for Sensor Networks. *Proceedings of the Sixth Annual International Conference on Mobile Computing and Networking*, August 2000.

[Johnson and Maltz 1996] Johnson DB and Maltz DA. Dynamic Source Routing in Ad Hoc Wireless Networks. *Mobile Computing,* edited by Tomasz Imielinski and Hank Korth, Chapter 5, pp. 153-181, Kluwer Academic Publishers, 1996.

[Ko and Vaidya 1998] Ko YB, Vaidya N. Location-Aided Routing (LAR) in Mobile Ad Hoc Networks. *Proceedings of ACM/IEEE MobiCom*, October 1998.

[Lee and Gerla 2001] Lee SJ, Gerla M. Split Multipath Routing with Maximally Disjoint Paths in Ad Hoc Networks. *IEEE International Conference on Communications*, 2001.

[Li et al. 2000] Li J, Jannotti J, De Couto DSJ, Karger D, Morris R. A Scalable Location Service for Geographic Ad Hoc Routing. *Proceedings of ACM MOBICOM.* August 2000.

[Marina and Das 2001] Marina MK, Das SR. On-Demand Multipath Distance Vector Routing in Ad Hoc Networks. *Proceedings of IEEE International Conference on Network Protocols (ICNP)*, 2001.

[Nasipuri and Das 1999] Nasipuri A, Das SR. On-Demand Multipath Routing for Mobile Ad Hoc Networks. *Proceedings of the IEEE International Conference on Computer Communication and Networks (ICCCN'99)*, October 1999.

[Nguyen et al. 2002] Nguyen N, Wang AI, Reiher P, Kuenning GH. Magnetic Field Routing for Ad Hoc Networks. Submitted to the *First Workshop on Hot Topics in Networks (HotNets-I)*, October 2002.

[Park and Corson 1997] Park VD, Corson MS. A Highly Adaptive Distributed Routing Algorithm for Mobile Wiresless Networks. *Proceedings of IEEE INFOCOM*, April 1997.

[Pearlman and Haas 1998] Improving the Performance of Query-Based Routing Protocols Through "Diversity Injection." *IEEE Wireless Communications and Networking Conference 1999 (WCNC'99)*, September 1999.

[Pei et al. 2000] Pei G, Gerla M, Hong X, LANMAR: Landmark Routing for Large Scale Wireless Ad Hoc Networks with Group Mobility. *Proceedings of IEEE/ACM MobiHOC 2000*, pp. 11-18, August 2000.

[Perkins and Royer 1999] Perkins CE, Royer EM. Ad Hoc On Demand Distance Vector Routing. *Proceedings of IEEE WMCSA '99*, pp. 90-100, February 1999.

[Prier et al. 2002] Prier G, Schnaider M, Robinson M. SMR-GPS: Split Multipath Routing with GPS Based Route Selection. Unpublished manuscript, 2002.

[Royer and Toh 1999] Royer EM, Toh CK. A Review of Current Routing Protocols for Ad Hoc Mobile Wireless Networks. *IEEE Personal Communications*, April 1999.

Chapter 11

COMPETITIVE ANALYSIS OF HANDOFF REROUTING ALGORITHMS

Yigal Bejerano*, Israel Cidon[†] and Joseph (Seffi) Naor[†]
* *Bell Laboratories, Lucent Technologies*
† *Technion, Israel Institute of Technology*

11.1 INTRODUCTION

Personal Communication Systems (PCS) enable people and devices to communicate independently of their location and while moving from place to place. For providing continuous communication to mobile users every PCS network employs a *mobility management* composed of two components, *location management* and *handoff management*. In contrast to the telephone number in traditional telecommunication systems that specifies the location of the end user, the PCS subscriber number does not provide the location of the mobile user. Therefore, the system must maintain a *location management* mechanism for locating mobile users. This mechanism maps subscriber numbers to the current location of the requested users for call delivery operations. The *handoff management* enables the PCS network to maintain sessions with mobile users while they

* Portions reprinted, with permission, from Y. Bejerano, I. Cidon and J. Naor, Effi-cient Handoff Rerouting Algorithms: A Competitive On-Line Algorithmic Approach", Proceeding of INFOCOM 2000, Tel Aviv, April 2000. © 2002 IEEE.

* Portions reprinted, with permission, from Y. Bejerano, I. Cidon and J. Naor, "Effi-cient Handoff Rerouting Algorithms: A Competitive On-Line Algorithmic Approach", IEEE/ACM Trans. on Networking to appear in December 2002. © 2002 IEEE.

change their attachment points with the system's infrastructure. Such changes are called *handoff* or *handover* operations [8, 30]. In this chapter we consider only rerouting algorithms for supporting handoff operations. More general overviews of mobility management and handoff operations can be found in [6, 30, 33]. In our discussion we consider only inter-cell handoff operations that result from user movements to new cells[1], that include actions at both the wireless level and the network infrastructure. From the wireless perspective, the handoff mechanism determines the next serving base station and allocates new wireless resources for the session. The network needs to ensure the continuity of the traffic flow without interruptions or disordering the packets.

Most modern communication systems that are used as an infrastructure for PCS networks [3, 13, 22, 23, 25, 27] are based on connection-based technologies such as telephone and ISDN technologies [28], Frame-Relay [21], ATM networks [21] and more recently MPLS Networks [12]. Such networks require the establishment of a *virtual channel* (VC) between the session participants and maintaining it during the life of the session. In PCS networks, each time a session participant performs a handoff operation, the session VC must be modified for maintaining end-to-end connectivity. Thus, using efficient handoff rerouting algorithms is important for an efficient management of PCS networks. The effectiveness of a handoff management is evaluated by its ability to use efficiently the network resources while ensuring minimal disruption to the ongoing sessions. We classified the criteria for evaluating the handoff algorithms into two components. The *setup cost* represents the cost associated with the handoff operations, in particular signaling cost and handoff latency, and the *hold cost* determines the expense related to the use of network resources held by the VC. The *overall cost* of a session is defined to be the sum of its setup and hold costs.

In this chapter we describe currently used handoff rerouting algorithms and discuss their efficiency in term of overall cost in comparison to an optimal solution. We show that the ratio between the overall session cost achieved by these algorithms and the cost of an optimal solution can be very large (to be defined precisely later) in the worst case. Then, we introduce new handoff rerouting algorithms for which we can bound this ratio in both the worst case and average case. We use analytical

---

[1]another type of handoff is intra-cell handoff that results from the deterioration of radio cannel quality beyond a determined threshold due to fading effects or reassignment of wireless cell resources.

proofs for the first result and simulation models for the latter result.

11.1.1   On-Line Algorithms and Competitive Analysis

Our goal, as already mentioned, is to reduce the overall session cost. Optimizing the VC routing for this long term goal, after each movement of a user, is a complicated task, since the handoff algorithm is typically not able to predict the session duration and the future movements of the users. Thus, the problem of minimizing the overall session cost is an on-line dynamic decision problem, where decisions are based on the current state of the network without knowledge of future events.

The difficulties that an algorithm faces are illustrated in the following example. Consider two static users that can be connected by one of two paths. Suppose that the hold cost of the first path is one unit per minute, while the hold cost of the second path is ten units for the entire session with no time limit. In this situation there is no single optimal path between the two users. VCs of short sessions should be routed over the first path while VCs of long ones should be routed over the second path. However, if the session duration is not known in advance, an optimal path cannot be chosen by the algorithm at the session initialization. Nevertheless, our algorithm has to make decisions without prior knowledge of the session duration.

A common way for measuring the performance of an on-line algorithm is *competitive analysis*. The idea is to compare the costs associated with an on-line algorithm with the costs spent by an optimal off-line algorithm that has complete knowledge of the future. The maximum ratio between their respective costs, taken over all possible input sequences, is called the *competitive ratio*. It guarantees an upper bound on the worst case performance with respect to an optimal off-line algorithm. For the above example, the on-line strategy for this problem that yields the best competitive ratio routes the session VC over the first path for a duration of ten minutes and then reroutes it over to the second path. This strategy guarantees that in the worst case the session cost will be at most twice the cost of an optimal off-line strategy. This example is known in the on-line literature as the *ski rental* problem [11, 17].

In recent years, competitive analysis was extensively used for analyzing the performance of various algorithms for different communication problems, such as call admission, circuit routing, scheduling and load balancing. Extensive surveys of this area are given in [11, 17]. An alternative approach to measuring the quality of on-line algorithms is through *average case* analysis, which relies on some hypothesis on the distribution of the input. Each of the two approaches has clear advantages as well as limitations, and the reader is referred to [11] for a related discussion.

11.1.2 Competitive Handoff Rerouting Algorithms

This chapter provides a worst-case analysis of handoff rerouting algorithms for general communication networks. We distinguish between two types of algorithms: *handoff anticipating* versus *handoff non-anticipating* rerouting algorithms. Algorithms of The first type are permitted to hold unused resources anticipating that these resources will be used later on, while algorithms of the latter type are not allowed to do so. By slightly increasing the session hold cost, the handoff anticipating algorithms may significantly reduce the setup cost, as we demonstrate in the following example. Consider a session in which one of the participants frequently moves between two adjacent radio cells. As a result, after each such movement the session VC is modified, and some of the previous VC resources become unused until the next movement. If the link setup costs is significantly higher than their hold costs[2] then the session overall cost can be reduced by keeping the two VC that end at each radio cell, rather than releasing and allocating the same resources again and again. However, holding unused resources may hurt the network utilization and increase the call blocking probability.

Each link e in the network is associated with two independent costs, the setup cost, s_e, and the hold cost, h_e. Initially, for the case of arbitrary graphs, we present a lower bound of $\Omega(\log n)$ on the competitive ratio of any handoff algorithm, where n is the number of nodes in the graph. Then, for the handoff-anticipating model, we consider several important cases where competitive on-line algorithms exist. We present a 2-competitive on-line algorithm for managing a session in a tree network. The difficulty in this case is how to avoid repeated allocations of the same resources which may increase the session cost with respect to the cost achieved by an optimal off-line algorithm. We also provide a matching lower bound on the competitive ratio. We next consider a ring and present a 4-competitive algorithm. In this case the algorithm

---

[2]We should take into account also the movement frequency for comparing these costs.

is required to determine both the VC path at each time step during the session and when to release unused link resources.

In the following section we turn our attention to the non-anticipating model and show that if there is a correlation between the setup and the hold costs of the links, then efficient competitive algorithms do exist. We assume that the ratio between the hold cost and the setup cost is bounded by two constants c_1 and c_2, where $c_1 \leq c_2$. With these restrictions, we present two $(\frac{c_2}{c_1} + 2)$-competitive algorithms. When $c_1 = c_2$, the algorithms are 3-competitive. This case is of practical importance since it reflects environments of current mobile networks.

11.1.3 The Chapter Organization

The chapter is organized as follows. Section 11.2 describes the network model and a short survey of different known handoff rerouting algorithms is given in section 11.3. In Section 11.4 we present both lower bounds and competitive algorithms for the handoff-anticipating model. Competitive algorithms for the non-anticipating model are Given in Section 11.5. We strengthen our analytical results by performing simulations and present the results in Section 11.6. Finally, we conclude this study in Section 11.7. For clarity of exposition, some of the proofs are omitted and others are deferred to the appendix.

11.2 THE NETWORK MODEL

We assume an arbitrary connection-oriented network modeled by an undirected graph $G(V, E)$, where the nodes and edges represent communication switches and full duplex links respectively. Users are attached to the nodes and can be either static or mobile. A mobile user may move and change its attachment node. In our discussion we do not consider any specific network architecture and we assume that a switch can serve a single base station or a group of base stations. The set of base stations serviced by a single node (switch) is called the *node coverage area*[3]. We consider only inter-node handoff operations, i.e., a movement of a mobile user from the coverage area of one node to the coverage area of another.

---

[3]Recall that large PCS networks have a hierarchical architecture. Each network switch connects a group of base station controllers (BSC) to the infrastructure network and each BSC serves a group of base stations.

A session between two users requires the establishment of a *virtual channel* (VC) between the corresponding nodes and holding it during the session. This means allocating resources at each edge over the VC path and holding them during the session. Each edge $e \in E$ is associated with a *linear cost function*, $f_e(\tau) = s_e + h_e \cdot \tau$, that defines the cost of using edge e for a duration of τ time units. The *setup cost*, $s_e \geq 0$, is the cost of allocating resources over edge e, and the *hold cost*, $h_e \geq 0$, is the cost of holding these resources for a single time unit. The hold time, τ, is measured from the time the edge resources are allocated until they are released. Hence, the entire cost of a VC session of duration τ, which is routed over a path p, is given by $f_p(\tau) = \sum_{e \in p} f_e(\tau)$. We use cost as a general term, and it can capture delay, dollar cost, handoff latency, signaling cost, or an aggregation of several measures.

Now, consider a session σ between two users that starts at time zero and terminates at time τ. The session is defined by a sequence of m triplets, $\sigma = \{(u_i, v_i, t_i)\}_{i=0}^{m}$, where the i-th triplet represents a movement of a user from node u_i to node v_i at a time t_i. We also consider the session initialization and the termination as movements. We assume that before a session starts, both users are attached to node u_0 and at time zero one of them moves to node v_0. Similarly, the session terminates at time t_m, when one of the users moves to the node to which the other attaches, and they both do not change their attachment node anymore.

The session cost depends on the used handoff rerouting algorithm. This cost is composed of the overall setup cost and the overall hold cost. For a given handoff algorithm $\mathcal{A}$ and a session σ, the first term is denoted by $Setup\_Cost_{\mathcal{A}}(\sigma)$ and the second term by $Hold\_Cost_{\mathcal{A}}(\sigma)$. Thus, the overall cost of a session σ results by algorithm $\mathcal{A}$ is,

$$Cost_{\mathcal{A}}(\sigma) = Setup\_Cost_{\mathcal{A}}(\sigma) + Hold\_Cost_{\mathcal{A}}(\sigma)$$

11.3 EXISTING HANDOFF REROUTING SCHEMES

We present the main handoff rerouting algorithms that are described in the literature and evaluate their performance using competitive analysis. Recall that this is a worst case analysis comparing the session cost of the algorithm under discussion with the session cost of an optimal off-line algorithm.

11.3.1   The Algorithm Description

As mentioned above, the objective of handoff rerouting algorithms is to maintain the ongoing sessions with mobile users while they change their attachment points with the system's infrastructure. They should minimize the consumption of network resources while ensuring minimal disruption to the traffic flow. A handoff algorithm consumes network resources by employing signaling and session rerouting each time a session's participant moves to a new place and a handoff operation occurs. In addition, the required bandwidth is reserved at all the links along the session VC. The current existing handoff rerouting algorithms can be broadly divided into the following four categories:

- Connection reestablishment.

- Path extension.

- Connection modification.

- Multicast Connection Rerouting (Handoff anticipation).

These algorithms differ in their resource consumption and their interference with the session flow. A summary of handoff rerouting algorithms and comparisons is given in [8].

The Connection Reestablishment Algorithm

The *connection reestablishment* is a simple handoff algorithm that establishes a completely new VC between the users at each handoff operation and releases the previous VC [18]. This algorithm optimizes the network utilization at the expense of high signaling cost and high handoff latency, making it inefficient in the case of small cell sizes or when the users are distinct.

The Path Extension Algorithm

The *path extension algorithm* [2, 3, 31], also called the *chaining handoff approach* [26], allocates a new segment between the old and the new attachment points and connects it to the existing VC. It never tears off any established VC segment with the exception of cycles. In practical implementations, optimizations such as detection and releasing of routing loops are usually performed. This is a simple handoff algorithm with low handoff latency and low signaling cost that also preserves the packets' order. The latter eliminates the need for large buffers and packet reorder mechanisms that is required in other schemes. As a result, the setup cost and the service interruption of this scheme are minimal, at the expense of possibly highly stretched routes that increase the session latency and reduce the network available capacity for other connections. This scheme was especially tailored for Wireless ATM-LAN environment where the base stations are directly connected to the ATM-LAN network. The proponents of these schemes claim that in such environments the additional delay that results from the link chaining is negligible, and the increasing hold cost is insignificant relative to the large bandwidth capacity available in the system infrastructure.

The Connection Modification Category

The *connection modification* (CM) category, also called the *partial rerouting* method, contains a large group of algorithms that they are all employ the following approach. When a handoff occurs, a CM algorithm uses part of the existing session VC and establish a new path between the user's new location and a *crossover switch* (COS). Algorithms using this approach differ by the way the COS is selected.

The Fixed Anchor Rerouting Scheme [5, 7, 24] - In this scheme a single node in the VC is selected as an anchor and only the path to the anchor is modified. The scheme is based on the assumption that during the time-life of a session the mobile user usually remains in the switch coverage area. Thus, both the signaling cost and the session interference are kept low and the VC route is near optimal. A similar scheme is also used by GSM networks [13, 23].

The Last Divergent Rerouting Algorithm [19, 20] - This algorithm selects as the COS the first node on the shortest path between the users that is also included in the existing VC. This approach can be viewed as an improvement of the connection reestablishment scheme and it is targeted to minimize the network resources held by the session. Like connection reestablishment scheme, this algorithm gives the best performance in terms of resource utilization. However, the handoff signaling as well as the session disruption may be large when the variation between the old and the new VCs is significant.

The Minimal Path Update (MPU) Algorithm [29, 19, 4] - This approach, also called the *nearest common node rerouting* (NCNR), selects as a COS the node in the existing VC that is the closest one to the user new location. This algorithm is an improvement of the path extension approach and it minimizes the rerouting cost and the latency of the handoff operations.

All the previous methods use either a *hard* or a *soft* handoff procedure. In a hard handoff procedure, the scheme finds the COS and then releases the VP path between the COS and the previous attachment point before establishing the new VC with the user current attachment point. In a soft handoff procedure, the new VC is allocated before releasing unnecessary resources. Thus, soft handoff reduces the handoff latency experienced by the user. In the following we present a handoff predicting approach that aims to further reduce the handoff latency.

The Multicast Connection Rerouting

The *multicast connection rerouting* category, also called *handoff anticipation* approach establishes a multi-point VC to several adjacent nodes in anticipation of a possible handoff. This approach reduces the handoff latency at the expense of processing cost and network bandwidth utilization. Several variations to this approach have been proposed in the literature.

The Static Virtual Connection Tree Rerouting Scheme (VCT) [1] - In this algorithm, during the connection initialization, a tree structure is established between a root node and a set of nodes in the vicinity of the user current attachment point (usually this tree is established between a switch and the set of base stations in its coverage area). Each packet designated to the user is duplicated and sent to all the leaves of the tree. Thus, while the user moves in the coverage area of the tree its information is always available. Each time the user leaves the coverage area of the current VCT, the algorithm performs inter-VCT handoff where the tree root and its structure are modified.

The Multicast Rerouting Algorithm (MR) [15] - In this method a a set of connections is established from an anchor node to the current attachment point of the mobile user and all its neighboring nodes. When a packet arrives to the anchor node it is duplicated and sent to all the VC-s. Thus, when the user moves to another attachment point its data is already available. The handoff algorithm updates the set of connections by adding a VC to new neighboring nodes and releasing VC-s that are no longer needed. This approach avoids the need to perform inter-VCT handoffs.

The Dynamic Virtual Connection Tree Rerouting Algorithm (DVCT) [32] - Here, the handoff algorithm combines the multicast methods used by the VTC and MR algorithms. It constructs a dynamic virtual tree that spans the user current attachment point and its neighboring nodes. However, network resources are allocated only along the path with the user current location, which is called the active connection. When the user moves to another node, the VC that ends at this node become active and the required resources are allocated along this path.

Summary of the Handoff Algorithms' Properties

In general, most of the above algorithms attempt to optimize the session cost only from the perspective of a single criterion, either the setup cost or the hold cost, at the expense of other criteria. The path extension is a simple algorithm that aims to reduce the setup cost, while the connection reestablishment scheme guarantees the optimal hold cost. The multicast connection rerouting algorithms use a handoff anticipating approach whose goal is to reduce the handoff latency at the expense of high usage of network resources. Connection modification algorithms attempt to reduce the overall session cost, however they cannot guarantee low cost relative to the optimal cost. The properties of these schemes are summarized in Table 11.1, (See also [6]).

11.3.2   Competitive Analysis of the Existing Algorithms

We turn to calculate lower bounds on the competitive ratios of the handoff rerouting algorithms described above. We classify the algorithms

The Schemes	Advanages	Disadvanages
Connection Reestablishment & Last Divergent	Optimal route. Minimal hold cost. Simple.	High handoff latency. High setup cost.
Path Extension and Minimal Path Update	Low handoff latency. Keeps packet order. Simple.	Inefficient connection route. High hold Cost.
Connection Modiflcation - Fixed Anchor	Reduced resource usage (setup & hold). Moderate handoff latency.	Complicated.
Multicast Connection	Low handoff latency. Keeps packet order.	Waste of net resources. Very high hold cost. Complicated.

Table 11.1: Comparison of Handoff Rerouting Algorithms.

into two groups. This first group contains *Hold-cost-minimization* algorithms that minimize the session hold cost and ignore its setup cost. For instance, this group contains the connection reestablishment algorithm [18] and the connection modification algorithm presented in [19], where the selected COS is the first node on the shortest path between the users that is also included in the existing VC. The second group includes *Setup-cost-minimization* algorithms that minimize the session setup cost without considering its hold cost. Such algorithms are the path extension algorithm [2],[3], the minimal path update algorithm [29] and the anchor rerouting algorithm [5]. This separation enables us to show that any algorithm that minimizes the cost of only one component may result with high cost of the second component.

Theorem 1: *The competitive ratio of any hold-cost-minimization algorithm is at least $n/2$, where n is the number of nodes in the network, even if the setup cost and hold cost are correlated.*

Proof: Consider the network $G(V, E)$ that is described in Figure 11.1, where for every edge $e \in E$, $s_e = h_e$. For every $i \in [2 \cdots n]$, let $S_{1,i} = (i - 1)$ and for every $i \in [3 \cdots n]$, let $S_{i-1,i} = 1 + \epsilon$. We assume a session σ between two users that are located in node 1. Immediately after the session initialization, one of the users moves from node 1 to node n

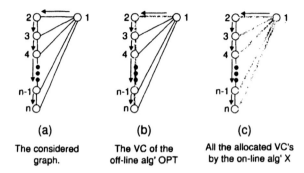

(a)	(b)	(c)
The considered graph.	The VC of the off-line alg' OPT	All the allocated VC's by the on-line alg' X

Figure 11.1: The graph $G(V, E)$ used by Theorem 1.

through the path $1, 2, 3, \cdots, n$. As soon as the user reaches node n the session terminates. We assume that the mobile user moves fast enough so that the session duration and hold cost are negligible. Now, consider the best hold-cost-minimization algorithm, denoted by $\mathcal{X}$. After each movement from node $i - 1$ to node i the algorithm releases the current VC and establishes a new one between the nodes 1 and i. The cost of this setup operation is $(i - 1)$, thus the setup cost of the algorithm is $Setup\_Cost_{\mathcal{X}}(\sigma) = \frac{n \cdot (n-1)}{2}$. However, the optimal off-line algorithm $\mathcal{OPT}$ extend the current VC after each movement. Thus, its setup cost is $Setup\_Cost_{\mathcal{OPT}}(\sigma) = (n - 1) \cdot (1 + \epsilon)$. As a result,

$$\frac{Cost_{\mathcal{X}}(\sigma)}{Cost_{\mathcal{OPT}}(\sigma)} = \frac{n \cdot (n - 1)/2}{(n - 1) \cdot (1 + \epsilon)} = \frac{n/2}{1 + \epsilon}$$

Since ϵ can be any small positive value, the lower bound is at least $n/2$.
□

Corollary 1: *The competitive ratio of the connection reestablishment algorithm is at least $n/2$, where n is the number of nodes in the network.*

Proof: The connection reestablishment algorithm will make the same routing decisions that the best setup-cost-minimization algorithm has made in the example given in the proof of Theorem 1.   □

Theorem 2: *The competitive ratio of any hold-cost-minimization algorithm is at least n, where n is the number of nodes in the network, even if the setup cost and hold cost are correlated.*

Proof: Consider a ring network $G(V, E)$ as depicted in Figure 11.4, with even number of nodes n and for every edge $e \in E$, $s_e = h_e$. Let $S_{1,n} = 1 + \epsilon$ and let the setup cost of the other edges $e \in E - \{(1, 2)\}$ be $S_e = 1$. We assume a session σ between two users that are located at the nodes 1 and $n/2 + 1$. Immediately after the session initialization, the user from node $n/2 + 1$ moves to node 2 through its left neighbor. We assume that the mobile user moves fast enough so that movement time from node $n/2 + 1$ to node 2 is negligible and the user stays at node 2 for a very long time. Recall that the setup cost of the edges between nodes 1 and $n/2 + 1$ along the left side of the ring (the path $1, 2, \cdots, n/2 + 1$) is $n/2 + \epsilon$ while the cost of the path along the right side is $n/2$. Now, consider the best setup-cost-minimization algorithm, denoted by $\mathcal{X}$. At the beginning, it establish the VC a long the right side of the ring, and after each movement from node $i - 1$ to node i, $i \in [n/2 + 2, n]$, the algorithm adds the edge $(i - 1, i)$ to the existing VC. Thus, $Setup\_Cost_{\mathcal{X}}(\sigma, t) = n - 1$ and the hold cost until time t is $Hold\_Cost_{\mathcal{X}}(\sigma) = (n - 1) \cdot t$. Let us tern to describe the routing decisions of the off-line algorithm $\mathcal{OPT}$. Algorithm $\mathcal{OPT}$ routes the VC along the left side of the ring, and after each movement it removed the unused edge. Thus, $Setup\_Cost_{\mathcal{OPT}}(\sigma) = n/2 + \epsilon$ and the hold cost until time t is $Hold\_Cost_{\mathcal{OPT}}(\sigma, t) = (1 + \epsilon) \cdot t$. Consequentially,

$$\frac{Cost_{\mathcal{X}}(\sigma, t)}{Cost_{\mathcal{OPT}}(\sigma, t)} = \frac{(n - 1) + (n - 1) \cdot t}{n/2 + \epsilon + (1 + \epsilon) \cdot t} \simeq n$$

Thus, for large t and small ϵ the ratio is approximately n.   □

Corollary 2 : *The competitive ratios of the path extension, the minimal path update, and the anchor rerouting algorithms are at least n, where n is the number of nodes in the network.*

Proof: These algorithms will make the same routing decisions that the best setup-cost-minimization algorithm has made in the example given in the proof of Theorem 2.   □

The above theorems prove that the described handoff algorithms may make poor routing decisions that yield high session cost with respect to the cost of an optimal algorithm. In the following we focus on competitive handoff algorithms that balance between the session setup and hold costs for obtaining constant competitive ratios.

11.4   HANDOFF ANTICIPATING ALGORITHMS

In this section we consider handoff anticipating competitive algorithms that are allowed to allocate and hold unused edge resources. As we have shown in Section 11.1.1, these resources may be used for reducing the overall session cost if it is known that these resources will be used later on during the session. It is clear that this ability gives a considerable advantage to the off-line algorithm that knows all the future movements in advance. Therefore, as we prove in Theorem 3, there is a lower bound of $\Omega(\log n)$ for any on-line algorithm for general graphs, where n is the number of nodes in the network. This is not a surprising result, since the connection management problem can be shown to be related to the on-line Steiner tree problem [16] that has a similar lower bound, by viewing each movement of a user as corresponding to a node leaving the tree and another node joining it.

Theorem 3: *Consider a general graph, mobile users, and arbitrary edge cost functions with positive setup cost. Then, the competitive ratio of the best on-line algorithm is at least $\Omega(\log n)$, where n is the number of nodes in the network.*

We defer the proof of Theorem 3 to Appendix 11.A In the sequel, we consider two special cases where constant competitive on-line algorithms do exist. In Section11.4.1, we consider a tree topology and Section 11.4.2 intrudes a competitive algorithm for a ring topology. To this end, let us first examine what are the properties of an optimal off-line algorithm. A connection management algorithm is called *lazy* if it changes the VC route only as a response to a movement of a mobile user and at the time of the movement.

Theorem 4: *If the edge cost functions are linear, then there is a lazy optimal off-line algorithm.*

Proof: Suppose in contrast that such lazy optimal algorithm does not exist. Consider a non-lazy optimal off-line algorithm and define a lazy algorithm that makes the same rerouting operations but postpones their execution until one of the user moves. Thus, the setup cost of the lazy algorithm is as most as the setup cost of the optimal algorithm. Since the lazy algorithm allocates the used resources after the optimal algorithm, its hold cost is no more than the hold cost of the optimal Algorithm. Hence, the lazy algorithm yield the same cost or better than the non-lazy off-line optimal algorithm.   □

Note that Theorem 4 is satisfied also when considering non-anticipating competitive algorithms.

11.4.1   An On-Line Algorithm for a Tree Topology

Managing a session in a graph with a tree topology may not seem a difficult task, since there is a unique path between every pair of nodes. However, the following example shows that session resources need to be released judiciously, otherwise the competitive factor is not bounded.

Example 1: Consider a graph with two nodes u and v and a single edge e between them. Now refer to a session between two users. One of them is static and is attached to node u. The other is mobile, and it moves as follows: At the session initialization it is located at node v. Each time a VC is established over edge e it moves to node u. When the VC resources are released the mobile user returns to node v.

The session algorithm may use one of the following strategies: (a) release the edge resources whenever they are not in use; (b) maintain unused resources in case the mobile user returns to node v. It is not hard to see that both strategies are not competitive.

Our algorithm, referred to as Algorithm $\mathcal{T}$, uses the *postponement principle* for determining the time for releasing unused resources. The release of VC resources over an edge e is postponed by a time period of s_e/h_e with respect to the last time they were in use. This period is called the *postponed period*, and the edge is called a *postponed edge*. If, during that period, the edge resources are required again, then there is no setup cost. The only cost incurred is the cost of maintaining the resources during the time they are not used. Thus, the edge resources are released precisely when the maintenance cost is equal to the setup cost. In this case, the postponed period is called *redundant*. In the following, we use the term *path activation* for establishing a VC over a given path p, which may include postponed edges as well as edges without any allocated resources. The activation cost of each edge $e \in p$ at a given time t is calculated as follows. Let $\tau_e(t)$ be the period that has elapsed since the last time edge e was in use until time t, or $\tau_e(t) = \infty$ if it was never used before. If e is a postponed edge then its activation cost is $h_e \cdot \tau_e(t)$. Otherwise, the activation cost is the sum of two components: the edge setup cost s_e and the cost of the redundant postponed period before the edge resources are released. The cost of this period is $h_e \cdot (s_e/h_e) = s_e$. As a result, the cost of activating path p at time t is:

$$Active\_Cost(p, t) = \sum_{e \in p} \begin{cases} h_e \cdot \tau_e(t) & \tau_e \le s_e/h_e \\ 2 \cdot s_e & \text{otherwise} \end{cases} \tag{11.1}$$

The hold cost of each edge remains unchanged.

Theorem 5: *Algorithm $\mathcal{T}$ is 2-competitive.*

<u>Proof:</u> Consider a session σ between mobile users that are controlled by a cruel adversary. The adversary increases the session cost achieved by $\mathcal{T}$ (with minimal effect on the cost of OPT) by forcing the on-line algorithm to repeatedly allocate VC resources immediately following their release. When VC resources become active, the adversary moves the mobile users to new nodes where these resources are not required any more (as described in Example 1). We claim that this is the worst-case scenario from the perspective of Algorithm $\mathcal{T}$. Let us bound the contribution to the cost of the session in both OPT and Algorithm $\mathcal{T}$ of every edge $e \in E$. Suppose that edge e is activated n_e times, and let the total period of time in which edge e is used be θ_e. Algorithm $\mathcal{T}$ incurs a cost of $2 \cdot s_e$ for each activation of edge e, where this cost includes a setup cost and a postponed period cost. Concerning OPT, when edge resources become unused, OPT has two options to choose from. It can either maintain the edge resources until the next activation, i.e., for a period of s_e/h_e, or release them immediately when they become unused, and reallocate them at a later point of time. The activation cost of both options is s_e. Hence,

$$Cost_A(\sigma) = \sum_{e \in E} (n_e \cdot 2 \cdot s_e + \theta_e \cdot h_e) \le$$

$$\le 2 \cdot \sum_{e \in E} (n_e \cdot s_e + \theta_e \cdot h_e) = 2 \cdot Cost_{OPT}(\sigma)$$

$\square$

In [9], it is shown that the competitive ratio of any on-line algorithm for a tree topology is at least 2. Thus, Algorithm $\mathcal{T}$ is the optimal on-line algorithm for trees.

11.4.2   An On-Line Algorithm for a Ring Topology

In this section we present a 4-competitive algorithm, referred to as Algorithm $\mathcal{R}$, for session management in a ring topology. In such a graph, at every step, the algorithm is required to determine both the VC path and when to release unused edge resources. The proposed algorithm uses the *postponement principle* for handling unused resources, as described in Section 11.4.1, where a path activation cost is defined by Equation 11.1. For determining the VC path during the session the algorithm uses a retrospective approach [17]. It keeps track of the past and routes the VC in accordance with the routing decisions made by OPT.

Our algorithm uses Theorem 4 for estimating the possible decisions of algorithm OPT. This theorem states that there is a lazy off-line algorithm OPT for optimal session management. Moreover, OPT knows the future movements of the users, and when edge resources become unused it can determine whether to release them immediately or to maintain them as postponed resources until they are needed again. Hence, the activation cost of a path p at time t according to OPT is

$$Active\_Cost^*(p,t) = \sum_{e \in p} \begin{cases} h_e \cdot \tau_e(t) & \tau_e \le s_e/h_e \\ s_e & Otherwise \end{cases} \qquad (11.2)$$

Algorithm $\mathcal{R}$ maintains a *decision tree* that represents all possible routing options and their cost. A *routing option* is a sequence of routing decisions that defines the VC path after the movement of each participant until a given time t. Thus, each possible routing option is a path in the decision tree from the root to a leaf. At the session initialization the tree contains only two routing options. After each movement of a user, each path in the decision tree splits into two new paths which correspond to the two new routing options (as described in Example 2). For every routing option z, let $p(z,t)$ be its VC path at time t, and let $c(z,t)$ be its accumulated cost until time t, calculated according to Equation 11.2. A routing option is called *optimal* at time t if it has the minimal accumulated cost until that time. We denote by $\bar{z}(t)$ the optimal routing option at time t, and let $\bar{p}(t)$ and $\bar{c}(t)$ be its VC path and its accumulated cost at time t correspondingly. At each given time t, algorithm $\mathcal{R}$ routes the session VC over the path $\bar{p}(t)$ defined by optimal routing option. It is clear from this description, that Algorithm $\mathcal{R}$ may follow several routing options during a session, adapting the behavior of new routing option when it becomes optimal, as describe by the next example.

We note that in practice, the decision tree is required to represent only routing options that may become optimal in the future. Thus, there is no need to maintain all possible routing options.

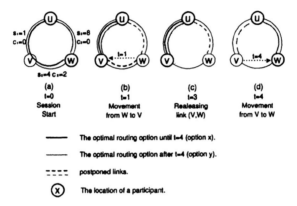

Figure 11.2: An example of a session in a ring.

Example 2: In this example, a routing option becomes optimal due to an activation of a postponed edge. Consider a ring with three nodes as depicted in Figure 11.2, where the setup cost and hold cost are denoted over the edges. Now, refer to a session between two users. One of them is static and it is attached to node u. The other is mobile, and it is located at node w at time $t_0 = 0$ when the session starts. At time $t_1 = 1$, the mobile user moves to node v, and at time $t_2 = 4$ it returns to node w. It is clear from the decision tree in Figure 11.3, that establishing a VC over edge (u, w) at time t_0, and a VC over edge (u, v) at time t_1 is the best routing option (option y), but this is revealed only at time $t_2 = 4$ when the mobile users returns to node w.

Example 2 shows us that two different routing options may be both optimal, one before time t and the other after time t, even if they both route the VC over the same path just before time t. This results from using two different VC paths in the ring in the past.

Theorem 6: *Algorithm $\mathcal{R}$ is 4-competitive.*

The proof of Theorem 6 is given in Appendix 11.B.

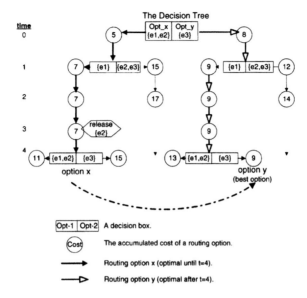

Figure 11.3: The decision tree of the session.

11.5   HANDOFF NON-ANTICIPATING SCHEMES

We turn to the class of *handoff non-anticipating competitive algorithms*. The latter algorithms are considered more practical than the handoff anticipating algorithms from two main reasons. First, the uncertainty of predicting the future movement of users, results in holding unused resources that increases the session hold cost without ensuring any reduction of the setup cost. Second, holding unused resources may hurt the network potential utilization and increase the call blocking probability.

We start by showing lower bounds for this problem. Although, in this case, the offline algorithm is more restricted than the handoff anticipating offline algorithm, the competitive ratio of the best on-line algorithm for general graph is still at least $\Omega(\log n)$, where n is the number of nodes in the graph. Then we restrict our discussion to the case where there is a correlation between the setup cost and the hold cost. For each link $e \in E$, the ratio between the hold cost and the setup cost is bounded by two constants c_1 and c_2, such that $c_1 \leq h_e/s_e \leq c_2$. With these restrictions, we present two $(\frac{c_2}{c_1} + 2)$-competitive algorithms. When $c_1 = c_2$, the algorithms are 3-competitive. This case is of practical importance since it reflects environments of current mobile networks.

11.5.1 Lower Bounds for Non-Anticipating Algorithms

First, we consider the case of general graphs where there are no restrictions on the setup and hold costs of a link.

Theorem 7: *Consider a general graph and linear cost functions. Then, the competitive ratio of the best on-line algorithm is at least $\Omega(\log n)$, where n is the number of nodes in the network.*

The proof of Theorem 7 can be found in [10]. This bound holds even if all edges have the same setup cost and the mobile users are allowed to move only between adjacent nodes. The proof strongly uses the independence between the setup and hold costs of each edge. Therefore, we restrict our study to a *correlated cost model* where the two costs of each edge are correlated. The graph is associated with two positive constants, c_1 and c_2, that bound the ratio h_e/s_e for every edge $e \in E$, i.e., $c_1 \le h_e/s_e \le c_2$. Now, let us also show the lower bound of the correlated cost model.

Theorem 8: *If the setup cost and hold cost are correlated, then the competitive ratio of any on-line algorithm is at least 2.*

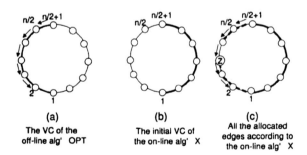

Figure 11.4: The ring for the lower bound proof

Proof: We assume in this proof that $c_2 = c_1$. Suppose, in contrast, that there is an on-line Algorithm $\mathcal{X}$ with a competitive ratio $\gamma < 2$. Consider a ring with n nodes, where n is an even number greater then $\frac{2}{2-\gamma}$, and both the setup and hold costs of each edge are 1. Clearly, $\frac{2\cdot(n-1)}{n} > \gamma$. Let σ be a session between a mobile and a static user which are initially at distance $n/2$ away from each other, e.g., at nodes $n/2+1$ and 1, respectively, as depicted in Figure 11.4. During the session initialization, Algorithm $\mathcal{X}$ establishes a VC connecting the two users, and suppose that the VC is routed through the right side of the ring. Immediately after the VC setup the mobile user moves to node $n/2$, and continues its movement until it reach node 2 through its left neighbor. We assume that the mobile user moves fast enough so that the session duration and hold cost are negligible.

Let us turn to calculate the setup cost of Algorithm $\mathcal{X}$. Initially, it uses the path-extension method until the mobile user reaches some node z, where it decides to reroute the session VC through the left side of the ring. Thus, its setup cost is at least $(n-1)$. Note that node z may be any node in the left side of the ring including nodes 2 and $n/2$. Concerning the off-line algorithm OPT, since it knows the mobile user movements in advance, it routes the session VC through the left side of the ring. After each movement it only releases unused VC resources without allocating new ones. Therefore, its setup cost is $n/2$. The competitive ratio of Algorithm $\mathcal{X}$ is $\frac{Cost_{\mathcal{X}}(\sigma)}{Cost_{OPT}(\sigma)} = \frac{n-1}{n/2} > \gamma$, in contrast to the above assumption. $\square$

11.5.2 Efficient Competitive Algorithms

In the sequel, we present two lazy on-line algorithms for arbitrary graphs and a correlated cost model with a competitive ratio of $1 + 2 \cdot \frac{c_2}{c_1}$. If $c_1 = c_2$, then the competitive ratio is 3. The two algorithms use different techniques for balancing between the setup and hold costs of a session. The first algorithm, denote by $\mathcal{A}$, is the simpler one and we also provide a full analysis of its competitive ratio. The second algorithm, termed Algorithm $\mathcal{B}$, employs more cost-effective methods, and although it achieves the same competitive ratio it yields better results in average as we show in Section 11.6.

Algorithm $\mathcal{A}$

The Algorithm's Description

> *Upon the session initialization between nodes u and v do:*
> Establish a VC over the path $p_{u,v}^*$
>
> *Upon a movement from node w to u when the second*
> *user is attached to node v do:*
> $p \leftarrow p_{w,v} \cup p_{w,u}^*$
> If $(s(p) \leq \alpha \cdot s^*(u,v))$ then
>   Establish a VC over the path $p_{w,u}^*$
>   Add $p_{w,u}^*$ to the VC path $p_{u,v}$
> Else
>   Release the current VC
>   Establish a new VC over the path $p_{u,v}^*$

Figure 11.5: A formal description of Algorithm $\mathcal{A}$.

The first algorithm, which we denote by $\mathcal{A}$, balances between the path extension and the connection reestablishment algorithms. In our description we use the following notation. For a path p, let $h(p) = \sum_{e \in p} h_e$ and $s(p) = \sum_{e \in p} s_e$ be the *hold cost* and the *setup cost* of path p, respectively. For every pair of nodes u and v, let the *shortest path* between them be the path which has *minimum setup cost*. Denote this path by $p_{u,v}^*$ and its setup cost by $s^*(u,v)$. In addition, let $h^*(u,v)$ be the *minimum hold cost* between nodes u and v. Note that the hold cost of the shortest path, $p_{u,v}^*$, may be more than $h^*(u,v)$. However, if the constants c_1 and c_2 are close, then the hold cost of the shortest path between a pair of nodes is close to the minimum hold cost.

Algorithm $\mathcal{A}$ works as follows. During the session initialization it establishes a VC over the shortest path between the users. Now, suppose that during a session one of the users moves from node w to node u, while the other user is attached to node v. The algorithm finds the path p which is obtained by concatenating the shortest path between nodes w and u, $p_{w,u}^*$, to the current VC. If the setup cost of p is not more than α times the setup cost of the shortest path between the two users, $s^*(u,v)$, then the path $p_{w,u}^*$ is established and it becomes part of the VC route.

Otherwise, the current VC is released and a new VC over the shortest path $p_{u,v}^*$ is established. A formal description of the algorithm is given in Figure 11.5, where $p_{w,v}$ is the VC path before the movement.

The algorithm uses the *credit principle*. It attempts to minimize the setup cost of each handoff operation under the constraint that the VC total setup cost is at most α times the setup cost of the shortest path between the users. We call α the *credit parameter*. The credit principle guarantees that the hold cost of $\mathcal{A}$ will not exceed $\frac{c_2}{c_1} \cdot \alpha$ times the hold cost of OPT. In the sequel we show that a proper selection of the parameter α yields a small competitive ratio.

The Algorithm's Competitive Ratio

We turn to prove that the competitive ratio of the algorithm is $1 + 2 \cdot \frac{c_2}{c_1}$. Consider a session σ that starts at time zero and is defined by a sequence of m triplets, $\sigma = \{(w_i, u_i, t_i)\}_{i=0}^{m}$, where the i-th triplet represents a movement of a mobile user from node w_i to node u_i at time t_i, as described in Section 11.2.

Lemma 1: *For every session σ,*
$Hold\_Cost_{\mathcal{A}}(\sigma) \leq \frac{c_2}{c_1} \cdot \alpha \cdot Hold\_Cost_{OPT}(\sigma).$

Proof: For every edge $e \in E$, $c_1 \leq h_e/s_e \leq c_2$. Therefore, for every pair u and v, $s^*(u,v) \leq h^*(u,v)/c_1$. In addition, for every path p, $h(p) \leq c_2 \cdot s(p)$. By the credit principle, at any time during the session, the VC route, p, satisfies $s(p) \leq \alpha \cdot s^*(u,v)$, where the users are attached to nodes u and v. Hence, $h(p) \leq c_2 \cdot s(p) \leq c_2 \cdot \alpha \cdot s^*(u,v) \leq \frac{c_2}{c_1} \cdot \alpha \cdot h^*(u,v)$ proving the lemma.   □

Consider the i-th movement from node w_i to node u_i. Let $s(M_i) = s^*(w_i, u_i)$ be the setup cost of the shortest path between these nodes, called the *movement cost*, and let $s(M) = \sum_{i=0}^{m} s(M_i)$.

Lemma 2: *For every session σ, $Setup\_Cost_{OPT}(\sigma) \geq \frac{s(M)}{2}$.*

Proof: First, assume that OPT pays for allocating the resources of an edge in two installments: the first half is paid for at the time of allocation, and the second half is paid for when the edge resources are released. Now consider a user's movement from node w_i to node u_i, and let us bound its contribution to the total setup cost of OPT. As a result of this movement, part of the VC route between node w_i and some node x is released, and a new path is established between nodes x and u_i. Hence, the setup cost of this movement is at least $(s^*(w_i, x) + s^*(x, u_i))/2$. By the triangle inequality, we get that $s^*(w_i, x) + s^*(x, u_i) \geq s^*(w_i, u_i)$ Therefore, the total cost is

$$Setup\_Cost_{OPT}(\sigma) \geq \sum_{i=0}^{m} \frac{s^*(w_i, u_i)}{2} = \frac{s(M)}{2}$$

$\square$

Lemma 3: *For every session* σ, $Setup\_Cost_A(\sigma) \leq \frac{\alpha}{\alpha-1} \cdot s(M)$.

Proof: We partition the session into phases so as to calculate the total setup cost of the session. The first phase, called phase 0, begins at the session initialization. A new phase begins each time the algorithm decides to release the current VC and to establish a new one over the shortest path. Suppose that the session contains K connection reestablishment operations ($K + 1$ phases). Let $s(p_k)$ be the setup cost of the VC path that is established at the beginning of phase k, and let $s(M_k)$ be the sum of all the movement costs that are made during phase k. Note that $s(p_0) = 0$, since we consider the session initialization as a movement. According to the credit principle,

$$
\begin{aligned}
s(p_k) &\leq \frac{1}{\alpha} \cdot [s(M_{k-1}) + s(p_{k-1})] \\
&\leq \frac{1}{\alpha} \cdot s(M_{k-1}) + \frac{1}{\alpha^2} \cdot [s(M_{k-2}) + s(p_{k-2})] \\
&\leq \frac{1}{\alpha} \cdot s(M_{k-1}) + \frac{1}{\alpha^2} \cdot s(M_{k-2}) + \frac{1}{\alpha^3} \cdot [s(M_{k-3}) + s(p_{k-3})] \\
&\leq \frac{1}{\alpha} \cdot s(M_{k-1}) + \frac{1}{\alpha^2} \cdot s(M_{k-2}) + \frac{1}{\alpha^3} \cdot s(M_{k-3}) + \cdots + \frac{1}{\alpha^k} \cdot s(M_0) \\
&\leq \sum_{j=0}^{k-1} \frac{1}{\alpha^{k-j}} \cdot s(M_j)
\end{aligned}
$$

Thus, the total cost of the VC's that are established at the connection reestablishment operations is

$$\sum_{k=1}^{K} s(p_k) = \sum_{k=1}^{K} \sum_{j=0}^{k-1} \frac{1}{\alpha^{k-j}} \cdot s(M_j) = \sum_{j=0}^{K-1} s(M_j) \cdot \sum_{k=1}^{K-j} \frac{1}{\alpha^k} \leq$$

$$\leq \left(\sum_{j=0}^{K-1} s(M_j)\right) \cdot \left(\sum_{k=1}^{\infty} \frac{1}{\alpha^k}\right) \leq \frac{1}{\alpha - 1} \cdot s(M)$$

Hence, the total setup cost is

$$Setup\_Cost_{\mathcal{A}}(\sigma) \leq \sum_{k=0}^{K}[s(p_k) + s(M_k)] \leq \sum_{k=0}^{K} s(p_k) + \sum_{k=0}^{K} s(M_k) \leq$$

$$\leq \frac{1}{\alpha - 1} \cdot s(M) + s(M) \leq \frac{\alpha}{\alpha - 1} \cdot s(M)$$

$\square$

Theorem 9: *Algorithm* $\mathcal{A}$ *is* $(2 + \frac{c_2}{c_1})$-*competitive for* $\alpha = 1 + 2 \cdot \frac{c_1}{c_2}$.

Proof: The total cost of a session σ is the sum of two components, the setup cost and the hold cost. According to Lemma 1, the competitive ratio of the hold cost is

$$\frac{Hold\_Cost_{\mathcal{A}}(\sigma)}{Hold\_Cost_{OPT}(\sigma)} \leq \frac{c_2}{c_1} \cdot \alpha$$

According to Lemma 2 and Lemma 3, the competitive ratio of the setup cost is

$$\frac{Setup\_Cost_{\mathcal{A}}(\sigma)}{Setup\_Cost_{OPT}(\sigma)} \leq \frac{\frac{\alpha}{\alpha-1} \cdot s(M)}{\frac{1}{2} \cdot s(M)} = \frac{2 \cdot \alpha}{\alpha - 1}$$

The value of α that minimizes the competitive ratio of both components is obtained from the equation $\frac{2 \cdot \alpha}{\alpha - 1} = \frac{c_2}{c_1} \cdot \alpha$. Hence, the best competitive ratio is obtained by setting $\alpha = 1 + 2 \cdot \frac{c_1}{c_2}$, and its value is $2 + \frac{c_2}{c_1}$. $\square$

Corollary 3: *If* $c_2 = c_1$, *then Algorithm* $\mathcal{A}$ *is 3-competitive (for* $\alpha = 3$) *for general graphs*.

Algorithm $\mathcal{B}$

The second algorithm, which we denote by $\mathcal{B}$, uses the connection modification approach for improving the first algorithm in terms of the session cost. It is based on the following two improvements in the selection of the crossover switch (COS) at each handoff operation.

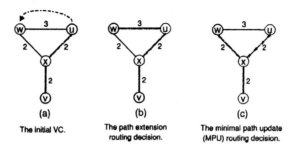

Figure 11.6: Selecting the VC route according to the path extension and the minimal path update algorithms.

The first improvement is achieved by selecting the COS according to *minimal path update (MPU)*. The selected COS is the node on the existing VC which is the closest to the user new attachment node. This is the cheapest modification of the existing VC and Figure 11.6 demonstrates that such a selection always yields lower setup and hold costs compared with the path extension algorithm.

The second improvement is achieved by removing exceptionally "heavy" segments from an MPU VC. A segment $p_{a,b}$ between nodes a and b is called an *exceptional segment* if its cost is at least α times the setup cost of the shortest path, i.e., $s(p_{a,b}) \geq \alpha \cdot s^*(a,b)$. If this happens, then the algorithm checks if there are exceptional segments that contain the selected COS, and replaces the most expensive exceptional segment by a shortest path, called a *shortcut*. This improvement considerably reduces the VC cost by establishing low cost shortcuts and releasing unused resources. It is especially useful for users that have local movement patterns, as described by Figure 11.7. In this figure the users are initially attached to the nodes u and v. After the session initialization, the mobile user from node u moves around node u, and creates exceptional segments along its way. Each time such a segment is detected it is replaced in the VC by a shortcut. Thus the VC length during the session is kept close to optimal.

So far we have described the two improvements as two separate steps which are employed sequentially. However, a better COS that further reduces the VC cost is as follows. For each handoff operation we allocate a budget equal to the cost of the allocated paths according to both of the

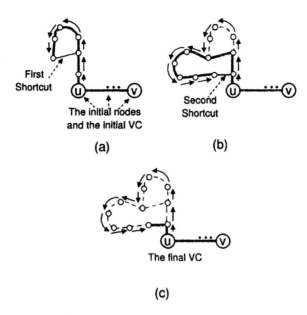

Figure 11.7: An example of session with exceptional segment removal operations.

above improvements. This budget upper bounds the cost of the allocated path. It is used for finding the COS that further reduces the VC cost as much as possible while satisfying the budget constraint. Note that if an exceptional segment is not found, then the selected path is the same as the path selected by an MPU decision. Otherwise, if an exceptional segment exists after the MPU decision, then it is removed at the end of this stage.

The final algorithm works as follows. At the session initialization, it establishes a VC over the shortest path between the session users. Now, suppose that one of the users moves from node w to node u, while the other user is attached to node v, and let $p_{w,v}$ be the VC path before the movement. The algorithm uses three steps for calculating the new path for the VC, $\tilde{p}_{u,v}$. In *Step 1*, it finds the node $x \in p_{w,v}$ which is the closest one to node u, $x = arg\_min_{x \in p_{w,v}}\{s^*(u,x)\}$. This function computes the value of x that minimizes $s^*(u,x)$. Let $\tilde{p}_{u,v}$ be the path that is obtained by concatenating the VC segment between nodes v and

---

Upon a session initialization between nodes u and v do:
  Establish a VC over the path $p_{u,v}^*$

Upon a movement from node w to u when the second
  user is attached to node v do:
  $x \leftarrow arg\_min_{x \in p_{w,v}} \{s^*(u,x)\}$
  $\tilde{p}_{u,v} \leftarrow p_{u,x}^* \cup p_{x,v}.$
  $(a,b) \leftarrow arg\_max_{a \in p_{u,x}^*, b \in p_{x,v}, s(\tilde{p}_{a,b}) \geq \alpha \cdot s^*(a,b)} \{s(\tilde{p}_{a,b})\}$
  If $(\tilde{p}_{a,b} \neq \emptyset)$ then
      $B \leftarrow s^*(u,a) + s^*(a,b)$
      $c \leftarrow arg\_min_{c \in p_{x,v}, s^*(u,c) \leq B} \{s^*(u,c) + s(p_{c,v})\}$
      $\tilde{p}_{u,v} \leftarrow p_{u,c}^* \cup p_{c,v}$
  Allocate the missing edge resources in $\tilde{p}_{u,v}$.
  Release unused edge resources.
  Route the VC over the path $\tilde{p}_{u,v}$.

Figure 11.8: A formal description of Algorithm $\mathcal{B}$.

x with the shortest paths between nodes x and u, $\tilde{p}_{u,v} = p_{u,x}^* \cup p_{x,v}$. In *Step 2*, the algorithm finds the most expensive exceptional segment in the path $\tilde{p}_{u,v}$ that includes node x, denoted by $\tilde{p}_{a,b}$. In *Step 3*, the algorithm selects a COS. If such an exceptional segment was found, then let the handoff budget be $B = s^*(u,a) + s^*(a,b)$. the algorithm selects a COS, c, that minimizes the VC total cost, $s^*(u,c) + s(p_{c,v})$, under the budget constrain, $s^*(u,c) \leq B$. The received path is $\tilde{p}_{u,v} = p_{u,c} \cup p_{c,v}^*$. Otherwise, node x remains the COS. Finally, the algorithm allocates the missing edge resources in the received path $\tilde{p}_{u,v}$, routes the session VC over this path, and releases unused resources. A formal description of Algorithm $\mathcal{B}$ is given in Figure 11.8.

Figure 11.9 provides an example of handoff rerouting operation according to Algorithm $\mathcal{B}$ with $\alpha = 3$. Initially, the users are attached to nodes w and v. Then, the user from node w moves to node u (Figure 11.9-a). In Step 1, the algorithm adds the path (x,z,a,u) to the existing VC and let $\tilde{p}$ be the resulting path (Figure 11.9-b). This path contains two exceptional segments, $\tilde{p}_{y,z}$ and $\tilde{p}_{a,b}$, where $s(\tilde{p}_{y,z}) = 120$ and $s(\tilde{p}_{a,b}) = 200$ (Figure 11.9-c). Note that node u by itself is *not* included in any legitimate exceptional segment. The most expensive segment is

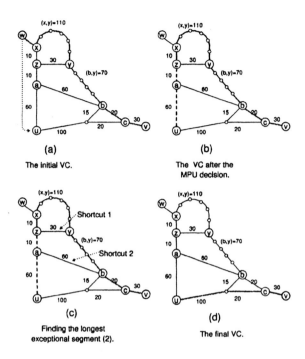

(a)

The initial VC.

(b)

The VC after the
MPU decision.

(c)

Finding the longest
exceptional segment (2).

(d)

The final VC.

Figure 11.9: An example of handoff operation according to Alg. $\mathcal{B}$.

$\tilde{p}_{a,b}$. Therefore, the handoff operation budget is $B = s^*(u, a) + s^*(a, b) = 60 + 60 = 120$. The node that minimizes the VC cost under the budget constraint is node c and thus the final VC cost is 150 (Figure 11.9-d).

Theorem 10: *Let* $\alpha = 1 + 2 \cdot \frac{c_1}{c_2}$. *Then, Algorithm* $\mathcal{B}$ *is* $(2 + \frac{c_2}{c_1})$-*competitive.*

Corollary 4: *For general graphs with* $c_1 = c_2$, *Algorithm* $\mathcal{B}$ *is* 3-*competitive for* $\alpha = 3$.

The competitive analysis of Algorithm $\mathcal{B}$ can be found in [10]. Moreover, our simulations show that Algorithm $\mathcal{B}$ achieves better results on the average than Algorithm $\mathcal{A}$.

11.6   SIMULATION RESULTS

We compare by simulations the performance of the proposed competitive handoff rerouting algorithms with respect to the other schemes in average sense. The evaluated algorithms are the connection reestablishment algorithm [18], the path extension algorithm [2],[3] and two connection modification algorithms: the minimal path update algorithm [29] and the anchor rerouting algorithm [5]. For the latter, the initial location of the mobile user is selected as an anchor (a permanent COS) and only the path to the anchor is modified. We also evaluated the performance of algorithms $\mathcal{A}$ and $\mathcal{B}$, that are described in Section 11.5. Our simulations considered different networks, different roaming distances and various initial distances between the users. For algorithms $\mathcal{A}$ and $\mathcal{B}$ we also evaluated the affect of different credit parameter values, α, on their performance.

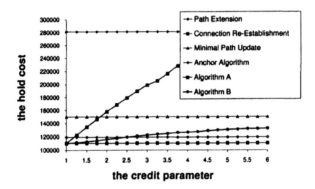

Figure 11.10: The effect of the credit parameter, α, on the hold cost.

Selected typical results from our experiments are described in in Figures 11.10-11.13. In this example, the tested communication network is a grid graph, where both the setup cost s_e, and the hold cost, h_e, of each edge $e \in E$ are 1. We evaluated the average setup and hold costs of sessions with the following characteristics. The initial distance between the users is 200 edges, where one of them is static and the other is mobile. The mobile user moves along a random path with 500 nodes. Its handoff rate is one handoff operation per time unit and the movement range is limited to a square of 100×100 nodes for achieving local movement

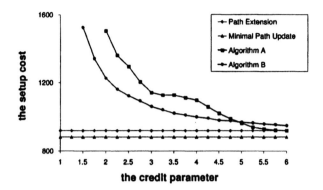

Figure 11.11: The effect of the credit parameter, α, on the setup cost.

effect. The session overall cost is calculated by the following equation,

$$Total\_Cost = \beta \cdot Setup\_Cost + (1 - \beta) \cdot Hold\_Cost$$

where $0 \leq \beta \leq 1$ determines the effect of the setup and the hold costs on the session overall cost.

Our experimental results show that the connection reestablishment algorithm yields the lowest hold cost with the expense of high setup cost, while the minimal path update algorithm achieved the lowest setup cost with high hold cost. The anchor rerouting algorithm and our proposed algorithms balance between the setup and the hold cost of the session, where Algorithm B yields a relatively low setup cost and hold cost. Figure 11.10 describes the effect of α on the hold cost and Figure 11.11 shows its effect on the setup cost. In the latter figure the results of the connection reestablishment and the anchor rerouting algorithms were omitted due to their high setup cost relative to the other algorithms (110365 and 19331 respectively). Figures 11.12 and 11.13 demonstrate the effect of α over the session overall cost for the cases where β equal 0.5 and 0.99 respectively. In the first case the hold cost is the dominant component of the session total cost. Therefore, $\alpha = 1.25$ yields the lowest total cost for both algorithms A and B. Moreover, Algorithm B produces the minimal session cost for every α[4]. In the second case the setup and the hold cost have a similar effect over the session total cost. Here, for every $\alpha \geq 2.25$,

---

[4]The evaluated α is in the range $[1, 6]$.

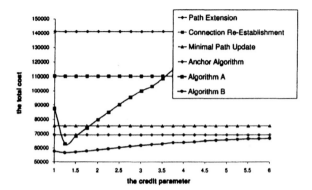

Figure 11.12: The effect of the credit parameter, α, on the session total cost with $\beta = 0.5$.

Algorithm B produces the minimal cost. We see that the credit parameter, α, can be used for balancing between the setup cost and the hold costs. If the hold cost is high relative to the setup cost, then selecting a low value to α guarantees low overall session cost. If the setup cost is the dominant component, then a high value to α is preferable.

11.7   CONCLUSIONS AND FUTURE WORK

Efficient handoff rerouting algorithms are essential for maintaining continuous connections between mobile users. These algorithms are required to meet two main goals, the first is reducing the network resources used by the connections (both signaling and allocated bandwidth) and the second is minimizing the handoff operation time. For capturing the different goals of the connection management problem, we introduce a network model where each link is associated with setup and hold costs. In this model the different needs can be consolidated into a single objective function and the goal is to reduce the connection overall cost. By reducing this cost, the system efficiently uses the network resources with moderate handoff time. Since, the connection times as well as the movements of the mobile users are not known in advance, we use competitive analysis for evaluating the performance of the various handoff algorithms. Initially, we have shown that most of the proposed handoff rerouting

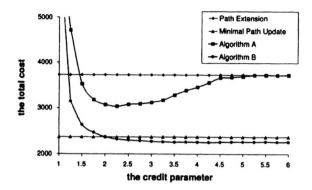

Figure 11.13: The effect of the credit parameter, α, on the session total cost with $\beta = 0.99$.

algorithms perform very poorly in the worst case with respect to the optimal session cost. We have also proved a lower bound of $\Omega(\log n)$ on the competitive ratio of any handoff algorithm for general graphs, where n is the number of nodes in the graph. Then we turned our attention to describe novel handoff algorithms that ensure constant competitive rations in two models. In the case of handoff-anticipating model, we presented a 2-competitive algorithm for a tree topology, and a 4-competitive for a ring topology. For the non-anticipating model we considered the case of correlated costs, where for every edge e of the network graph the ratio of its setup cost, s_e, and hold cost, h_e, bounded by two positive constants c_1 and c_2, i.e., $c_1 \leq h_e/s_e \leq c_2$. In this case, we described two new handoff rerouting algorithms with a competitive ratio of $(2 + \frac{c_2}{c_1})$. These algorithms balance between setup and hold costs by using a credit parameter α. Our simulation results demonstrated that the proposed algorithms yield low overall cost also in the average sense relative to the best algorithms described in the literature. These experiments show that selecting a proper value of α is essential for minimizing the session cost. This value depends on the network parameters as well as on the users movement characteristics. Finding the best α that optimizes the algorithms performance is still an open question.

References

[1] A. Acampora and M. Naghshineh. An Architecture and Methodology for Mobile Executed Handoff in Cellular ATM networks. IEEE J. on selected area in communication (JSAC). Vol. 12 No. 8. October 1994.

[2] A. Acharya, S. Biswas, L. French and D. Raychaudhuri. Handoff and Location Management in Mobile ATM Networks. Proc. of the 3rd Int. Conf. on Mobile Multimedia Communication (MoMu-C3), September 1996.

[3] P. Agrawal, E. Hyden, P. Krzyzanowski, P. Mishra, M. B. Srivastava and J. A. Trotter. "SWAN: A Mobile Multimedia Wireless Network", IEEE personal communication, April 1996.

[4] M. Ajmone Marson, C. F. Chiasserini, A. Fumagalli, R. Lo Cigno and M. Munafo. "Local and Global Handovers Based on In-Band Signaling in Wireless ATM networks", Wireless Networks, Vol. 7, No. 4, pp-425-436, 2001.

[5] I. F. Akyildiz, J. S. M. Ho and M. Ulema. "Performance Analysis of the Anchor Radio System Handover Method for Personal Access Communications system", Proc. IEEE INFOCOM 96, 1996.

[6] I. F. Akyildiz, J. Y. McNair, J. S. M. Ho, H. Uzunalioglu, and W. Wang, "Mobility Management in Next Generation Wireless Systems", IEEE Proceedings Journal, Vol. 87, No. 8, pp. 1347-1385, August 1999.

[7] B. Akyol and D. Cox. "Re-routing for Handover in a Wireless ATM Network"., IEEE Personal Communication, October 1996.

[8] B. A. J. Banh, G. J. Anido and E. Dutkiewicz. "Handover Rerouting Schemes for Connection Oriented Services in Mobile ATM Networks", Proc. IEEE INFOCOM 98, 1998.

[9] Y. Bejerano, I. Cidon, and J. Naor. "Dynamic Session Management: A Competitive On-Line Algorithmic Approach", Proc. of the Dial-M for Mobility workshop. August 2000.

[10] Y. Bejerano, I. Cidon, and J. Naor. "Efficient Handoff Rerouting Algorithms: A Competitive On-Line Algorithmic Approach", Proc. IEEE INFOCOM 98, 1998. To appears in the IEEE Transaction on Networking.

[11] A. Borodin and R. El-Yaniv. "Online Computation And Competitive Analysis", *Cambridge University Press*, 1998.

[12] J. Davie and Y. Rekhter, "MPLS: Technology and Applications", San Francisco: Morgan Kaufman Publishers, 2000.

[13] J. Eberspacher, H. J. Vogel and C. Bettstetter. GSM Switching, Services and Protocols", 2nd edition. John Wiley and Sons, 1999.

[14] A. Festag, T. Assimakaopoulos, L. Westerhoff and A. Wolisz, "Rerouting for Handover in mobile Networks with Connection-Oriented Backbones: An Experimental Testbad", Proc. of ICATM 2000, June 2000.

[15] R. Ghai and S. Singh. "An Architecture and Communication Protocol for Pico-cellular Networks", IEEE Personal Communication Magazine, Vol. 1 No. 3. 1994.

[16] M. Imaze and B. M. Waxman. "Dynamic steiner tree problem", SIAM Journal on Discrete Mathematics 4:369-384, 1991.

[17] S. Irani and A. R. Karlin. On Online Computation. Chapter 13 in "Approximation Algorithms for NP-Hard Problems", Edited by D. S. Hochbaum, *PWS Publishing Company*, 1996.

[18] K. Keeton, B. A. Mah, S. Seshan, R. H. Katz and D. Ferrari. "Providing Connection-Oriented Network Services to Mobile Hosts", Proc. of the USENIX Symp. On Mobile and Location-Independent Computing, Cambridge Massachusetts, August 1993.

[19] J. Li, R. Yates and D. Raychaudhuri. "Performance Analysis on Path Rerouting Algorithms for Handoff Control in Mobile ATM Networks", Proc. IEEE INFOCOM 99, 1999.

[20] J. Li, R. Yates and D. Raychaudhuri. "Mobile ATM: A Generic and Flexible Network Infrastructure for 3G Mobile Services", . Journal of Communications and Networks, March 2000.

[21] D. E. McDysan and D. L. Spohn. "ATM Theory and Application". McGraw-Hill, 1994.

[22] J. Mikkonen, C. Corrado, C. Evci and M. Progler. "Emerging Wireless Broadband Networks", IEEE Communication Magazine, February 1998.

[23] M. Mouly and M. B. Pautet. "The GSM System For Mobile Communication", M. Mouly, 49 rue Louise Bruneau, Palaiseau, France 1992.

[24] J. Naylon, D. Gimurray, J. Porter and A. Hopper. "Low-Latency Handover in Wireless ATM LAN", IEEE J. on selected area in communication (JSAC). Vol. 16 No. 6. August 1998.

[25] T. Ojanpera and R. Prasad. "An Overview of Third-Generation Wireless Personal Communication: A European Perspective", IEEE Personal Communication, December 1998.

[26] R. Ramjee, T. F. La Porta, J. Kurose and D. Towsley. "Performance Evaluation of Connection Rerouting Schemes for ATM-based Wireless Networks", , IEEE/ACM Transaction on Networking, Vol. 6. No. 3. June 1998.

[27] D. Raychaudhuri. "Wireless ATM: An Enabling Technology for Multimedia Personal Communication", Wireless Networks, Vol. 2, pp 163-171, 1996.

[28] W. Stallings. "ISDN and Broadband ISDN", 2nd edition, Macmillan Publishing Company, 1992.

[29] C. K. Toh. "Performance Evaluation of Crossover Switch Discovery Algorithms for Wireless ATM LAN". Proc. IEEE INFOCOM 96, 1996.

[30] N. D. Tripathi, J. H. Reed and H. F. VanLandingham "Handoff in Cellular Systems" Performance Evaluation of Crossover Switch Discovery Algorithms for Wireless ATM LAN. IEEE Personal Communication. December 1998.

[31] V. W. S. Wong and V. C. M. Leung, "A path optimization signaling protocol for inter-switch handoff in wireless ATM networks", Computer Networks, Vol. 31, No. 9-10, pp. 975-984, May 1999.

[32] O. Wu and V. C. M. Leung, "Connection Architecture and Protocols to support efficient handoff over ATM/BISDN personal communication networks", Wireless Networks, Vol. 1, No. 2, 1996.

[33] V. W. S.Wong and V. C. M. Leung, "Location Management for Next-Generation Personal Communications Networks", IEEE Networks, Vol. 14, No. 5. pp. 18-24, Sep/Oct 2000.

11.A   THE PROOF OF THEOREM 3

Theorem 11 : *Consider a general graph, mobile users, and arbitrary edge cost functions with positive setup cost. Then, the competitive ratio of the best on-line algorithm is at least $\Omega(\log n)$, where n is the number of nodes in the network.*

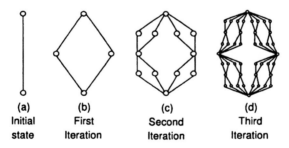

(a)   (b)   (c)   (d)
Initial   First   Second   Third
state   Iteration   Iteration   Iteration

Figure 11.14: The layered graph for proving the lower bound.

Proof: The proof of this theorem is similar to the one presented in [16]. We show a scenario where the cost of every on-line algorithm is at least

$\Omega(\log n)$ times the cost of an optimal off-line algorithm OPT. Let Algorithm $\mathcal{X}$ be an on-line algorithm. We use a layered graph which is defined recursively. We start with a graph that contains two nodes and a single edge connecting them, as depicted in figure 11.14-(a). At each iteration, each edge is replaced by a diamond shape subgraph that contains two new nodes and four edges. Figure 11.14 shows the layered graph obtained after three iterations. Consider a layered graph which is generated after K iterations. This graph contains $n = \frac{2 \cdot (4^K - 1)}{3} + 2$ nodes in $2^K + 1$ layers, hence, $n \le 4^K$. The layers are numbered in increasing order from 0 till 2^K, from top to bottom. Each node is associated with two indices: the *layer index* and the *iteration index*. The latter index specifies the iteration in which it was generated. Each edge is associated with a cost function. For simplicity we assume that the cost function of all the edges have a setup component which is equal to one unit. We say that node u is the *k-upper-neighbor* of node v if it is the nearest node to v with a higher layer index and its iteration index is at most k. In a similar way, a node u is the *k-lower-neighbor* of node v if it is the nearest node to v with a lower layer index and its iteration index is at most k.

Now, consider a session σ between two users. One is static and is located at the single node at layer 2^K, while the second is mobile. The movements of the mobile user are determined by a cruel adversary, whose goal is to increase the session cost of Algorithm $\mathcal{X}$ without increasing the cost of OPT. The adversary starts a session as described in Figure 11.15. The session contains $K+1$ phases numbered from 0 to K. At phase 0 the mobile user is located at the single node at layer 0 (see figure 11.15-(a)). As soon as the session VC is established, phase 1 starts and the mobile user moves to a new node. At phase j, the mobile user visits a single node from each layer whose iteration index is j, from top to bottom. A node is selected only if it is not included in an established VC, but its $(j-1)$-th upper and lower neighbors were visited by the mobile user. As soon as a VC between the selected node and the static user is established, the mobile user moves to a new node. This process continues until a node from each layer is visited. Such a session for $K = 3$ is described in Figure 11.15. At the given example, we assume without loss of generality that algorithm $\mathcal{X}$ prefers to route the VC over the right most possible path. Note that the session duration is negligible and therefore the cost of the session is attributed to the VC setup operations. Furthermore, all the visited nodes form a path with 2^K edges that goes through all the graph layers.

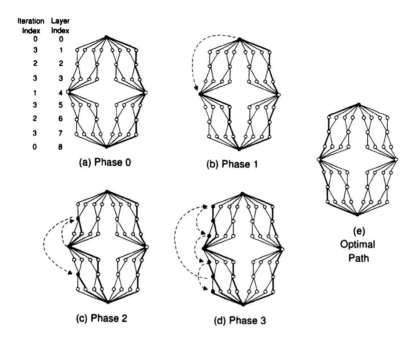

Figure 11.15: The movement pattern for proving the lower bound.

Next, we compare the session cost according to $\mathcal{X}$ and OPT. During the session initialization both algorithms are required to establish a VC that contains 2^K edges. However, while this was the only VC established by OPT, algorithm $\mathcal{X}$ is required to establish more VCs so as to respond to the movements of the mobile user. At each phase j, algorithm $\mathcal{X}$ establishes 2^{j-1} VCs, where each one contains 2^{K-j} edges. Hence, the setup cost of $\mathcal{X}$ at each phase $j \geq 1$ is 2^{K-1}. As a result, the following lower bound is obtained.

$$\frac{Cost_\mathcal{X}(\sigma)}{Cost_{OPT}(\sigma)} = \frac{2^K + K \cdot 2^{K-1}}{2^K} =$$

$$= \frac{K+2}{2} \geq \frac{\log_4 n + 2}{2} \geq \frac{\log_4 n}{2} = \Omega(\log n)$$

$\square$

11.B   THE COMPETITIVE RATIO OF ALGORITHM $\mathcal{R}$

In the following we prove that algorithm $\mathcal{R}$ is 4-competitive. This proof contains two steps. First we assume that $\mathcal{R}$ is not required to pay for redundant postponed periods. Hence, the cost of activating a path is given by Equation 11.2. Later on, we calculate the contribution of these redundant postponed periods to the session cost.

Let T be a sequence of time intervals that specifies when an edge e was used during a given session. We denote by $Cost_T^*(e)$ the accumulated cost of using edge e during these time intervals, where the activation cost is defined by Equation 11.2.

Lemma 4: *Let T_1 and T_2 be two sequences of time intervals such that $T_1 \subseteq T_2$, then for every edge $e \in E$, $Cost_{T_1}^*(e) \leq Cost_{T_2}^*(e)$*

<u>Proof</u>: The correctness of this Lemma is derived from the fact that both the activation cost and the hold cost functions of edge e are continuous and monotonic non-decreasing functions.   □

Consider a session σ that starts at time zero. We denote by $Cost_{\mathcal{R}}^*(\sigma, t)$ the session cost with respect to algorithm $\mathcal{R}$ until time t, when the activation cost of a path is calculated according to Equation 11.2.

Lemma 5: *For every session σ and a given time $t_{end} \geq 0$, $Cost_{\mathcal{R}}^*(\sigma, t_{end}) \leq 2 \cdot \tilde{c}(t_{end})$.*

<u>Proof</u>: First, let us divide the session duration into phases, and suppose without loss of generality that the session contains K phases from its initialization time until time t_{end}. Each phase $k \in \{1 \ldots K\}$ starts at time t_{k-1} and ends at time t_k, when $t_0 = 0$ and $t_K = t_{end}$. In phase k the session VC is the same as the VC path of a single routing option, denoted by z_k, which is optimal at time t_k. Hence, the VC path of algorithm $\mathcal{R}$ at time t, $p(\mathcal{R}, t) = p(z_k, t)$ for each time $t \in (t_{k-1}, t_k]$. Such a phase partition is obtained by defining the phases in reverse order. The routing option z_K is the one that is optimal at time $t_K = t_{end}$, and phase K starts just after the latest time t' such that $p(\mathcal{R}, t') \neq p(z_K, t')$. Time t' defines also the time t_{K-1} when phase $K - 1$ terminates. Let z_{K-1} be the routing option that is optimal at that time, and continue

the phase partition process as described above until all the phase are defined. From the phase definition we get that $p(z_k, t_k) \neq p(z_{k+1}, t_k)$. Hence, $p(z_k, t_k) \cup p(z_{k+1}, t_k) = E$.

Let $\Delta c(z_k, t, t')$ be the cost of the routing option z_k during the time interval (t, t'), given that at time t the session resources are allocated over all the ring edges. Note that $\Delta c(z_k, t, t') \leq c(z_k, t') - c(z_k, t)$.

The cost of each phase k is the sum of two components. The first is the accumulated cost of the routing option z_k during the phase period. This cost is $c(z_1, t_1)$ for the first phase, and $\Delta c(z_k, t_{k-1}, t_k)$ for each phase $k > 1$. The second component is the transition cost from routing option z_k to z_{k+1} at time t_k. This transition is performed by allocating all resources that routing option z_{k+1} has at that time. According to Lemma 4 the transition cost is bounded by the cost of the routing option z_{k+1} during the period of phase k, which is $c(z_2, t_1)$ for the first phase, and $\Delta c(z_{k+1}, t_{k-1}, t_k)$ for $k > 1$. Therefore,

$$Cost_\mathcal{R}^*(\sigma, t_{end}) \leq c(z_1, t_1) + c(z_2, t_1) + \sum_{k=2}^{K-1} \{\Delta c(z_k, t_{k-1}, t_k) +$$

$$+ \Delta c(z_{k+1}, t_{k-1}, t_k)\} + \Delta c(z_K, t_{K-1}, t_K)$$

Now, let us separate the contribution of the odd and even routing options.

$$Cost_\mathcal{R}^*(\sigma, t_{end}) \leq c(z_1, t_1) + \sum_{k=3, odd}^{K} \{\Delta c(z_k, t_{k-2}, t_{k-1}) +$$

$$+ \Delta c(z_k, t_{k-1}, t_k)\} + c(z_2, t_1) + \Delta c(z_2, t_1, t_2) +$$

$$+ \sum_{k=4, even}^{K} \{\Delta c(z_k, t_{k-2}, t_{k-1}) + \Delta c(z_k, t_{k-1}, t_k)\}$$

$$\leq c(z_1, t_1) + \{\sum_{k=3, odd}^{K} \Delta c(z_k, t_{k-2}, t_k)\} +$$

$$+ c(z_2, t_1) + \Delta c(z_2, t_1, t_2) + \sum_{k=4, even}^{K} \Delta c(z_k, t_{k-2}, t_k)$$

Since $\Delta c(z_k, t_{k-2}, t_k) \leq c(z_k, t_k) - c(z_k, t_{k-2})$ follows that

$$Cost_\mathcal{R}^*(\sigma, t_{end}) \leq c(z_1, t_1) + \sum_{k=3, odd}^{K} \{c(z_k, t_k) - c(z_k, t_{k-2})\} +$$

$$+c(z_2, t_2) + \sum_{k=4,even}^{K} \{c(z_k, t_k) - c(z_k, t_{k-2})\}$$

By using $c(z_k, t_k) = \tilde{c}(t_k)$ and $c(z_k, t_{k-2}) \geq \tilde{c}(t_k)$ we receive,

$$Cost_{\mathcal{R}}^*(\sigma, t_{end}) \leq \tilde{c}(t_1) + \sum_{k=3,odd}^{K} \{\tilde{c}(t_k) - \tilde{c}(t_{k-2})\} +$$

$$+\tilde{c}(t_2) + \sum_{k=4,even}^{K} \{\tilde{c}(t_k) - \tilde{c}(t_{k-2})\}$$

These are two telescopic sums, therefore, $Cost_{\mathcal{R}}^*(\sigma, t_{end}) \leq 2 \cdot \tilde{c}(t_{end})$. $\square$

Lemma 6: *For every session σ and a given time $t_{end} \geq 0$,*
$Cost_{\mathcal{R}}(\sigma, t_{end}) \leq 4 \cdot \tilde{c}(t_{end})$.

<u>Proof</u>: According to Lemma 5, $Cost_{\mathcal{R}}^*(\sigma, t_{end}) \leq 2 \cdot \tilde{c}(t_{end})$, when we ignore the cost of redundant postponed periods. This cost contains both edge setup cost and edge hold cost. To include also the cost of redundant postponed periods in the total cost, we should count twice the cost of each edge setup operation, as defined by Equation 11.1. Hence, $Cost_{\mathcal{R}}(\sigma, t_{end}) \leq 2 \cdot Cost_{\mathcal{R}}^*(\sigma, t_{end})$ As a result, $Cost_{\mathcal{R}}(\sigma, t_{end}) \leq 4 \cdot \tilde{c}(t_{end})$. $\square$

Theorem 12: *For every session σ, $Cost_{\mathcal{R}}(\sigma) \leq 4 \cdot Cost_{OPT}(\sigma)$.*

<u>Proof</u>: On one hand, from Lemma 6, we get that $Cost_{\mathcal{R}}(\sigma, t_{end}) \leq 4 \cdot \tilde{c}(t_{end})$. On the other hand, the routing option of OPT is also included in the decision tree of the session. Therefore, for any given time $t \geq 0$, $Cost_{OPT}(\sigma, t) \geq \tilde{c}(t)$. As a result, $Cost_{\mathcal{R}}(\sigma) \leq 4 \cdot Cost_{OPT}(\sigma)$. $\square$

Chapter 12

CACHE-BASED COMPACTION
A Technique for Optimizing Wireless Data Transfer

Mun Choon Chan and Thomas Y.C. Woo
Networking Research Laboratory
Bell Laboratories, Lucent Technologies
{munchoon,woo}@research.bell-labs.com

12.1   Introduction

Despite the emergence of wide-area wireless data services, the bandwidth available over a wide-area wireless channel is still fairly limited and likely to remain so in the forseeable future due to the inherent limitation of wireless transmission in a limited spectrum. In the United States, new third generation technology such as the 3G1X offering speed up to 144Kbps is already being deployed and emerging high-speed wireless data standard such as the High Data Rate (HDR) [Bender et al., 2000] can eventually provide rates up to 2Mbps. However, since the bandwidth available over a wireless network are shared by a number of mobile users, the average bandwidth available to each user is usually much smaller.

From an end user's perspective, the primary measure of data transfer performance is response time or latency. Strictly speaking, latency is an end-to-end quantity, which consists of two main components, namely, *processing delay* and *transport delay*. The former refers to the processing time incurred in the end device, in the origin server and all the intermediate servers (see Figure 12.1), while the latter refers to the time spent in traversing all the interconnecting links. While processing power of end devices and servers have grown signf}cantly in the past and is expected to continue its growth, the relative growth in wireless bandwidth lags behind signfcantly. Trading-off computation for bandwidth is one way to exploit this drastic difference between the cost of computation and bandwidth.

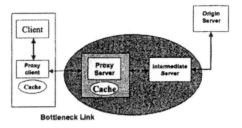

Figure 12.1. System Model

In this work, we consider the case where the transport delay is dominated by the delay incurred in the last (or first depending on perspective) hop. Our objective is to reduce the overall latency by reducing the transport delay, or specifically the last hop delay.

The key innovation behind our cache-based compaction technique is as follows. Instead of "coding" the requested object on its own, a more compact encoding is performed by leveraging other objects that are already available in the client's possession. In particular, if a client already possesses "similar" objects in its cache, then those objects (called *reference* objects) are used as an extended "dictionary" based on which the newly requested object may be coded. The more "similar" the reference objects are to the requested object and the more such "similar" reference objects are available in the client's possession, the smaller is the resulting transfer.

Our approach is a compression technique in that it uses a dictionary-based compression technique. However, unlike standard compression techniques such as *gzip*, a set of "similar" objects, not just the requested object itself, is used as the compression dictionary.

Our approach can also be viewed as a differential transfer technique in that it compares objects, and transfers mainly the differences. However, unlike existing differential transfer techniques, comparison is not restricted to just objects from earlier versions. Our approach can potentially leverage multiple *unrelated* objects in a client's cache.

Our proposed compaction technique is lossless. Therefore, it is most applicable to transfer of documents (e.g., Web pages, emails, text files). For graphics, lossy compression techniques are more effective, since the resolution and color of the graphical objects can often be reduced while the objects retain most of their "visible" quality.

To be accurate, our cache-based compaction technique represents a general approach rather than a specific algorithm. At a high-level, it consists of two key components: (1) a *selection* algorithm for choosing reference objects, and (2) an *encoding/decoding* algorithm that encode and decode a new object using a collection of reference objects. A specific instantiation of the compaction technique is obtained by providing concrete implementations of the selection and encoding/decoding algorithms.

Apart from its use for optimizing Web and email transfer, the cache-based compaction idea can also be viewed as a systematic foundation that ties together the three most-used, yet completely decoupled, techniques — caching, compression, and prefetching — for improving wireless data applications.

To demonstrate our compaction idea, we have applied it to two of the most popular (wireless) network applications, namely, Web browsing and email. In the former, we use compaction to optimize HTTP-based Web page transfer. In the latter, we use compaction to reduce the transfer size of email messages. In both cases, we provide experimental results showing significant improvement using our compaction technique over existing ones.

The balance of the chapter is organized as follows. In the next section, we review related work addressing similar problems. In Section 12.3, we lay out a system model for our compaction approach followed by descriptions of the encoding and decoding algorithms, which are common to both the Web and email applications. In Section 12.4, we present the details of the selection algorithm for Web transfer and support it with extensive experimental results. In Section 12.5, we present the selection algorithm for email transfer and results of its evaluation. Finally, we conclude in Section 12.6.

12.2   Related Work

The major techniques used for optimizing data transfer over a slow link are compression, caching, differencing and prefetching.

Compression can be divided into lossy and lossless. Lossy compression is usually applied to graphical and audio objects, and lossless compression is applied to text and binary objects. The benefits of using lossless data compression algorithms such as *gzip* (which is based on LZ77 [Ziv and Lempel, 1977]) and *vdelta* [Hunt et al., 1998] to compress non-video and non-audio objects is studied in [Mogul et al., 1997]. The use of data-specific technique for reducing object size is described in [Fox and Brewer, 1996]. Reduction was achieved via lossy compression, for example by reducing resolution and/or color of a graphical object.

Caching is frequently used to improve the performance of distributed systems. Traditional caching algorithms search for identical object or the most recent version of an object. The use of caching to enhance Web browsing has been studied extensively in the literature, see for example

[Dingle and Partl, 1997] and [Wessels and Claffy, 1998]. However, the utility of traditional caching applied to Web pages is limited by a number of factors, including the access patterns, valid lifetime of a Web page, and the number of static (cacheable) pages vs. the number of dynamic (non-cacheable) pages. The limits of latency reduction from caching, based on searching for objects with the same URL is studied in [Kroeger et al., 1997]. In this work, storage available is assumed to be very large and it is found that latency could be reduced by at best 26%. Some other limits of caching be found in [Mogul, 2000], which also provides a good survey of various caching, differencing and prefetching techniques used in optimizing transfer of Web objects.

Differencing compares an earlier version of an object to the current version. Usually, only two objects of the same URL or output of CGI script with different parameters are considered. Some of the differencing algorithm used are UNIX *diff* and *vdelta* [Hunt et al., 1998]. Delta coding is one form of differencing and is studied in [Mogul et al., 1997]. The authors found that differencing worked for 10% of all "status 200" response at the proxy level. IETF has attempted to standardize the use of delta encoding in HTTP [Mogul et al., 2002]. The *WebExpress* system described in [Housel and Lindquist, 1996] also describes a form differencing technique. Object comparison was based on the object's URL as well as a digital signature of the object. Differencing was applied mainly to output of CGI scripts.

Prefetching can be used when the knowledge of an object's future access pattern is known with some probability. Such techniques have been applied to file transfer [Griggioen and Appleton, 1996] and Web transfer [Fan et al., 1999]. Many of these techniques rely on actively transferring possibly unwanted data to the client during idle periods. Thus, access latency is reduced by increasing the overall bandwidth usage. However, due to the shared nature and the (often adopted) usage-based pricing of wireless data network, trading bandwidth for latency may not be a good choice.

Our proposed compaction scheme is unique in that unlike caching, compression and differencing, unrelated objects can be used to optimize transfer. Compaction operates on much smaller unit of short strings rather than an entire data object. This allows compaction to be much more flexible in its ability to extract similarity among different objects.

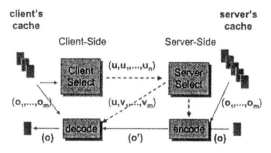

Figure 12.2. System Overview

12.3 Our Cache-based Compaction Technique

In this section, we first give an overview of the compaction technique. Then in Sections 12.3.2, we present the encoding and decoding algorithms that are common to both the Web and email applications.

12.3.1 Overview

Let C be a client and S a server.[1] Let C contains a set of objects denoted by $C.cache$, and that C would like to obtain a new object o from S. Let S contains a set of objects denoted by $S.cache$, where $o \in S.cache$.[2]

Processing starts when the client requests an object o from the server by sending the name of the request object, u, together with the *names* of the set of reference objects, $u_1, \ldots, u_n$ to S (Figure 12.2).

The server, S, looks into its cache and obtains the objects $o_1, \ldots, o_m$ with names $v_1, \ldots, v_m$ such that $v_1, \ldots, v_m$ is a subset of $u_1, \ldots, u_n$.

The server then computes a new object o' using o and $o_1, \ldots, o_m$ and sends o' together with the names of the reference objects used $(v_1, \ldots, v_m)$ to C. On receiving o', C recovers o from o' and $o_1, \ldots, o_m$ given $v_1, \ldots, v_m$. The computation of o' by S is the encoding step, while the reconstruction of o by C is the decoding step. The algorithms used in the encoding and decoding steps satisfy the following relationship:

$$o = \text{decode}(\text{encode}(o, o_1, \ldots, o_n), o_1, \ldots, o_n)$$

When $n = 0$, it reduces to essentially a compression technique. When $n = 1$ and o_1 is an earlier version of o, it reduces to previously studied delta encoding technique. In other words, our compaction technique

---

[1]C and S can be any of the entities in the process chain shown in Figure 12.1. C and S need not even be adjacent if some form of "tunneling" is available.
[2]The requirement that $o \in S.cache$ is a simplification. S can fetch o on demand if necessary.

subsumes most existing proposals and is most interesting when $n > 1$ and $o_1, \ldots, o_n$ are not simply variations of o.

In this work, we use a dictionary-based compression scheme as our encoding and decoding algorithms. Specifically, we uses $o_1, \ldots, o_n$ as "extended" dictionaries for compression of o. The objects $o_1, \ldots, o_n$ are determined via a *selection* algorithm which tries to identify objects that are "similar" to o. The measure of similarity is the number and length of common substrings.

Obviously, saving is possible with our compaction technique if and only if

$$t_{\text{select}} + t_{\text{encode}} + t_{\text{decode}} + \frac{|o'|}{s} < \frac{|o|}{s}$$

where $|o|$ denotes the size of an object o and s is the bit transfer rate on the link between C and S. A necessary condition for this is $|o'| < |o|$, and the absolute reduction in latency is proportional to the size of o and inversely proportional to s. Hence, our compaction scheme will make most sense when transferring data over the last hop, where o is of reasonable size and s is typically small.

The underlying observation is that dictionary-based compression scheme (the most well known being the LZ77 [Ziv and Lempel, 1977] and LZ78 [Ziv and Lempel, 1978] family) works because of the recurrence of common sub-strings within a document. The basic idea in our proposal is to exploit this notion of *similarity* among multiple documents for reducing transfer. If a number of similar documents have already been transferred from the web server to the client, transfer of the next similar document can be done in an efficient manner.

There are two sets of algorithms used in the compaction technique. A selection algorithm is used by the client and server to select *similar* objects and a pair of encoding/decoding algorithms for data transfer. The selection algorithms are specific to each application and we present one for each application in Sections 12.4 and 12.5. The encoding/decoding algorithms, on the other hand, are generic and we present them next.

12.3.2   Encoding and Decoding Algorithms

The encoding and decoding algorithms are based on the universal compression algorithm described in [Ziv and Lempel, 1977].

The encoding algorithm is shown in Figure 12.3. An object o is essentially viewed as a byte array, with $o[i]$ denoting the i-th byte of o (we start counting from 1), $o[..i]$ denoting the part of o from the beginning up to and including the i-th byte, and $o[i..]$ denoting the part of o beginning at i-th byte of o to the end.

```
function encode ( o, o₁, o₂, ..., oₙ, threshold ) {
    i = 1;
    o' = empty string;
    while ( o[i..] is non-empty ) {
        o₀ = o[..i − 1];
(1)     CS = {(ℓ, j) | ℓ is a prefix of o[i..] and ℓ occurs in oⱼ, 0 ≤ j ≤ n};
        if ( CS = ∅ )
            L = o[i];
        else
(2)         pick (L, k) ∈ CS such that ∀(ℓ, j) ∈ CS : |L| ≥ |ℓ|;
        if ( |L| ≤ threshold )
            append to o' the character token L;
        else
            append to o' the triplet token ( k, starting position of L in oₖ, |L| );
        i = i + |L|;
    }
    return o';
}
```

Figure 12.3. Encoding Algorithm

```
function decode ( o', o₁, o₂, ..., oₙ ) {
    o = empty string;
    while ( o' is non-empty ) {
        remove first token t from o';
        if ( t is a character token )
            append t to o;
        else {
            /* t must be a triplet token */
            let t = (k, pos, l);
            append to o the substring in oₖ starting at position pos of length l;
        }
    }
    return o;
}
```

Figure 12.4. Decoding Algorithm

The parameter threshold should be set to at least the encoded size of a triplet (whose size is at least 1) to ensure that the size of the compressed result is smaller than the original.

The steps (1) and (2) represent the searching of the longest common substring between the part of o currently being processed and the part of o that has been processed (o_0 specifically) together with the n reference objects o_1, ..., o_n. The more similar the reference objects are to o, the more common substrings there are, and the better is the compression.

This is the most time consuming part of the encoding algorithm, and is implemented using hash tables in our case.

In general, compression gets better with larger n, though the marginal improvement diminishes. The case when $n = 0$ is basically the *LZ77* algorithm. In that case, an ordered pair token, instead of a triplet, is sufficient.

Decoding is straightforward and is comparatively much faster. Its detail is shown in Figure 12.4, and should be self-explanatory.

It should be clear that the above encoding algorithm is lossless. Though it works for any objects, it is most applicable to text (e.g., plain ascii, HTML) objects. For graphical objects that are already in compressed form (e.g., GIF, JPEG), the amount of non-trivial similarity among objects is minimal. Lossy compression techniques can drastically reduce the size of a graphical object while retaining most of its "visible" quality. In the sequel, we consider the use of compaction on text objects only.

12.4   Applying Compaction to Web Transfer

12.4.1   Selection Algorithm

In order to obtain good compression result, the selection algorithm needs to be able to pick a set of reference objects that are similar to the requested object. While examining the content of the objects is the only sure way of deciding if they are similar, this process is too computationally expensive, especially when the collection of objects is large (e.g., all the objects in a cache). Therefore, we are left with using heuristics based on other attributes of the objects.

A natural choice for selection parameter is the name or the Uniform Resource Locator (URL) of the object. Generally, the URL does not tell much about an object's content. We argue though that the structure of a URL may provide good enough hints.

By treating URL as path name, a collection of objects can be viewed as leaves in a forest, with all objects from the same site represented in a distinct tree. We observe that a large majority of sites tend to follow a consistent design style, which translates into the use of similar structure and formating instructions. Additionally, the hierarchy is often structured in terms of related topics, and objects pertaining to similar topics tend to share common content [Chan and Woo, 1999b].

In summary, we conjecture that Web documents that are "close" together in the hierarchy formed by their URLs tend to be more similar than those that are "far apart." A degenerate case of this is used in the differencing scheme described in [Banga et al., 1997, Housel and

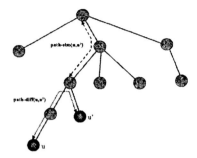

Figure 12.5. Path Similarity and Path Difference

Lindquist, 1996, Mogul et al., 1997], where an older version of a document with the same URL is used to compute the delta for transferring a newer version of the document.

To precisely specify our heuristics, we first introduce some notations.

Notations. Let u and u' be two URLs.[3] They are written respectively as $h/p_1/p_2/\ldots/p_n$ and $h'/q_1/q_2/\ldots/q_m$. Each of the h, h', p_i and q_i is called a *segment*. The *length* of an URL is defined as the number of segments in the URL. Thus, $|u| = n + 1$. We define an *enumerator* function $[\cdot]$ for URL as follows: $u[0] = h$, and for $1 \leq i \leq n$, $u[i] = p_i$. Additionally, $u[i..j]$ is the path $u[i]/\ldots/u[j]$ p is a *prefix* of u if for all $0 \leq i \leq |p|$, $p[i] = u[i]$. p is a *common prefix* of u and u' if p is a prefix of both u and u'. □

We define the *path similarity* between two URLs u and u' as *path-sim*(u, u') be the length of the longest common prefix of u and u' and their *path difference* as *path-diff*$(u, u') = |u| + |u'| - 2 *$ *path-sim*(u, u').

These definitions are graphically illustrated in Figure 12.5. As an example, the URLs http://www.cnn.com/US/news/abc and http://www.cnn.com/US/def have a path similarity of 2 and a path difference of 3.

When the path similarity is 0, it means that the two URLs refer to objects from different Web sites. When path similarity is at least 1, the path difference indicates the number of "hops" it takes to go from one URL to the other in the tree hierarchy. In particular, if the path difference is 2, it means that the two URLs belong to the same directory.

---

[3]Since we consider only HTTP URLs, for ease of our disposition, we assume the protocol part has been omitted.

```
function select ( u, C, n ) {
    /* filter out objects from different sites */
(1) C = C - {u' ∈ C | path-sim(u, u') = 0};
    S = ∅;
    while ( |S| < min(n, |C|) ) {
        T = subset of C - S with minimum path difference with u;
        S = S + subset of T with maximum path similarity with u;
    }
    return S;
}
```

Figure 12.6. Selection Algorithm for Web Pages

With the above, we are now ready define a similarity relationship. Given a URL u, the *similarity ordering* relative to u, denoted by $\sqsupseteq_u$, is defined as follows:

$$u_1 \sqsupseteq_u u_2 \quad \text{iff} \quad \begin{array}{l} path\text{-}sim(u, u_1) \geq path\text{-}sim(u, u_2) \\ \text{and} \\ path\text{-}diff(u, u_1) \leq path\text{-}diff(u, u_2) \end{array}$$

This essentially says that a URL u_1 is considered more *similar* to URL u than another URL u_2 if u and u_1 share a longer common prefix and it takes fewer hops to go from u to u_1 than from u to u_2. In other words, u and u_1 share more common path and fewer disjoint hops. It is easy to see that $\sqsupseteq_u$ is a partial ordering.

A precise selection policy requires a total ordering. Thus, we extend $\sqsupseteq_u$ to a total ordering $\sqsupseteq_u^t$ as follows. If u_1 is related to u_2 under $\sqsupseteq_u$, then they are related in the same way under $\sqsupseteq_u^t$. Otherwise, $u_1 \sqsupseteq_u^t u_2$ iff $path\text{-}diff(u, u_1)$ is at most $path\text{-}diff(u, u_2)$. Clearly, there are multiple ways to extend a partial ordering to a total ordering. This particular definition gives priority to the path difference, and is the one studied in this work.

Once $\sqsupseteq_u^t$ is defined, the selection algorithm is straightforward. It basically will select the top n most similar URLs from all the URLs available for selection. The pseudo code for the selection algorithm is given in Figure 12.6. We note that C is the set of URLs that are available for selection,[4] and n is the maximum number of most similar URLs needed. In summary, the selection heuristic tries to minimize the path difference, while maximizing the path similarity. The filtering step (step (1)) removes all the URLs that do not belong to the same site as u.

---

[4]This is typically a subset of the cache.

12.4.2   Experimental Results

In this section, we present the results of our experimental evaluation of our proposed compaction technique. Our experiments are broken down into 3 sets, each of which is intended to establish a distinct claim.

Set 1. In the first set of experiments (Section 12.4.2.1), we examine if the similarity ordering $\sqsupseteq_u^t$ introduced in Section 12.4.1 (or equivalently the selection algorithm shown in Figure 12.6) does actually pick out "good" reference objects that are useful in the encoding procedure (Figure 12.3). In other words, we would like to verify our conjecture that similarity in URL implies certain degree of similarity in their content.

For comparison purposes, we perform the same experiments with a standard compression scheme, namely, *gzip* [Mogul et al., 1997], and a standard differencing scheme, namely, *diff -e | gzip* (abbreviated as *diff* in the sequel) [Banga et al., 1997, Housel and Lindquist, 1996, Mogul et al., 1997]. *gzip* is like compaction with $n = 0$, but with a number of additional optimizations. *diff* is like compaction with $n = 1$. Strictly speaking, most existing differencing schemes apply only to objects with the same URL.

Set 2. From experiments in Set 1, we demonstrate that objects high in the similarity ordering serves better as reference objects than those low in the similarity ordering. The remaining question to ask is, in a real-life browsing session, how "high" in the similarity ordering can the selection algorithm typically find objects at. In order words, we study the actual distribution of path difference and path similarity in a typical browsing session.

We perform this set of experiments (Section 12.4.2.2) using actual client-side access log. Our objective is to demonstrate that typical browsing patterns of actual users contain sufficient locality such that reference objects with high content similarity (as defined by the similarity ordering) are frequently available in the client cache, and hence can be selected.

Set 3. Finally, in the last set of experiments (Section 12.4.2.3), we "follow" real user access trace to perform actual downloading of Web objects using compaction. We compute the actual savings and compare that with the results of identical downloading under *gzip* and *diff*. We also compute and compare the average latency incurred by compaction, *gzip* and *diff* under various link bandwidth.

In the following, we refer to our compaction scheme as $npact(n)$, where n is the number of reference objects used. While we have experimented with many different values of n, all experiments presented below used

Section	mean δ	mean σ	Selected subset of Ω
12.4.2.1.1	0	≥ 1	traces from periodic downloading
12.4.2.1.2	2	≥ 1	traces from CGI output
12.4.2.1.3	2	≥ 1	random selection from Web sites
12.4.2.1.4	> 2	≥ 1	random selection from Web sites
12.4.2.1.5	≥ 0	≥ 0	random selection from Web sites

Figure 12.7. Summary of Experiments Performed

$n = 3$. This value is selected because the performance of $npact(3)$ is noticeably better than $npact(1)$, while $npact(7)$ is only slightly better than $npact(3)$. Furthermore, because of the length limitation, we can only present results from selected sites and Web traces.

12.4.2.1   Set 1: Usefulness of Similarity Ordering.   We like to study the performance of $npact$ when reference objects of different path differences and similarities are used. Different groups of experiments are performed, they are broken down by path differences and path similarities (see Figure 12.7 for a summary). To precisely state our results, we first introduce some notations.

Notations. Let Ω be a set of objects available for selection, and $\{u_1, \ldots, u_n\} \subseteq \Omega$. Define:

$$path\text{-}sim(u, \{u_1, \ldots, u_n\}) = \sum_{i=1}^{n} path\text{-}sim(u, u_i)$$

$$path\text{-}diff(u, \{u_1, \ldots, u_n\}) = \sum_{i=1}^{n} path\text{-}diff(u, u_i)$$

For brevity, we denote by σ the aggregate *path-sim* and δ the aggregate *path-diff* as defined above. We often refer to their mean values which is formally defined as:

$$\text{mean } \sigma(u, \{u_1, \ldots, u_n\}) = \frac{1}{n} path\text{-}sim(u, \{u_1, \ldots, u_n\})$$

$$\text{mean } \delta(u, \{u_1, \ldots, u_n\}) = \frac{1}{n} path\text{-}diff(u, \{u_1, \ldots, u_n\})$$

In the sequel, the parameters $u, u_1, \ldots, u_n$ will be omitted unless otherwise noted.   □

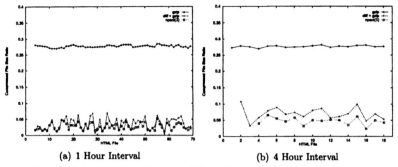

(a) 1 Hour Interval (b) 4 Hour Interval

Figure 12.8. Different Versions of http://www.abcnews.com/index.html

20 Web sites were used in the experiments.[5] 6 of these sites were ranked in the top 25 most visited sites, and 14 were ranked in the top 500 sites.[6] The rest of the sites were chosen to include various categories. The category breakdown is 3 news sites, 3 information sites, 7 commercial sites, 3 technical sites, 2 academic sites and 2 government sites. For the files collected, all binary, graphics and audio files were removed. Also, only files with size between 1K and 64K were considered.

The experiments in this set operate as follows: For each site studied, we first pick a random object from the site. Then we try to simulate the transfer of the chosen object using compaction by selecting n other objects (from same or different sites) to be used as reference objects. We compute the size of the encoded object, and tally this by path difference and path similarity values.

12.4.2.1.1   Objects with same URL (mean $\delta = 0$, mean $\sigma \geq$ 1). For brevity, we present our results only for a representative site, www.abcnews.com. In this experiment, we collected objects from the Web site www.abcnews.com every hour, over a period of 5 days (from the May 23 1998 to May 28 1998). Different versions of objects with the same URL were grouped together and sorted in chronological order. For each sequence of objects, we apply *gzip*, *diff* (between the current and the last version), and *npact* (the 3 most recent versions) to determine the size

---

[5]www.abcnews.com (news) www.aol.com (information) www.bofa.com (commercial) www.cisco.com (commercial) www.cnet.com (techncial) www.columbia.edu (academic) www.edmund.com (commercial) sportszone.espn.com (news) www.fcc.org (government) www.ibm.com (commercial) www.javasoft.com (technical) www.lucent.com (commercial) www.microsoft.com (commercial) www.netscape.com (commercial) www.nycvisit.com (information) www.techweb.com (technical) www.tripod.com (information) www.umass.edu (academic) www.usatoday.com (news) www.ustreas.gov (government)
[6]Source: MediaMetix (www.mediametix.com)

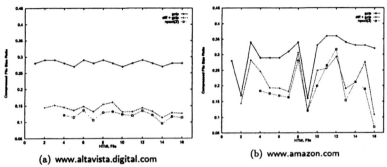

Figure 12.9. Objects from CGI Scripts with Different Parameters

of transfer. While comparisons were performed for a number of URLs, only the URL http://www.abcnews.com/index.html, which generated a total of 69 different objects, will be described here. Other URLs exhibit similar trends.

Figure 12.8 shows the ratio of the encoded and original size for all 69 objects.

The results show that for objects with same URL, which tend to have similar content, *diff* and *npact* performed much better than *gzip*. In addition, Figure 12.8(b) shows that *npact* is better in capturing similarity among less similar objects. For the set of objects selected every hour, the mean compression ratios are 0.2772 for *gzip*, 0.0358 for *diff* and 0.02718 for *npact*. When the set of objects is selected every 4 hour, the mean compression ratios are 0.2768 for *gzip*, 0.0703 for *diff* and 0.0473 for *npact*.

12.4.2.1.2   Objects from CGI scripts with different parameters (mean $\delta = 2$, mean $\sigma \geq 1$). We submitted a number of queries to the search engine www.altavista.digital.com with different query strings. Figure 12.9 shows the output for 16 pages, the first 8 pages are for the query string *java* and the next 8 pages for the query string *network*.

The results show that both *diff* and *npact* perform very well (mean compression ratio of 15%), while *gzip* performs much poorer (mean compression ration of 30%). An interesting observation was that there was no significant difference in result when responses from different query strings were used for referencing. This implied that most of the similarity came from formating. Similar results were also obtained from requests to online shopping sites like www.amazon.com.

In general, HTML pages that are being updated continuously (stock quote, sports scoreboard, newspaper headlines, weather, movie showtimes, etc.) can be transferred very efficiently under *npact*.

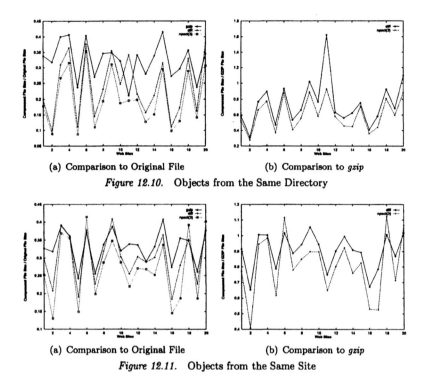

(a) Comparison to Original File   (b) Comparison to *gzip*

Figure 12.10.   Objects from the Same Directory

(a) Comparison to Original File   (b) Comparison to *gzip*

Figure 12.11.   Objects from the Same Site

12.4.2.1.3   Objects from the same directory (mean $\delta = 2$, mean $\sigma \geq 1$).   For each of the 20 Web sites, the objects collected were filtered such that only objects in directories with 4 or more objects were extracted. After this filtering, the minimum number of objects left per site was 61 (www.columbia.edu), maximum was 3,964 (sportszone.espn.com) and the average was 1,110.

The mean compression ratio for each Web site is plotted in Figure 12.10.

For all 20 Web sites, *npact* performs better than *gzip* on the average. Of the 20 sites, *npact* performs more than 50% better for 6 sites, 10% to 50% better for 13 sites, and 10% better or less for only 1 site. The performance of *diff* tracked that of *npact*, though not in all cases. In 3 out of 20 sites, *diff* performed worse than even *gzip*. The mean compression ratios over all 20 site were 0.2012 for *npact*, 0.2419 for *diff*, and 0.3242 for *gzip*.

12.4.2.1.4   Objects from the same Web site but different directories (mean $\delta > 2$, mean $\sigma \geq 1$).   In our experiments, the minimum, maximum, and average number of files per site was 250 (www.nycvisit.com), 5,550 (sportszone.espn.com), and 1,594.

Figure 12.11 shows the mean compression ratio for each of 20 sites. As expected, the results show that objects chosen randomly from the same Web site had a smaller amount of similarity. Nevertheless, out of the 20 sites, relative to *gzip*, *npact* performed 50% better in 2 sites, 10% to 50% better in 12 sites, less than 10% better in 6 sites.

In 3 sites though, *npact* performed worse than *gzip*. Two of them were academic sites and the third was a government site. Both academic sites contained a large number of objects from different departments and (probably) prepared by different people, with little common in style and formating. The government site contained a large number of plain text file with very minimum formatting. Since the current implementation of *npact* performed worse than *gzip* if used purely as a compression scheme, *npact* thus performed worse in these cases.

Note that when objects are very different, the mean compression ratios of *gzip* and *diff* will be very close because *diff* will simply output the requested object plus some overhead. Therefore, the observation that the mean compression ratios of *gzip* and *diff* were approximately the same for these 3 sites provides further evidence that the reference objects share little similarity with the requested object.

Over all 20 sites, the mean compression ratios were 0.2726 for *npact*, 0.3013 for *diff* and 0.3317 for *gzip*.

12.4.2.1.5   Objects from different Web sites (mean $\delta > 0$, mean $\sigma > 0$).   In this case, *npact* did not perform as well as both *gzip* and *diff*. The mean compression ratios were 0.3145 for both *gzip* and *diff*, and 0.3455 for *npact*. This confirms that similarity among randomly selected objects is low.

12.4.2.1.6   Compression ratio with respect to δ and σ.   The previous experiments showed that *npact* performed well on the average for specific ranges of δ and σ values. Due to space limitations, we cannot include the measurement plots. Instead, we will highlight observations drawn from these experiments.

1 The use of δ and σ as selection parameters correctly selected objects with *similar* contents for the case of small δ and large σ. For example, choosing reference objects from the same directory generated encoded objects that were smaller than *gzip* consistently (on the average) for all sites studied. Therefore, the conjecture that "closeness" of URLs implies similarity in content was true for the case of small δ and large σ, but not true for large δ and small σ.

2 The parameter δ was better in predicting good performance for small δ, as in the case of $\delta = 0$ (same URL) and $\delta = 2$ (same directory). However, for larger values, σ may be a better indicator of good performance than δ.

3 While the compression is good for small δ and large σ, it shows no clear trend when only one of the dimensions (δ or σ) is varied.

12.4.2.2   Set 2: Distribution of Similarity Ordering in Actual Traces.   Results from Section 12.4.2.1 show that *npact* performed significantly better than *gzip* and *diff* if objects with high similarity ordering are used as references. The objective of this section is to show that in an actual user browsing session, our proposed selection algorithm is able to pick up reference objects with high similarity ordering most of the time.

To verify our claim, we made use of an actual user trace. We had two requirements for the trace. First, the requested URL must be retained in the trace in order to compute δ and σ. Second, the trace should record the behavior of the actual client making the request so that per-client statistics could be collected. The first requirement ruled out the use of most publicly available HTTP logs (e.g., UC Berkeley Home IP Web Traces[7] and Digital's Web Proxy Trace[8]) because the URLs had been anonymized for privacy reason. The second requirement ruled out the NLANR[9] cache access logs because the log entries were highly aggregated. With these limitations, we can only find an older log from Boston University [Cunha et al., 1995] which satisfied our requirements.

The Boston University trace contains 762 unique users, and after removing URLs with extensions that indicated that they may be non-HTML or non-text objects (e.g., those with extension GIF, JPEG, etc.), 197,004 URLs were left. The maximum number of URLs per user is 4,412 and the minimum number of accesses per user is 16. 448 users have 100 or more URLs accesses. For each user's access log, we used the selection algorithm to select n reference objects, whose aggregate path differences and path similarities are recorded as a (δ, σ) tuple. To simulate the effect of caching, we used a moving window size of w, where w represents the cache size.

Figure 12.12 shows the density function of the (δ, σ) tuples for the case of $w = 64$. Of all the requests, 78% found 3 or more URLs from the same site. Among these requests, 37.7% found 3 or more URLs with δ

---

[7]http://www.cs.berkeley.edu/~gribble/traces/index.html
[8]ftp://ftp.digital.com/pub/DEC/traces/webtraces.html
[9]http://ircache.nlanr.net

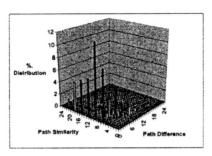

Figure 12.12. Distribution of Path Difference and Path Similarity in Boston University Trace using $w = 64$

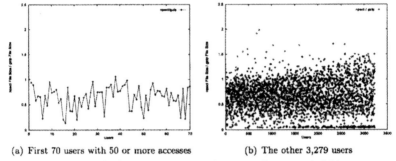

(a) First 70 users with 50 or more accesses   (b) The other 3,279 users

Figure 12.13. Performance of *npact* using traces from www.bell-labs.com

$= 0$ (same name), 76.7% had $\delta \leq 6$, and 90.7% had $\delta \leq 10$. When w is increased to 1024, the improvement only improves slightly.

The distribution of (δ, σ) was heavily concentrated in the regions of $\delta \leq 10$ and 80% of the tuples had $\delta \leq 6$ (i.e., mean $\delta = 2$). Earlier results (Section 12.4.2.1) demonstrated that the region with mean $\delta = 2$ corresponds to region of high object similarity. With $\delta = 6$, the 3 reference objects must either be all from the same directory (Section 12.4.2.1.3) or have at least one object with the same name (Section 12.4.2.1.1). From the results of Section 12.4.2.1, the first case has a 38% improvement over *gzip* and 17% improvement over *diff*, while the second case can have a 90% improvement over *gzip* and a 23% improvement over *diff*.

In summary, the distribution of (δ, σ) in this trace contains a significant amount of reference locality such that our selection heuristics can find highly similar URLs. In fact, at least 80% of the accesses would benefit from the use of *npact*.

12.4.2.3   Set 3: Performance of *npact* based on Actual Traces.

n	0	1
t_{select} (ms)	0.00	0.17
encode (kB/s)	900	722
decode (kB/s)	2,830	6,011

n	3	7
t_{select} (ms)	0.20	0.23
encode (kB/s)	380	176
decode (kB/s)	6,255	6,750

Figure 12.14. Select/Encode/Decode Processing Times

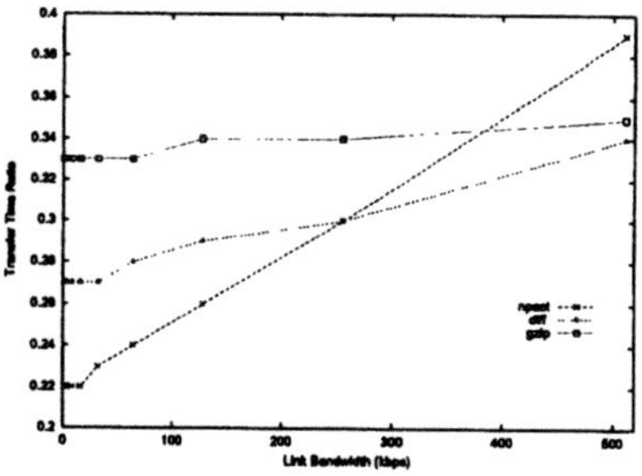

Figure 12.15. Transfer Time Ratio vs. Link Bandwidth

12.4.2.3.1   Object transfer size reduction.   In the final set of experiments, we compute the actual amount of savings using *npact* by performing *npact* (i.e., selection, encoding and decoding) for actual access traces. The Boston University trace we used in Section 12.4.2.2 could not be used here because the age of the logs meant that many of the URLs were outdated, and could no longer be fetched.

What we chose to do instead was to take a specific multi-day server log, divide it into per-user access traces, perform *npact* for each such trace, and compute the mean compression ratio. In the following, we presented our results based on the site www.bell-labs.com, where we had access to the detailed server log.

Specifically, we obtained the server log of www.bell-labs.com for 7 days from June 20 1998 to June 26 1998. In this log, we were able to extract 3,296 user access trace (based on unique IP addresses) that we "followed" using *npact*, *gzip* and *diff*.

Figure 12.13 plots the performance of *npact* relative to *gzip*. Logs were ordered by their number of access. Figure 12.13 shows the ratio for the first 70 users with the most number of accesses and Among the remaining 3,226 users, 2,831 out of 3,297 users had ratios smaller than 1 (*npact* outperforms *gzip*), and 69 out of first 70 users had ratio smaller than 1. The average ratio of *npact* over *gzip* for all 3,279 users was 0.6693.

In summary, 86% of all users would benefit from the use of *npact*. The larger the number of accesses, the more likely that *npact* would perform better. It can also be observed from Figure 12.13(b) that substantial savings were possible even for users with very few accesses.

The comparison with *diff* is similar. The compression ratio of *npact* relative to *diff* over all users was 0.8170 on the average.

12.4.2.3.2   Latency reduction.   Latency reduction is achieved
in our scheme if $t_{select} + t_{encode} + t_{decode} + \frac{|\sigma'|}{s} < \frac{|\sigma|}{s}$

In order to quantify the reduction in latency, the selection, encoding
and decoding speed of *npact*, *gzip* and *diff* were measured. Figure 12.14
shows the average execution time of the *npact* with respect to n, averaging over 1,000 files. The measurements were done on a SUN Ultra2.
Note that the comparison is to *npact*'s disadvantage since the code for
gzip and *diff* is highly optimized and very efficient while *npact*'s code
is not. Nevertheless, the comparison provides a useful gauge. The compression speed of *gzip* is 2,545 kbyte/s and *gunzip* decompresses at 12,270
kbyte/s. The encoding speed of *diff* -e is 961 kbyte/s.

We use the compression ratio obtained in Section 12.4.2.3.1 for various
algorithms to compare latency reduction. The compression ratios of *gzip*,
diff and *npact* are 0.33, 0.27 and 0.22 respectively.

Latency reduction is evaluated by computing the time it takes to
transfer the same amount of data using *gzip*, *diff* and *npact* over a wireless
link relative to using no compression. This ratio is called the *transfer
time ratio*. For example, if *gzip* has a transfer time ratio of 0.5, and it
takes 1 second to transfer a file without compression, it will take *gzip*
0.5 second to transfer the same file.

Figure 12.15 shows how the transfer time ratio for various compression
schemes with link bandwidth ranges from 1 kbps to 512 kbps. With link
bandwidth below 64kbps, transport time dominates the overall latency
and *npact* performs much better than *diff* and *gzip*. For link bandwidth
between 64kbps and 256kbps, computation overhead of *npact* becomes an
increasingly dominant component and reduces the utility of *npact* substantially. Beyond 256kbps, *diff* outperforms *npact* and beyond 400kbps,
gzip outperforms *npact*. For link bandwidth greater than 8Mbps, plain
transfer without compression incurs the least latency.

It is important to note that as computation becomes faster, the range
of bandwidth in which *npact* is useful will increase.

12.5   Applying Compaction to Email Transfer

Email transfer occurs in two directions, from the server to the client
(client reads an email) and from the client to the server (client composes
and sends an email).

In the server-to-client direction, when a new email arrives for a client,
a notification containing, among others, the *sender name, date, message
length* and *subject* is sent to the client. To retrieve an email, the client
sends back to the server the "name" of the email as well as that of a set of
reference messages chosen among the set of locally cached messages using

a selection algorithm.[10] On receiving the request, the server encodes the requested email using the set of reference messages and sent the result to the client. Finally, the client receives and decodes the encoded object using locally cached reference messages.

The client-to-server direction is similar except that the roles of compaction client and server are reversed. The email client selects and encodes while the server decodes.

The cache-based compaction technique can be easily integrated into the IMAP4 protocol as an extended capability. A compaction-enabled client can check if compaction is available on the server by using the IMAP *CAPABILITY* command.

12.5.1   Selection Algorithm

The selection algorithm partitions email messages into two components, the message header, and the message body. For server-to-client transfer, the selection algorithm looks at both the header and the body. For client-to-server transfer, selection looks only at the body since the header is usually very small (recipient and subject only) [Chan and Woo, 1999a].

Let *edit distance* between 2 strings be the minimum number of copy, replace, delete and insert operations required to transform the first string to the second string. For example, the edit distance between the strings *compaction* and *Re: compaction* is 4 (all deletes) and the edit distance between *munchoon@dnrc* and *woo@dnrc* is 6 (5 deletes and 1 replace).

In the selection process, the following similarity indexes are defined. Let M_i and M_j be two email messages:

- *sender-similarity*$(M_i, M_j) = \text{edit\_distance}(N_i, N_j)$, where N_i and N_j are sender names of M_i and M_j

- *subject-similarity*$(M_i, M_j) = \text{edit\_distance}(S_i, S_j)$, where S_i and S_j are message subjects of M_i and M_j

The selection algorithm uses these similarity indexes as heuristic for choosing similar objects and is presented in Figure 12.16. The algorithm takes four parameters, the email requested (only the sender, subject and date are needed), the set of message headers cached, the set of message bodies cached, and the maximum total size of all the reference objects selected.

---

[10]In an actual implementation, these "names" are typically unique interger identifiers generated by the server, e.g., the UIDs in IMAP4 [Crispin, 1994].

```
function select ( u, C_H, C_B, N ) {
    /* u is the message required
    C_H is the set of message header cached
    C_B is the set of message body cached
    N is the total size of the reference objects */
    S = ∅;
    while ( |S| < min(N, |C_H| + |C_B|) ) {
        H = header in C_H − S with maximum sender-similarity with u, using
            subject-similarity then date as tie-breaker
        S += H
        B = body in C_B − S with maximum subject-similarity with u, using
            sender-similarity then date as tie-breaker
        S += B
    }
    return S;
}
```

Figure 12.16. Selection Algorithm for Email Messages

User	No. Msg.	Av. Size(B)	gzip	compaction
1	879	5817	0.4639	0.3139
2	267	4359	0.4656	0.3621
3	280	4062	0.4817	0.3496
4	174	3554	0.4800	0.3466
5	268	2760	0.4763	0.3367

Figure 12.17. Summary of Experiments Performed

The selection criteria for message header is different from the selection criteria for message body. For message header, the selection algorithm looks for message headers from similar senders using the heuristic that these message headers tend to share similar paths, context types etc. Among the message headers from the most similar senders, the header with the most similar subject is selected.

For message body, the selection algorithm looks for message bodies with similar subjects using the heuristic that message bodies with similar subjects tend to contain similar content. Among the message bodies with the most similar subjects, the message body from the most similar sender is selected. The last tie breaker in both cases is the date; the more recent messages are preferred.

In this selection algorithm, a threshold for the maximum size of all reference objects is used to terminate the selection instead of the number of reference objects as is the case in *npact* [Chan and Woo, 1999b].

12.5.2   Experimental Results

In this set of experiments, the objective is to verify if the selection algorithm introduced in Section 12.5.1 does actually pick out "good" reference objects that are useful in the encoding procedure.

In order to limit the processing time of the selection algorithm, which runs a dynamic program to compute the edit distance, the maximum length of the subject and sender string is limited to 32 bytes.

In encoding, all reference objects are first concatenated into a single object, o_1, before it is given to the encoder. In other words, the encode function is invoked as $o' = $ encode$(o, o_1, threshold)$. The advantage of this approach is that it reduces the encoding overhead and allows multiple email messages, which are usually relative short, to be treated as a single reference object. In addition, the encoder will utilize only up to N bytes for encoding even though the size of o_1 can be larger than N. In the experiments, N is set to 64KB. Similarly, decoding is invoked by calling $o = $ decode(o', o_1).

We performed two sets of experiments, the first set for server-to-client transfer and the second set for client-to-server transfer. In both cases, given a set of chronologically ordered emails, emails starting from the 65th message are retrieved, using the previous 64 emails as cache. Email messages larger than 64KB are ignored. They are not transferred and are not considered as possible reference objects.

For server-to-client transfer, we measure the reduction in transfer size for 5 clients receiving emails from the server. The compressed size of the email messages using *gzip* is used for comparison.

The measurement inputs and results are summarized in Figure 12.17. Column 2 shows the total number of messages in the input data set from each user and column 3 shows the average size of these messages in byte. Column 4 shows the average compression ration achieved using *gzip* and the last column shows the average compression ratio achieved using compaction. The results show that the selection algorithm is able to choose reference messages that allow the encoding algorithm to achieve substantial reduction in transfer size. On the average, compaction is about 30% better than *gzip*.

The size of the email message also influences the compression ratio achieved and compaction achieves better compression ratio for small messages in this experiment. Figure 12.18 shows the distribution of message size and Figure 12.19 compares the average compression ratio achieved by *gzip* and compaction over different message sizes for all 1549 messages compressed. It can be seen that the sizes of most email mes-

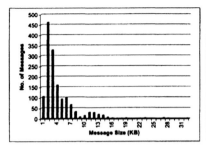 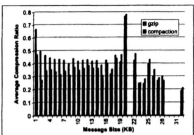

Figure 12.18. Distribution of Message Size   *Figure 12.19.* Comparison of Compression Ratio with respect to Message Size

sages are less than 16KB and for messages of these sizes our compaction scheme performs significantly better than *gzip*.

The result for client-to-server transfer is similar. For a user composing and sending 50 messages to the server, the compression ratio using *gzip* is 0.6776 and the compression ratio for compaction is 0.4379. The average message size is 1279 bytes. The compression ratio is expected to be lower in this direction because of the absence of message headers which are highly redundant.

While our compaction scheme achieves substantial size reduction, additional improvements are possible. First, note that the email retrieval being studied is performed selectively and is on-demand. Therefore, the selection algorithm needs to be efficient so as not to offset the decrease in latency achieved through transfer size reduction. However, for email delivery using Post Office Protocol 3 (POP3) [Myers and Rose, 1996], retrieval can be performed using bulk transfer and update can be performed periodically. In such a protocol, the selection algorithm used can be very computationally intense so as to obtain a even better reduction in transfer size. For example, in the selection algorithm described in Figure 12.16, after reducing the potential set of references objects to the message headers (bodies) from (with) the most similar senders (subjects), a brute force comparison can be used to find the set of reference objects that achieves the best compression ratio.

Next, message body is treated as a single entity by the same encoding/decoding algorithm. However, for larger messages, it is likely that these message are multi-part messages and should be treated as such. Finally, since the content types of the individual parts are known, content specific encoding/decoding algorithms can be used to improve the compression ratio.

12.6 Conclusion

We presented a technique which we call *cache-based compaction* for reducing the size (optimizing the latency) of transferring data over a wireless link. The two key ideas behind our technique is: (1) an efficient selection heuristic, and (2) the use of an extended dictionary (specifically the client cache) for compression. From our experiments, we observe that our compaction technique provides significant improvement over previously proposed techniques for real-life user accesses.

Looking to the future, we note that a few trends will enhance the effectiveness of our technique. First, with the proliferation of popular Web design tools like Microsoft's Frontpage and the tendency to enforce commonalities among pages from the same Web site, the amount of similarity among Web pages will increase. Next, as the cost of storage continues to decrease, caches will be large enough to capture most of the benefits of caching static content. Additional performance improvement must employ new techniques. Finally, as Web content becomes more dynamic, the utility of static caching will decrease. Our compaction technique's ability to leverage unrelated objects will become critical.

While prefetching has not been investigated in this work, we believe it can be enhanced using our compaction technique. Specifically, given past access patterns of Web or email objects, a server can use the set of most popular objects to precompute reference objects containing the most used tokens. By prefetching these reference objects, subsequent transfer of requested objects using our compaction algorithm can be improved.

References

[Banga et al., 1997] Banga, G., Douglis, F., and Rabinovich, M. (1997). Optimistic deltas for WWW latency reduction. *USENIX*.

[Bender et al., 2000] Bender, P., Black, P., Grob, M., Padovani, R., Sindhushayana, N., and Viterbi, A. (2000). CDMA/HDR: a bandwidth-efficient high-speed wireless data service for nomadic users. *IEEE Communications Magazine*, 38(7).

[Chan and Woo, 1999a] Chan, M. C. and Woo, T. Y. (1999a). Application of compaction technique to optimizing wireless email transfer. *Proceedings of IEEE Wireless Communications and Networking Conference*.

[Chan and Woo, 1999b] Chan, M. C. and Woo, T. Y. (1999b). Cached-base compaction: A new technique for optimizing web transfer. *Proceedings of the IEEE INFOCOM*.

[Crispin, 1994] Crispin, M. (1994). Internet message access protocol - version 4. *Intnernet Engineering Task Force, RFC 1730*.

[Cunha et al., 1995] Cunha, C. R., Bestavros, A., and Crovella, M. E. (1995). Characteristics of WWW client-based traces. Technical Report BU-CS-95-010, Department of Computer Science, Boston University.

[Dingle and Partl, 1997] Dingle, A. and Partl, T. (1997). Web cache coherence. *Fifth International World Wide Web Conference*.

[Fan et al., 1999] Fan, L., Cao, P., Lin, W., and Jacobson, Q. (1999). Web prefetching between low-bandwidth clients and proxies: Potential and performance. In *Measurement and Modeling of Computer Systems*, pages 178–187.

[Fox and Brewer, 1996] Fox, A. and Brewer, E. (1996). Reducing WWW latency and bandwidth requirements by real-time distillation. In *Fifth International World Wide Web Conference*.

[Griggioen and Appleton, 1996] Griggioen, J. and Appleton, R. (1996). The design, implementation, and evaluation of a predictive caching file system. Technical Report CS-264-96, Department of Computer Science, University of Kentucky, Lexington, KY.

[Housel and Lindquist, 1996] Housel, B. C. and Lindquist, D. B. (1996). Webexpress: A system for optimizing web browsing in a wireless environment. *Proceedings of the Second Annual International Conference on Mobile Computing and Networking*, pages 108–116.

[Hunt et al., 1998] Hunt, J. J., Vo, K.-P., and Tichy, W. F. (1998). Delta algorithms: an empirical analysis. *ACM Transactions on software Engineering and Methodlogy*, 7(2):192–214.

[Kroeger et al., 1997] Kroeger, T. M., Long, D. D., and Mogul, J. C. (1997). Exploring the bounds of web latency reduction from caching and prefetching. *USENIX Symposium on Internet Technologies and Systems*.

[Mogul, 2000] Mogul, J. (2000). Squeezing more bits out of HTTP caches. *IEEE Networks*.

[Mogul et al., 2002] Mogul, J., Krishnamurthy, B., , Douglis, F., Feldmann, A., Goland, Y., van Hoff, A., and Hellerstein, D. (2002). Delta encoding in HTTP. *RFC 3229*.

[Mogul et al., 1997] Mogul, J. C., Douglis, F., Feldmann, A., and Krishnamurthy, B. (1997). Potential benefits of delta encoding and data compression for http. In *Proceedings of the ACM SIGCOMM*, pages 181–194.

[Myers and Rose, 1996] Myers, J. and Rose, M. (1996). Post office protocol - version 3. *Intnernet Engineering Task Force, RFC 1939.*

[Wessels and Claffy, 1998] Wessels, D. and Claffy, K. (1998). ICP and the Squid Web Cache. *IEEE Journal on Selected Areas in Communication,* 16(3):345-357.

[Ziv and Lempel, 1977] Ziv, J. and Lempel, A. (1977). A universal algorithm for sequential data compression. *IEEE Transaction of Information Theory,* IT-23(3):337-343.

[Ziv and Lempel, 1978] Ziv, J. and Lempel, A. (1978). Compression of individual sequences via variable-rate coding. *IEEE Transaction of Information Theory,* IT-24(3):530-536.

Chapter 13

PERFORMANCE IMPROVEMENTS IN MULTI-TIER CELLULAR NETWORKS

Vijoy Pandey
vijoy@nortelnetworks.com
Nortel Networks, 4655 Great America Parkway, Santa Clara, CA 95054

Dipak Ghosal
ghosal@cs.ucdavis.edu
Department of Computer Science, University of California, Davis, CA 95616

Biswanath Mukherjee
mukherje@cs.ucdavis.edu
Department of Computer Science, University of California, Davis, CA 95616

13.1.   Introduction

The goal of a Personal Communication System (PCS) is to provide a wide range of services, to multiple user types, at any location, and over multiple environments. Future cellular networks will be a central part of this goal and will therefore require improved quality of service (QoS), higher capacity, and a larger coverage area than existing networks. QoS can be improved if the system can achieve a lower new-call blocking probability and ensure that calls which are admitted into the system have lower failure rate (i.e., lower dropping probability). In order to increase the cellular network's capacity, we can employ a finer mesh of smaller cells (i.e., microcells) over areas with a large population of users in order to achieve higher channel reuse. On the other hand, to be able to cover a larger area and serve a large number of highly-mobile hosts, we should increase the cell size.

A multi-tier cellular network architecture can satisfy these various – and sometimes conflicting – requirements easily. In this architecture, a cell in a particular layer covers a number of smaller cells in the next lower layer. In a two-tier architecture, a lower layer of microcells is covered by an upper layer of macro (umbrella) cells. Microcells are used to achieve higher capacity, while macrocells provide a larger coverage area and reduce overheads due to handoffs.

The channel allocation problem has been studied extensively in the context of a single-tier cellular network architecture [Katzela and Naghshineh, 1996]. The key idea is to optimize an objective function that includes both the new-call blocking and the handoff blocking events. Some allocation schemes, in particular the guard channel scheme [Ramjee et al., 1997], employ reservation mechanisms in which certain number of channels are reserved exclusively for handoff calls.

The channel allocation problem is much more complex in a multi-tier architecture since it provides a mechanism to obtain better performance if various mobility and traffic types (such as voice and data) can be differentiated and assigned to different tiers. Some preliminary studies on channel allocation for multi-tier architectures have been carried out in which the microcells can use channels which are not in use at the macrocell layer [Worsham and Avery, 1993], or the microcells can reuse channels which are being used by the macrocells [Kinoshita et al., 1993, Kojima et al., 1998, Ho et al., 2001].

The design of cellular networks with consideration of various economic aspects has been addressed in [Gavish and Sridhar, 1995]. The authors study the interaction between the various costs associated with the setup and maintenance of cellular networks and the revenue obtained from subscribers. They develop a model which determines the optimal network configuration (in terms of number and size of cells and number of channels used) that maximizes net revenue. Economic aspects of configuring two-tier cellular networks have been studied in [Ganz et al., 1997]. Our study is mainly concerned with the performance aspects of configuring two-tier cellular networks.

In this work, we first study the channel allocation problem for a two-tier cellular network, for two different call types – voice (with short call holding time) and data (with long call holding time) – when there are no overheads associated with handoff requests. Given a limited number of channels, we examine different channel assignments between the two layers such that the blocking and the dropping probabilities are minimized. We compare the impact of two call admission algorithms on the network performance. The first algorithm treats all calls identically and first attempts to assign them to the microcell layer. The second algorithm first attempts to assign calls with long call holding time to the macrocell layer. Subsequently, we introduce queueing overheads by defining a *handoff controller* which models the Base Station Controller (BSC)/Mobile Switching Center (MSC) signaling system and serves handoff requests from a number of macrocells and their underlying microcells. The impact of handoff overheads on the blocking and the dropping probabilities is then studied for the cases outlined above.

We have the following main contributions. First, we show that, if no handoff overheads are present (i.e., there is no waiting time between a handoff request and a handoff completion), it is advantageous to allocate most of the channels to the microcell layer. Second, when handoff overheads are present and the load to the handoff controller is significant, there exists an optimal orthogonal assignment of channels between the two tiers that minimizes the blocking and the dropping probabilities. If the load at the handoff controller is low, allocating more channels to the microcell layer minimizes the blocking and the dropping probabilities. The load at the handoff controller can increase due to an increase in any or all of these three factors: (1) the mean service time of the handoff controller; (2) the average mobility of the users in the system; and (3) the total number of cells (macrocells and microcells) handled by the controller. This result suggests that a hierarchical design of the BSC/MSC controllers will help in minimizing the effects of handoff overheads on the network performance. Finally, we show that it is always better to assign all calls to the microcell layer on call admission, and not distinguish them based on their call holding times. Specifically, we show that assigning both data and voice calls first to the microcell layer on admission leads to better performance than a call admission policy which assigns data calls first to the macrocell layer.

The remainder of the paper is organized as follows. In the next section, we describe the network architecture as well as the call admission algorithms analyzed in this study. In Section 3, we state the assumptions that have been used to carry out the mathematical analysis and simulation. In Section 13.4, we develop an analytical model to compute the blocking probability for a two-tier network with just one type of call and no handoff overheads. The validation of the analytical model is discussed in this section. In Section 13.5, we examine the performance of a two-tier cellular network with both single- and two-class traffic and study the impact of two different call admission algorithms with no handoff overheads. In Section 13.6, we study the network when significant overhead costs are associated with handoffs. We examine single-class traffic for low and high mobility, and then compare the two call admission policies for a mix of both voice and data calls. In Section 13.7, we discuss the sensitivity of the results to variations in the parameter values. All the results in Sections 13.5 through 13.7 are obtained via simulation experiments. Section 13.8 outlines some related work. Finally, in Section 13.9, we present our conclusions and list some topics for future investigation.

336

13.2.   Network Architecture and Problem Statement

A cell is defined by the geographical region over which a mobile terminal can receive transmission from a particular base station. When a mobile terminal moves out of the coverage area of a base station (A) into the coverage area of another base station (B), a handoff is said to have occurred from base station A to base station B. In cellular systems, different cells may simultaneously use the same frequency channels [Lee, 1995], thus improving spectrum efficiency. This concept is known as frequency reuse. Interference due to the same channel being used in different cells, known as co-channel interference, puts a limit on the minimum distance over which the same frequency can be reused. This distance, normalized to the radius of a cell, is known as the reuse distance ratio. Clearly, making the cell size smaller leads to greater frequency reuse and hence larger capacity, which implies that more users can be accommodated in the network. On the other hand, smaller cell sizes will cause more handoffs, resulting in greater overheads in servicing these handoffs as well as in keeping track of the users' locations. To improve network capacity as well as to reduce

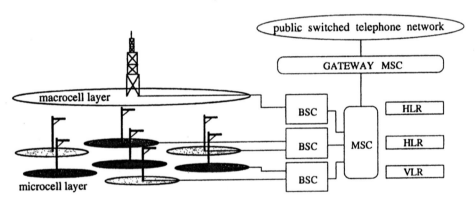

Figure 13.1.   A two-tier cellular network architecture.

the overheads due to handoffs, a two-tier cellular network architecture has been proposed in which a larger macrocell is overlayed over a number of smaller microcells. Such a two-tier cellular network is illustrated in Figure 13.1[1], where each macrocell overlays 7 microcells.

In this architecture, a group of cells (macrocells and microcells) is controlled by a Base Station Controller (BSC), which is responsible for channel assignment within these cells and the management of handoffs between neighbouring base stations. The Mobile Switching Center (MSC) is responsible for (1) set-

[1] The figure shows the GSM cellular network architecture [Redl et al., 1995], which is typical of most current cellular network architectures.

ting up, managing, and clearing connections; (2) routing incoming calls to the appropriate cell; and (3) managing inter-BSC and inter-MSC handoffs. The Home Location Register (HLR) is a database containing user specific data, such as user identity, subscribed services, and the current user location, for all subscribers registered in a geographical area. The Visitor Location Register (VLR) contains information about all users currently "visiting" its particular geographical area.

To incorporate the resource allocation and clearance overheads associated with a handoff procedure, we model the BSC/MSC signaling centers by a simple *handoff controller* which manages handoffs between a number of macrocells and their underlying microcells.

The two key factors which influence the performance of a two-tier cellular network are [Murata and Nakano, 1993]:

- *Channel assignment:* How should the given spectrum be shared between the two tiers so as to minimize the blocking and dropping probabilities? There are many different ways in which a given number of channels can be allocated to the two tiers and re-used [Worsham and Avery, 1993, Kinoshita et al., 1993]. In this study, we examine orthogonal sharing, where the channels are partitioned among the tiers without reuse *across* the tiers (i.e., microcells cannot reuse channels used by the macrocell layer and vice versa).

- *Call and handoff admission:* In which tier should an incoming new or handoff call be placed, and does such a placement improve the blocking and the dropping probabilities? We consider two different types of calls in the network – data calls and voice calls – the main difference between them being the call holding time which is significantly larger for the data calls. The manner in which these calls may be handled gives rise to two call admission algorithms which are discussed below.

Uniform Call Admission (UCA) Algorithm   The simplest call admission scheme is to treat all calls identically and always first attempt to admit them into the microcell layer. This leads to the following algorithm (Figure 13.2).

- A new call always enters a microcell. If all channels are busy, the call overflows to the corresponding macrocell. If all of the channels in the macrocell are also busy, the call gets blocked.

- When a call is handed-off, it is first attempted at the microcell layer. If there is no available channel, it overflows to the corresponding macrocell where it can get blocked if there is no available channel.

- When there is a handoff at the macrocell layer, the call is handed off to the appropriate microcell at the periphery of the adjacent macrocell.

338

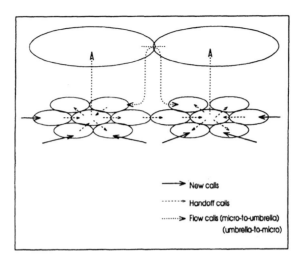

Figure 13.2. Uniform Call Admission (UCA) Algorithm.

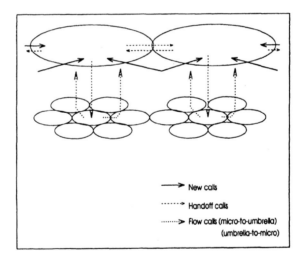

Figure 13.3. Non-uniform Call Admission (NCA) Algorithm.

Non-uniform Call Admission (NCA) Algorithm   In this algorithm we make a distinction between voice and data calls. Given that data calls have much longer call holding times and hence stay in the network for a longer time, they will generate more handoffs than voice calls if assigned to the microcell layer (as in the UCA algorithm). This might lead to higher blocking and dropping probabilities. On the other hand, if we assign data calls to the macrocell layer to reduce the number of handoffs produced by such a call, the handoff rate of the call will be reduced. Hence, the residence time of a data call in a cell will increase substantially, which might lead to larger blocking and dropping probabilities. To investigate this tradeoff, we consider a call admission algorithm where data calls are always first attempted at the macrocell layer. Details of this algorithm, for which we assume that it is possible to "sense" the type of the incoming call (voice or data), are presented below.

- Voice calls follow the UCA algorithm (i.e., they always first attempt the microcellular layer).

- New data calls always first attempt a macrocell. If no channel is available in the macrocell, the call overflows into its corresponding microcell under the macrocell. If all channels in this microcell are busy, the data call is blocked.

- When there is a data call handoff at the microcell layer, the data call attempts a handoff to the macrocell corresponding to the target microcell. In the macrocell layer, a handoff takes place to the adjacent macrocells.

- If no channel is available for the handoff data call in the macrocell, it overflows to its corresponding microcell. If no channel is available in this microcell as well, then the data call is blocked.

The above algorithms are depicted in Figures 13.2 and 13.3, respectively, each of which show two macrocells covering seven microcells each.

Problem Statement   Given a two-tier cellular network infrastructure with a limited number of channels, two types of calls (voice and data), and a single class of user mobility (either fast- or slow-moving users), the problem is to find an optimal orthogonal assignment such that the blocking and the dropping probabilities are minimized. While studying the problem, we investigate the effect of the two call admission algorithms (UCA and NCA) as well as handoff overheads on the blocking and the dropping probabilities.

13.3.   Modeling Assumptions and Performance Metrics

13.3.1   Assumptions

To study the channel partitioning algorithms via mathematical analysis and simulation, we make the following assumptions.

- Each macrocell in the two-tier network is overlayed over exactly N microcells. For simulation purposes, we assume that each handoff controller serves a total of M macrocells and their underlying microcells.

- We assume a total of C channels in the network. For the two-tier network, we allocate m channels to each microcell and c channels to each macrocell. Note that, $c = \lfloor (C - m \times R_m)/R_u \rfloor$, where R_u denotes the macrocell reuse distance ratio and R_m denotes the microcell layer reuse distance ratio.

- The total capacity (κ) of the network is defined to be the maximum number of calls the network can support at any given instant of time. The microcell layer can support a maximum of $M \times N \times m$ calls at any given instant. The macrocell layer can support a maximum of $M \times c$ calls. This leads to the following expression for κ:

$$\kappa = M \times (Nm + c) \qquad (13.1)$$

- The arrivals of new voice and data calls are drawn from a Poisson process with parameter λ (calls/sec). We use F and $1 - F$ to denote the fraction of data and voice calls in the network, respectively.

- Mobility is modeled using an uniform fluid-flow approximation. Under this model, the rate at which a mobile crosses a microcell boundary is given by:

$$\alpha = VL/\pi S \qquad (13.2)$$

where L is the perimeter of a microcell, V is the average velocity of the user, and S is the microcell area [Thomas et al., 1988]. The sojourn time of a mobile in a microcell is exponentially distributed with the mobility parameter α (cell crossings/sec). In macrocells, the mobility parameter turns out to be $\alpha_u = \alpha/\sqrt{N}$.

- The total call holding times have exponential distributions with means $1/\mu_v$ and $1/\mu_d$ (sec) for voice and data calls, respectively.

- The residence time of a voice call in a microcell is therefore exponentially distributed with parameter $\alpha + \mu_v$. There are two ways in which a call exits a microcell: (1) there is a handoff with probability $\theta = \alpha/(\alpha + \mu_v)$ and (2) the call completes with probability $1 - \theta$. The above holds for an macrocell as well, with α replaced by α_u, and also for data calls, with μ_v replaced by μ_d.

Default Parameter Values We have used the following default parameter values for the analytical and simulation results shown in this work. We have used a reuse distance ratio of 3 for the macrocells and microcells (i.e., $R_u = R_m = 3$) and the total number of channels available (C) to the network is 60. The mean user mobility, for a slow user, is assumed to be 3.5 kmph (2.25 mph) – the average walking speed of a human being. For a fast user, the mean mobility is assumed to be 35 mph (56 kmph) – the average mobile speed in a city business area. The microcell radius is assumed to be 200 m, and $N = 12$. Therefore, from Equation (13.2) we get $\alpha = 0.00367647$ (cell crossings/sec) for slow users, and $\alpha = 0.05718954$ (cell crossings/sec) for fast users. For voice calls, a mean call holding time of 2 minutes is used, while data calls are assumed to be 20 minutes long on average. Each simulation experiment was conducted for 10,000,000 new-call arrivals. The sensitivity of the results to variations in the default parameter values are discussed in Section 13.7.

13.3.2   Performance Metrics

To compare the various orthogonal assignments and call admission algorithms, we consider the following performance metrics.

- **Blocking Probability:** This metric can be further broken down into the voice and data blocking probabilities. The voice blocking probability is defined to be the ratio of the number of new voice calls blocked to the number of voice calls that tried to enter the system. Data blocking probability is defined similarly. Total blocking probability is defined as the ratio of new calls blocked (data and voice) to the total number of new-call attempts.

- **Dropping Probability:** Calls which have successfully entered the system can undergo handoffs due to user mobility. We define the total dropping probability as the ratio of the total number of calls (data and voice) that got blocked while attempting a handoff to the total number of calls that successfully entered the network. This blocking probability can also be broken down into voice and data dropping probabilities. This metric is a measure of the QoS provided by the network.

13.4.   Analysis of the Network with a Single Class of Traffic

In order to get a better understanding of the impact of the system parameters on the different blocking probabilities, we have developed a mathematical model of a two-tier network. The model currently applies only to a single class traffic with the additional simplification that there is no overhead associated with handoff requests[2]. The analysis extends the approach developed in [Meier-Hellstern, 1989] and [Lagrange and Godlewski, 1995, Lagrange and Godlewski, 1999], by including the effects due to mobility in the macrocell layer and by incorporating the admission control algorithms described in Section 13.2 into the model.

---

[2]This model is currently being extended to incorporate handoff overheads.

342

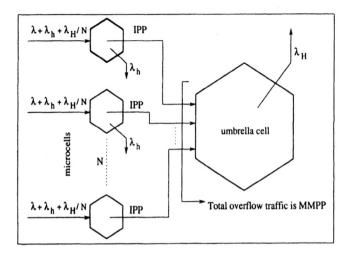

Figure 13.4. Analytical model of the two-tier architecture.

Figure 13.4 shows the model of the two-tier network architecture. Let λ_h and λ_H denote the mean handoff arrival rates at the microcell and macrocell layers, respectively. Both these handoff arrival processes have been approximated as Poisson processes for analytical tractability; the accuracy of this approximation is evaluated later in this section by comparing results from the analytical model with simulation.

Our analysis does not yield a closed-form solution for the blocking probability and we use an iterative approach. We assume initial values for λ_h and λ_H. Given λ, λ_h, and λ_H, each microcell is modeled as an M/M/m/m queue where m is the number of channels in each microcell. The arrival rate into each microcell due to macrocell handoffs is assumed to be λ_H/N. The handoff traffic from a macrocell enters only the microcells at the periphery of the adjacent macrocell. There is no direct impact of these handoff calls on microcells that are in the center of the macrocell, and hence, this approximation becomes less accurate when N becomes large. Once we calculate the total blocking probability at the microcell layer, we model the traffic overflowing into each macrocell as an Markov-Modulated Poisson Process (MMPP). From this formulation, we can calculate the blocking probability at the macrocell layer, which helps us in updating our estimate of the macrocell layer handoff rate λ_H. We can iterate through this procedure to calculate the total blocking probability to within a desired accuracy.

13.4.1   Analysis of the Microcell Layer

The total input traffic into a microcell is $\lambda_t = \lambda + \lambda_h + \lambda_H/N$. Let us define $\rho = \lambda_t/(\alpha + \mu)$. λ_h is calculated using the flow equilibrium equation given by [Foschini et al., 1993]:

$$\lambda_h = (\alpha/(\alpha + \mu))(\lambda_t)(1 - B(\rho, m)) \tag{13.3}$$

where $B(\rho, m)$ is the Erlang-B formula for offered load ρ and m servers. This flow equation simply states that the handoff rate into a microcell is a function of the incoming traffic to the microcell, the handoff rate from nearby microcells, and the blocking probability in the microcell. The generator matrix, Q, for each microcell can be written as:

$$Q = \begin{pmatrix} -\lambda_t & \lambda_t & \cdots & & \cdots & \\ \alpha + \mu & -\lambda_t - (\alpha + \mu) & \cdots & & \cdots & \\ \cdots & 2(\alpha + \mu) & \cdots & & \cdots & \\ \cdots & \cdots & \cdots & & \lambda_t & \\ \cdots & \cdots & \cdots & -\lambda_t - m(\alpha + \mu) \end{pmatrix} \tag{13.4}$$

The overflow traffic from such a queue is modeled as an MMPP which is defined by the generator matrix Q (Equation (13.4)) and an arrival matrix given by $\Lambda = diag(0, 0, 0, \ldots, \lambda_t)$. Both Q and Λ are $(m + 1) \times (m + 1)$ matrices.

Since there are N microcells in each macrocell, the cumulative *overflow traffic* into each macrocell is the superposition of N such MMPPs, assuming that they are statistically independent (Figure 13.4). Typically, N and m have values on the order of 19 and 20 which give us 13.25×10^{24} states. Since this state space is computationally very large, the overflow process can be approximated by an Interrupted Poisson Process (IPP) [Kuczura, 1973]. An IPP is a special case of the MMPP — it is a two-state MMPP with the arrival rate in one of the states being zero. For our case, the IPP can be represented by:

$$Q_{ipp} = \begin{pmatrix} -\sigma_1 & \sigma_1 \\ \sigma_2 & -\sigma_2 \end{pmatrix}, \quad and \quad \Lambda_{ipp} = \begin{pmatrix} \lambda_t & 0 \\ 0 & 0 \end{pmatrix} \tag{13.5}$$

i.e., we have a two-state Markov chain with transition probabilities σ_1 (from state 1 to state 2) and σ_2 (from state 2 to state 1). In state 1, the overflow traffic rate is $\lambda + \lambda_h + \lambda_H/N$, while in state 2, it is zero. To approximate the MMPP by an IPP, we match the first three noncentral moments of the instantaneous arrival rate and a time constant of the original MMPP to the IPP [Meier-Hellstern, 1989, Heffes, 1980]. To do so, we have to first calculate the stationary vector π of the underlying M/M/m/m process. Let $P_b = \pi_m$ represent the blocking probability for this process. Then, the mean η of the instantaneous arrival rate for the MMPP turns out to be $P_b(\lambda_t)$ and its variance ν becomes $(\lambda_t)^2(P_b - P_b^2)$. The time constant can be calculated by using:

$$\tau_c = \nu^{-1}[P_b(\lambda_t)(\bar{e}\bar{\pi} - Q)\Lambda\bar{e} - \eta^2] \qquad (13.6)$$

where $\bar{e} = (1, 1, 1, \ldots, 1)^T$. Having calculated these parameters, σ_1 and σ_2 can be obtained as:

$$\sigma_1 = \tau_c^{-1}(1 - P_b\bar{e}), \quad \text{and} \quad \sigma_2 = \tau_c^{-1}(P_b\bar{e}) \qquad (13.7)$$

Since all of the microcells are homogeneous, the resulting overflow to an macrocell is the superposition of N identical two-state MMPPs. The cumulative overflow traffic is an MMPP described by the matrices Q_{ovfl} and Λ_{ovfl} (see [Meier-Hellstern, 1989] for the corresponding mathematical details).

13.4.2   Analysis of the Macrocell Layer

Each macrocell can be represented by an MMPP/M/c/c queue. The $(c + 1)(N + 1) \times (c + 1)(N + 1)$ generator matrix, Q_{umb}, for each macrocell can be written as:

$$Q_{umb} = \begin{pmatrix} Q_{ovfl} - \Lambda_{ovfl} & \Lambda_{ovfl} & \cdots & \cdots \\ \mu I & Q_{ovfl} - \Lambda_{ovfl} - \mu I & \cdots & \cdots \\ \cdots & 2\mu I & \cdots & \cdots \\ \cdots & \cdots & \cdots & \cdots \\ \cdots & \cdots & \cdots & Q_{ovfl} - c\mu I \end{pmatrix} \quad (13.8)$$

where I is the identity matrix. The blocking probability can be calculated by solving for the stationary vector $\mathbf{\Pi}$, where

$$\mathbf{\Pi}Q_{umb} = 0 \quad \text{and} \quad \mathbf{\Pi}\bar{e} = 1 \qquad (13.9)$$

Note that $\mathbf{\Pi}$ consists of $c + 1$ vectors: $\mathbf{\Pi} = (\bar{\pi}^0, \bar{\pi}^1, \ldots, \bar{\pi}^c)$, where each $\bar{\pi}^i$ is of length $(N + 1)$. We can solve for these equations directly.

The probability that a call gets blocked in the macrocell, given that it has been blocked in the microcell tier, is given by:

$$P_B = \left(\sum_{i=0}^{N} i\bar{\pi}_i^c \right) \Big/ \left(\sum_{k=0}^{c} \sum_{i=0}^{N} i\bar{\pi}_i^k \right) \qquad (13.10)$$

Hence, the total blocking probability for the network is given by $P_B \times B(\rho, m)$.

Once we have calculated the total blocking probability (P_B) at the macrocell, we can update our estimate of λ_H by using a flow equation similar to Equation (13.3) given by:

$$\lambda_H = [1 - P_B][B(\rho, m)(\lambda + \lambda_h + \lambda_H/N)(N)][\alpha/(\alpha + \mu\sqrt{N})] \quad (13.11)$$

Having updated λ_H, we can iterate through the entire procedure to get a desired accuracy in the total blocking probability.

13.4.3 Validation of the Analytical Model

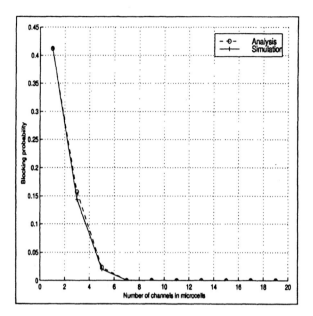

Figure 13.5. Total blocking probability for a single call type - comparison of analysis and simulation (N=12).

Figure 13.5 shows the total call blocking probability obtained via analysis for a call arrival rate of $\lambda = 0.035$ calls/sec and $N = 12$. Results obtained via simulation (for the same parameters) have also been shown in the figure. We observe that there is a close match between the analytical and simulation results, although the analysis is slightly optimistic in predicting the blocking probability. Such observations have been verified for higher values of N and λ as well, e.g., $N = 50$ and $\lambda = 0.065$. Discrepancies between the analytical and simulation results may be due to the following reasons. First, the analysis approximates handoffs into microcells and macrocells as Poisson processes. Second, the overflow traffic from a microcell is approximated by an IPP. Finally, microcells at the periphery of a cluster covered by a macrocell are likely to receive more macrocell layer handoffs than those at the center.

13.5. Two-Tier Cellular Network without Handoff Overheads

13.5.1 Voice Calls

Figures 13.6 through 13.9 show the performance characteristics of a two-tier cellular network for call arrival rates of $\lambda = 0.035, 0.045$ and 0.075 calls/sec, and

$N = 12$. In this section, we assume that handoff overheads are insignificant. We make the following key observations:

- The blocking and the dropping probabilities (shown in Figures 13.6[3] and 13.8) generally decrease as we allocate more channels to the microcell layer. There are two main reasons for this behaviour:

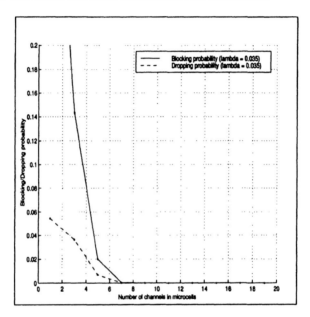

Figure 13.6. Blocking and dropping probabilities for voice calls (with 95% confidence intervals).

- As the number of channels in each microcell (m) is increased, the total capacity of the system (κ) increases. When $C = 60$, $M = 5$, and $m = 1$, we obtain $c = 19$. This implies that we can have a maximum of 155 calls[4] in the network at any instant ($\kappa = 155$ from Equation 13.1). On the other hand, when $m = 19$, the capacity becomes $\kappa = 1145$. This increase in capacity results from more reuse at the microcell layer. Since the offered load to the system is constant ($\lambda = 0.035$ calls/sec in this experiment), when there are fewer channels in each microcell, most new calls overflow to the macrocell which gets saturated, thereby resulting in higher blocking. When more channels are allocated to each microcell, the total

---

[3]95% confidence intervals are also shown in Figure 13.6, but notice that they are extremely tight and not very distinguishable; accordingly, we won't show confidence intervals in other performance graphs.
[4]$5 \times 12 \times 1 = 60$ calls in the microcells and $5 \times 19 = 95$ calls in the macrocells.

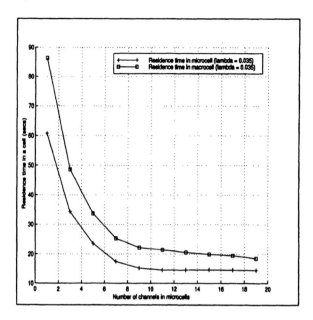

Figure 13.7. Residence time of voice calls in a cell.

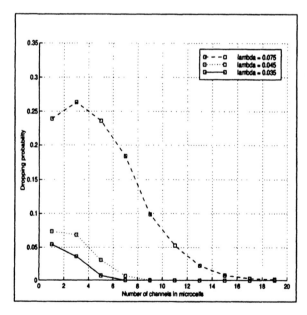

Figure 13.8. Dropping probabilities for voice calls.

capacity of the microcell layer increases, as a result of which most

348

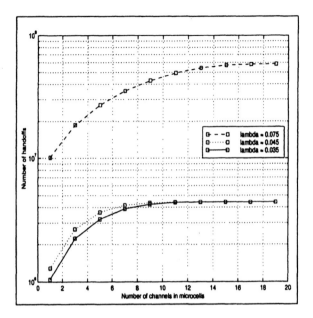

Figure 13.9. Number of handoffs generated for voice calls.

calls are satisfied at the microcell layer. Few calls overflow to the macrocell layer, hence, fewer calls get blocked.

– As we allocate more channels to the microcell layer, this layer carries most of the traffic. Since a microcell is smaller in size compared to a macrocell, a mobile terminal undergoes more handoffs at the microcell layer (as shown in Figure 13.9). As a result, the residence time of a call in a particular cell decreases as observed in Figure 13.7. Even though a voice call is assumed to have a call holding time of 120 secs., the residence time of a voice call in a macrocell or a microcell is much shorter due to a large handoff rate. This means that the call holds resources in a cell for a shorter amount of time which contributes in further lowering the blocking and the dropping probabilities as m is increased.

■ When we increase the call arrival rate, the dropping probability first increases with an increase in network capacity and then decreases, as shown in Figure 13.8 for $\lambda = 0.075$. This phenomenon can be explained by observing Figure 13.9, which shows the total number of handoffs as a function of the number of channels in a microcell[5]. When we have less

---

[5]The figure shows the number of handoffs generated over the entire simulation period. Each experiment was run for 10 million new call arrivals, and handoffs are generated by all calls which were successful.

system capacity, many new-call attempts get blocked. So there are very few *active* calls in the system, and therefore there are few handoffs. As we increase the capacity by increasing m, there are more active calls which cause more handoffs. When the network capacity becomes very large, the number of active calls in the system saturates, since the capacity is greater than the aggregate arrival rate. Thus, the number of handoffs generated also saturates which is observed in Figure 13.9. In the moderate-capacity region, the increase in the number of handoffs is greater than the increase in capacity. Therefore, we observe an increase in the dropping probability which explains the "hump" shape in Figure 13.8.

When the arrival rate is small, the new-call blocking probability is low even when the capacity of the network is small. Hence, there are a lot of active calls in the system and therefore a lot of droppings. When we increase the network capacity, the increase in the number of active calls is not as much as when the arrival rate was higher. The network is able to accommodate the resulting increase in the number of handoffs, and hence we do not observe a "hump" in the dropping probability curve when λ is small.

In conclusion, we observe that, in the absence of handoff overheads, it is best to operate a two-tier network with most of the channels allocated to the lower tier. However, when the arrival rate is high, there is a larger dropping probability in the moderate-capacity region, which might impact the QoS of some applications.

13.5.2   Two-Class Traffic: Voice Plus Data Calls

Figures 13.10 through 13.13 show the performance for the two call admission algorithms described in Section 13.2, with $N = 12$, $\lambda = 0.015$ calls/sec, and $F = 0.2$. Again we assume handoff overheads to be absent. We make the following observations.

- For both algorithms, the blocking and the dropping probabilities decrease as the capacity increases (see Figure 13.10). The reasons are similar to those presented in the previous section for voice calls. As we increase m, the capacity of the network increases. Moreover, as we can see from Figure 13.11, the residence time of a call (data or voice) in a cell decreases as m increases, for both algorithms. These two factors together result in the blocking and the dropping probabilities to decrease as m increases.

- The Non-uniform Call Admission (NCA) algorithm performs worse than the Uniform Call Admission (UCA) algorithm. The reason for this behaviour is as follows.

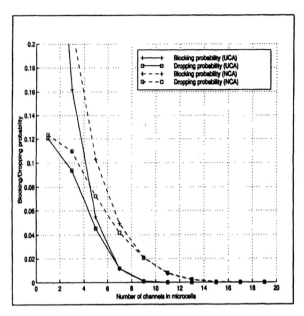

Figure 13.10. Blocking and dropping probabilities for combined voice and data calls.

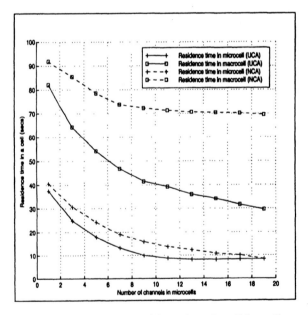

Figure 13.11. Average residence time of a call in a cell.

Figure 13.11 shows the residence time of a call in the macrocell and microcell layers for both algorithms. We observe that, in both the microcell

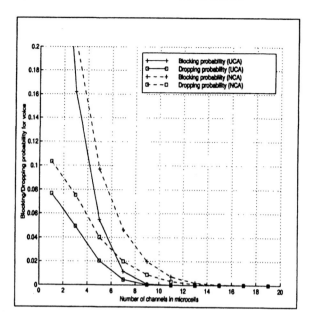

Figure 13.12. Voice blocking and dropping probabilities.

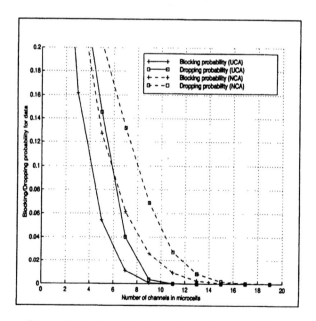

Figure 13.13. Data blocking and dropping probabilities.

and the macrocell layers, the residence time of a call is higher for NCA

352

than it is for UCA. This implies a higher call holding time, and hence higher blocking and dropping probabilities for the NCA algorithm.

In the NCA algorithm, a data call always first attempts the macrocell layer. Since macrocells are larger in size, a data call undergoes fewer handoffs. This is in contrast to the UCA algorithm where it first attempts the microcell layer. Coupled with a larger call holding time, a data call following the NCA algorithm has a significantly larger residence time in the macrocell layer, than a data call following the UCA algorithm (Figure 13.11). This results in a larger holding time for a data call in NCA which results in higher blocking and dropping probabilities.

- Figures 13.12 and 13.13 compare the impact of the two algorithms separately on voice and data calls, respectively. Since the longer call holding time of a data call in the macrocell layer in NCA affects both the voice and the data calls (due to the nature of the admission algorithms), we observe that UCA performs better separately for voice and data calls. The differences are more pronounced for data calls since voice calls always attempt the microcell layer first in both NCA as well as UCA.

In summary, we observe that the UCA algorithm outperforms the NCA algorithm. Moreover, UCA is easier to implement, since both data and voice calls are treated identically. Even for maintaining an acceptable QoS, UCA is better as it results in lower dropping probability.

13.6. Two-Tier Cellular Network with Handoff Overheads

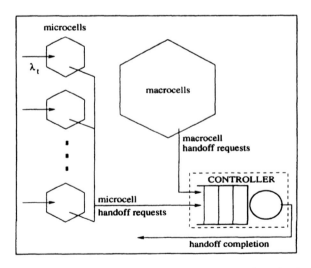

Figure 13.14. The model of the network with handoff overheads.

In the previous section, we did not consider the cost of handoff processing. In particular, we assumed that, when a user moves from one cell to the next, the resources in the old cell are instantly cleared for reuse by other new or handoff calls. This is not true in reality and protocol processing and resource clearing procedural costs are associated with handoffs. Furthermore, these costs may increase with the number of handoffs, with the net result that the channel holding time in a cell may increase resulting in higher blocking probabilities. To incorporate these handoff overheads, we model our cellular network architecture as shown in Figure 13.14. To simplify the modeling of the network, we have merged the BSC/MSC hierarchy seen in a typical cellular architecture (for example, the GSM cellular architecture shown in Figure 13.1) into a single controller for a group of macrocells. This controller is modeled as an infinite buffer queue with a single server with an exponentially-distributed service time with parameter μ_c. As stated earlier, our network consists of M macrocells and we assume that each controller handles handoff requests from these M macrocells as well as their underlying microcells. The calls requesting the handoffs occupy channels in their old cell throughout the period they spend in the controller queue. Once these handoff requests are serviced by the controller, the channels in the old cell are released for reuse by new and handoff calls. The infinite buffer approximation at the controller is used as a simplification to study the effects of the queueing delay on the blocking probability in the network.

The figures in this section have been plotted for the same parameters as outlined in Section 13.3. The mean controller service time is taken to be 200 ms [Brass and Fuhrmann, 1997] and it handles $M = 5$ macrocells. Each macrocell covers $N = 12$ microcells. In the following sections, we first consider slow user mobility and single-class traffic, followed by fast user mobility and single-class traffic. Finally, we compare the performance of the two call admission algorithms, NCA and UCA, for a mix of voice and data calls, and fast user mobility.

13.6.1   Slow Mobility and Single-Class Traffic

Figure 13.15 shows the total, new-call, and handoff blocking probabilities for a call arrival rate of $\lambda = 0.035$ calls/sec. UCA is used as the call admission strategy. Slow-moving mobiles have been considered with a mean speed of 2.25 mph.

We observe that the blocking and dropping probabilities are similar to the no-overhead case discussed in Section 13.5.1. The blocking probabilities decrease as more channels are assigned to the microcell layer, since the increase in the network capacity more than offsets the increase in handoff traffic. Further, since the mobility is slow, the handoff arrival rate at the controller is not large enough to cause significant delays. As a result, there is only a negligible increase in

354

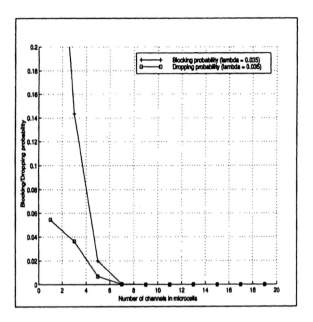

Figure 13.15. Voice blocking probability with handoff overheads - slow mobility.

the channel holding time due to the handoff processing overhead. Finally, for larger arrival rates, we observed that certain channel partitions in the moderate capacity region resulted in higher dropping probability, similar to the case with no handoff overheads. These results have not been presented to conserve space.

13.6.2 Fast Mobility and Single-Class Traffic

In this section, we examine the network behaviour for voice calls when mobiles move at a mean speed of 35 mph, and $\lambda = 0.06$ calls/sec. The rest of the system parameters remain the same as before. Figure 13.16 shows the blocking and dropping probabilities for voice calls, while Figure 13.17 shows the residence time of a voice call in a cell.

- As seen in Figure 13.16, there exists an optimal orthogonal assignment of channels between the two tiers that results in minimum blocking and dropping probabilities. There are two opposing factors that result in this optimization. As we allocate more channels to the microcell layer, the capacity of the network increases which decreases the blocking probabilities. Also, as stated earlier, the residence time of a call in a cell becomes smaller, reducing the blocking and the dropping probabilities further.

  On the other hand, as we allocate more channels to the microcell layer, the number of handoffs increases (Section 13.5.1), and therefore the rate of requests at the controller increases, which results in higher waiting times at the controller. This causes the channel holding time in the "old" cell to increase (Figure 13.7), resulting in higher blocking probability. The cumulative result of these two factors results in the optimization shown in Figure 13.16.

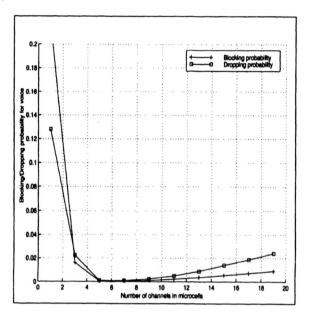

Figure 13.16. Voice blocking probability (handoff overheads, fast mobility).

- In the moderate and high capcity regions ($m = 3$ and above), the dropping probability is slightly larger than the blocking probability. This can be explained as follows. As the capacity of the network is increased, more calls are handled by the microcell layer. Since microcells are smaller in size, many more handoffs are generated when m is large. This increase in the number of handoffs has a two-fold effect. First, it loads the controller, increasing the holding time of a call in a cell, and hence increases the blocking probability. Therefore, fewer calls are successful in the large capacity region (e.g., $m = 17$) as compared to the moderate capacity region (e.g., at $m = 6$).

Second, this increase in the number of handoffs also implies an increase in the number of handoff drops. The rate of increase in the number of handoffs and the number of droppings, as m increases, is larger than the rate of increase in the number of new call blocks. Hence, by definition, the dropping probability becomes larger than the blocking probability as capacity increases.

356

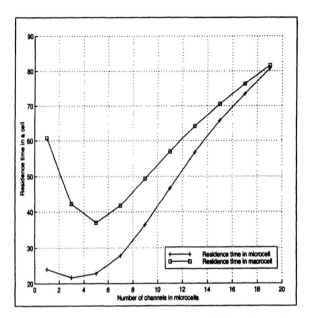

Figure 13.17. Residence time of a voice call in a cell (handoff overheads, fast mobility).

- For smaller values of M ($M = 1, 2$ or 3), i.e., when the handoff controller serves handoff requests from a small number of macrocells and microcells, the results obtained were similar to Figure 13.15, and we did not observe the optimization shown in Figure 13.16. Again, these results are not included here to conserve space.

On studying the results obtained in this section as well as those in the previous section, we find that if the handoff controller is lightly loaded, then allocating more channels to the microcell layer decreases the blocking and the dropping probabilities. When the load at the handoff controller becomes high, we observe an optimum in the blocking and the dropping probabilities.

There are three main factors which affect the load at the handoff controller. They are: (1) the mean controller service time ($1/\mu_c$); (2) the average rate of mobility of a user (α); and (3) the number of macrocells (M) and their underlying microcells being served by the controller. The average user mobility is not directly under our control. The mean controller service time can be reduced by employing faster processors at the BSC/MSC signaling centers. By following a hierarchical design of BSC/MSC controllers, where each handoff controller serves only a few macrocells and microcells, we can reduce the total handoff request traffic to the controller, thereby decreasing its load. Moreover, by having replicated controllers on this hierarchical structure, we can effectively reduce the mean controller service time.

To summarize, for fast mobility and single class traffic, we observe that the blocking and the dropping probabilities are optimized for certain channel partitions. Interestingly, both the blocking and the dropping probabilities are optimized for the same channel partition. Finally, our results suggest a hierarchical design of controllers to reduce the impact of handoff overheads on the blocking probabilities.

13.6.3 Fast Mobility and Two-Class Traffic

In this section, we compare and contrast the two different call admission algorithms (NCA and UCA) in the presence of handoff overheads. To do so, we have assumed that there are 20% data calls and 80% voice calls in the network (i.e., $F = 0.2$), as before. For both algorithms, the aggregate new-call arrival rate is assumed to be $\lambda = 0.06$ calls/sec. The mobiles move at a mean speed of 35 mph. The other network operating parameters are the same as before. Based on the results shown in Figures 13.18 through 13.21, we make the following observations.

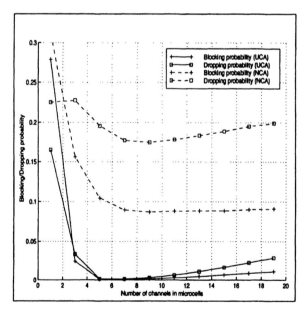

Figure 13.18. Blocking and dropping probabilities for combined data and voice calls.

- Even with two-class traffic, there exists an optimal orthogonal assignment of channels between the two tiers that minimizes the blocking and the

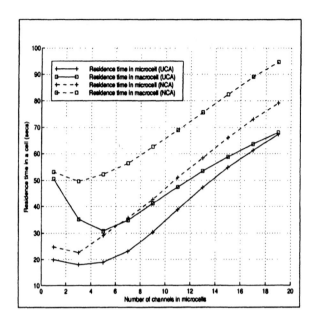

Figure 13.19. Average residence time of a call in a cell.

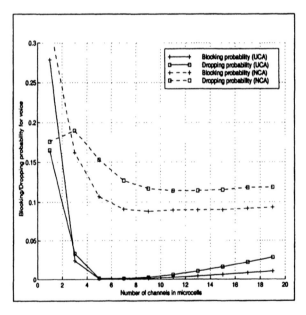

Figure 13.20. Voice blocking and dropping probabilities.

dropping probabilities. The reasons for this optimum are similar to those outlined in the previous section.

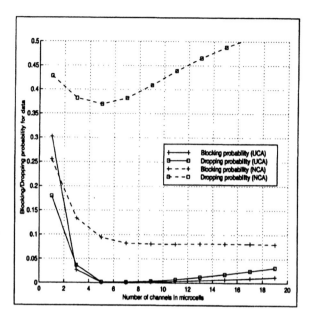

Figure 13.21. Data blocking and dropping probabilities.

- Similar to Section 13.5.2, we observe that the UCA algorithm outperforms the NCA algorithm, as can be seen in Figure 13.18. Figure 13.19 shows the average cell residence time of a call for the two different algorithms. As expected, the residence times for NCA are higher than those for UCA. Moreover, the residence time increases as capacity increases and more handoff requests are sent to the handoff controller.

- Figures 13.20 and 13.21 show the blocking and the dropping probabilities for voice and data calls separately. As mentioned before, the significantly longer residence time of a data call in NCA as compared to the UCA adversely affects both data and voice calls. This leads to higher blocking for voice as well as data calls in the NCA algorithm.

- In general, we find that treating voice and data calls identically, and letting them follow UCA leads to better network performance, irrespective of whether handoff overheads are significant or not. Moreover, since UCA does not have to "guess" the type of the call (data or voice) before admitting it into the network, it is easier to implement.

In summary, for fast mobiles, when both data and voice calls are present in the network, we observe that the UCA algorithm results in better network performance, a result that we have seen earlier in a network with no handoff overheads. Moreover, UCA is the simpler of the two call admission algorithms, and hence should be the preferred strategy while assigning calls to the two layers.

13.7.   Sensitivity of Results to Parameters

In Section 13.3, we mentioned a list of default parameters which have been used for the analytical and simulation experiments illustrated in this study. Though only a few results have been presented for clarity and conservation of space, we have varied the parameter values considerably to study its effects on our experiments. We observed the following.

- The call holding time, in the single-class traffic experiments, was varied from 90 secs. to 1200 secs. As the holding time increased, both the blocking and the dropping probabilities increased, as expected.

- In Section 13.5, when handoff overheads were assumed to be absent, increasing the user mobility, from 2.25 mph to 90 mph, led to a sharper fall in the blocking and the dropping probabilities. Since there was no penalty for handoffs, increased mobility implied shorter cell residence times, and hence lower blocking. On the other hand, the "hump" in the dropping probability curve (Figure 13.8) became more apparent with increasing mobility due to reasons presented in Section 13.5.1.

    Increasing user mobility in a network with significant handoff overheads resulted in a larger load at the handoff controller and hence larger blocking and dropping.

- By varying the reuse distance ratios (R_u and R_m) between 2 and 4, and the total number of channels (C) between 50 and 150, we studied the effects of network capacity on the blocking and the dropping probabilities. When handoff overheads were assumed to be absent, an increase in capacity always led to a decrease in blocking. When handoff overheads were significant, the blocking characteristics changed depending upon M. When M was small, the blocking and the dropping probabilities decreased with an increase in capacity. On the other hand, a larger M, coupled with a larger capacity increased the load at the handoff controller, thereby increasing the blocking and the dropping probabilities.

- By keeping the microcell radius constant and changing N from 7 to 50, we studied the effects of different macrocell sizes on blocking. An increase in the macrocell size implied an increase in coverage area, and hence an increase in the blocking and the dropping probabilities.

13.8.   Related Work

The idea of an overlayed and hierarchical cellular network architecture was proposed to combine the benefits of greater channel re-use by shrinking the cell size, while at the same time trying to minimize the number of mobile user handoffs due to the smaller cell size. Substantial effort is being placed in designing efficient load balancing and QoS improvement methodologies for multi-tier cellular networks.

In [Kojima et al., 1998], a macrocell/microcell cellular architceture has been studied with fixed channel allocation in the macrocell tier and dynamic channel allocation in the microcell tier. Through simulation, the authors show that the proposed system can minimize blocking probability under any traffic condition, including spatial nonuniform traffic. A soft frequency partitioning scheme between the hierarchical layers of a multi-tier network has been studied in [Ho et al., 2001]. The idea proposed demonstrates that a significant capacity gain can be achieved by sharing the frequency between the layers, without sacrificing the carrier-to-interference ratio performance of the macrocellular layer.

Frequency sectoring has been proposed in [Ho et al., 1999], which leads to greater system capacity, and also allows for the gradual deployment of a hierarchical cellular architecture. The idea of partitioning the macrocell tier into two or more sub-tiers is proposed in [Anpalagan and Katzela, 1999]. A call blocked in microcell tier can overflow to a macrocell sub-tier depending on the user's speed. It has been shown that this architecture has fewer handoffs as compared to a strict hierarchical architecture. In [Jabbari and Fuhrmann, 1997] the performance of a call admission scheme based on user speed has been studied. The network architecture also incorporates a "take-back" scheme, where calls are handed off to a different tier when no resources are available in the tier to which it has been assigned.

To utilize the spectrum of a multi-tier cellular network more efficiently, a fuzzy cell selection procedure based on a user's speed has been proposed in [Sung and Shum, 1999]. Cell selection through such a scheme lowers the call blocking probability while keeping the handoff rate low. Cell selection based on call-requested data rates has been studied in [Cimone et al., 1999]. Higher data-rate users are assigned to the microcell layer, while lower data-rate users are admitted into the macrocell layer. Cell selection based on both, the user's speed and the requested type of service, has been studied in [Ben-Othman and Valois, 1999].

An adaptive bandwidth reservation scheme based on a mobility pattern profile has been proposed for a two-tier cellular network in [Choi et al., 2001]. The authors also propose a call admission control based on this bandwidth reservation and "next-cell prediction" scheme using the mobility profile.

13.9.   Conclusion and Future Work

Multi-tier cellular networks can provide higher capacity as well as better QoS in terms of lower blocking and dropping probabilities in an environment which supports multiple classes of traffic and multiple types of user mobility. In order to achieve these advantages, it is necessary to address two key engineering problems: (1) how should channels be partitioned among multiple tiers? and (2) how should different calls be assigned to the different tiers? This study was an attempt to address these two important issues.

Our results show that, if the processing overhead of a handoff is neglected, then allocating more channels to the microcell layer yields lower blocking and dropping for both single- and two-class traffic. When the processing overhead of a handoff call is taken into consideration, we observe an optimal orthogonal assignment between the two layers which results in minimum blocking and dropping probabilities, when the load at the handoff controller is high. This optimization appears for voice and data calls, as well as a mix of both call types in the network. There are three main factors which contribute to the load at a handoff controller, namely, (1) the controller service time; (2) the average user mobility; and (3) the number of cells being served by the controller. When either or all of these factors are small, we do not observe any optimizations in the blocking and the dropping probabilities. This suggests that a hierarchical design of handoff controllers is preferred to mitigate the effects of handoff overheads on the network performance. Finally, we observe that a Uniform Call Admission (UCA) scheme is both simpler to implement and results in lower blocking and dropping probabilities as compared to the Non-uniform Call Admission (NCA) scheme which assigns calls with long call holding time to the macrocell layer.

References

[Anpalagan and Katzela, 1999] Anpalagan, A. S. and Katzela, I. (1999). Overlaid cellular system design with cell selection criteria for mobile wireless users. In *Proc., Canadian Conference on Electrical Computer Engineering*, volume 1, pages 24–28.

[Ben-Othman and Valois, 1999] Ben-Othman, J. and Valois, F. (1999). Multiservice allocation in hierarchical cellular networks (mahcn). In *Proc., IEEE International Symposium on Computers and Communications*, pages 80–86.

[Brass and Fuhrmann, 1997] Brass, V. and Fuhrmann, W. (1997). Traffic engineering experience from operating cellular networks. *IEEE Communications*, 35(8):66–71.

[Choi et al., 2001] Choi, C. H., Kim, M. I., Kim, T. J., and Kim, S. J. (2001). A call admission control mechanism using mpp and 2-tier cell structure for mobile multimedia computing. In *Proc., Conference on Computer Communications and Networks*, pages 581–584.

[Cimone et al., 1999] Cimone, G., Weerakoon, D. D., and Aghvami, A. H. (1999). Performance eveluation of a two layer hierarchical cellular system with variable mobility user using multiple class applications. In *Proc., IEEE Vehicular Technology Conference*, pages 2835–2839.

[Foschini et al., 1993] Foschini, G. J., Gopinath, B., and Miljanic, Z. (1993). Channel cost of mobility. *IEEE Transaction on Vehicular Technology*, 42(4).

[Ganz et al., 1997] Ganz, A. et al. (1997). On optimal design of multitier wireless systems. *IEEE Communications*, 35(2):88–93.

[Gavish and Sridhar, 1995] Gavish, B. and Sridhar, S. (1995). Economic aspects of configuring cellular networks. *ACM/Baltzer Wireless Networks (WINET)*, 1(1):115–128.

[Heffes, 1980] Heffes, H. (1980). A class of data traffic processes - covariance function characterization and related queueing results. *Bell System Technical Journal*, 59:897–929.

[Ho et al., 1999] Ho, C. J., Lea, C. T., and Stuber, G. L. (1999). Call admission control in hierarchical cellular networks. In *Proc., Vehicular Technology Conference*, volume 1, pages 320–324.

[Ho et al., 2001] Ho, C. J., Lea, C. T., and Stuber, G. L. (2001). Call admission control in the microcell/macrocell overlaying system. *IEEE Transactions on Vehicular Technology*, 50(4).

[Jabbari and Fuhrmann, 1997] Jabbari, B. and Fuhrmann, W. (1997). Teletraffic modelling and analysis of flexible hierarchical cellular networks with speed sensitive handoff strategy. *IEEE Journal on Selected Areas in Communications*, 15(8):1539–1548.

[Katzela and Naghshineh, 1996] Katzela, I. and Naghshineh, M. (1996). Channel assignment schemes for cellular mobile telecommunication systems: A comprehensive survey. *IEEE Personal Communications*, pages 10–31.

[Kinoshita et al., 1993] Kinoshita, Y., Tsuchiya, T., and Ohnuki, S. (1993). Common air interface between wide-area cordless telephone and urban cellular radio: Frequency channel doubly reused cellular system. *Transactions on IEICE*, J76-BII(6):487–495.

[Kojima et al., 1998] Kojima, F., Sampei, S., and Morinaga, N. (1998). An intelligent radio source management scheme for multi-tiered cellular system with different assigned bandwidth under nonuniform traffic conditions. In *Proc., International Conference on Universal Personal Communications*, volume 1, pages 157–161.

[Kuczura, 1973] Kuczura, A. (1973). The interrupted poisson process as an overflow process. *Bell System Technical Journal*, 52:473–448.

[Lagrange and Godlewski, 1995] Lagrange, X. and Godlewski, P. (1995). Teletraffic analysis of a hierarchical cellular network. In *IEEE Vehicular Technology Conference*, pages 882–886.

[Lagrange and Godlewski, 1999] Lagrange, X. and Godlewski, P. (1999). Performance of a hierarchical cellular network with mobility-dependent handover strategies. In *Proc., Vehicular Technology Conference*, volume 3, pages 1868–1872.

[Lee, 1995] Lee, W. (1995). *Mobile Cellular Telecommunications*. McGraw Hill, Inc., 2nd edition.

[Meier-Hellstern, 1989] Meier-Hellstern, K. S. (1989). The analysis of a queue arising in overflow models. *IEEE Transactions on Communication*, 37(4):367–372.

[Murata and Nakano, 1993] Murata, M. and Nakano, E. (1993). Enhancing the performance of mobile communication systems. In *IEEE International Conference on Universal Personal Communications*, pages 732–736.

[Ramjee et al., 1997] Ramjee, R., Towsley, D., and Nagarajan, R. (1997). On optimal call admission control in cellular networks. *ACM/Baltzer Wireless Networks*, 3(1):29–41.

[Redl et al., 1995] Redl, S. M., Weber, M. K., and Oliphant, M. W. (1995). *An Introduction to GSM*. Artech House Mobile Communications Series. Artech House.

[Sung and Shum, 1999] Sung, C. W. and Shum, K. W. (1999). Channel assignment and layer selection in hierarchical cellular system with fuzzy control. In *Proc., Vehicular Technology Conference*, volume 4, pages 2433–2477.

[Thomas et al., 1988] Thomas, R., Gilbert, H., and Maziotto, G. (1988). Influence of the moving of the mobile stations on the performance of a radio mobile cellular network. *Proceedings of the Third Nordic Seminar on Digital Land Mobile Radio Communications*.

[Worsham and Avery, 1993] Worsham, J. and Avery, J. (1993). A cellular band personal communications system. In *2nd Universal Personal Communications*, pages 254–257.

Chapter 14

TECHNOLOGY-INDEPENDENT LINK SENSING IN WIRELESS AD-HOC NETWORKS: BENEFITS AND CHALLENGES[1]

Lisa A. Shay* and Kenneth S. Vastola**
*Department of Electrical Engineering and Computer Science, United States Military Academy
**Deparmtent of Electrical, Computer, and Systems Engineering, Rensselaer Polytechnic Institute

14.1   INTRODUCTION

Wireless-connected computing and communication devices such as cell phones, pagers, hand-held PCs, and the latest "integrated" devices that incorporate a PC or PDA and a cell phone into one package are becoming increasingly more popular. Some of these devices have several network interface options, such as Bluetooth[2], 802.11 wireless LANs[3], or "2.5G" or "3G"[4] (GSM or GPRS-enabled) cellular service. As the devices become

---

[1] This work supported in part by MURI contract F49620-97-1-0382 funded through AFOSR.
[2] Bluetooth is a short-range (≈10m range) wireless network technology primarily designed to replace short-distance cable connections. See http://www.bluetooth.com for more information.
[3] The IEEE 802.11 standard was ratified in 1997 and specifies physical layer and medium-access layer behavior for wireless Local Area Networks. The 802.11HR standard is an amendment to the 802.11 standard to include higher data rates [13].
[4] 2.5G cellular networks are upgrades to the GSM cellular standard to provide higher data transfer rates. See http://www.gsmworld.com/technology/gprs/index.shtml for more information. The 3G or Third Generation wireless standards are part of the IMT-2000 standard of the International Telecommunications Union. Devices conforming to this standard will support data rates up to 2Mbps and are expected to be available in the US as early as 2003. See http://www.imt-2000.org/portal/index.asp for more information.

more sophisticated, it will be possible to automatically switch among the network interfaces, even within the same communication. For a device with multiple network interfaces, it would be useful to have a single method of determining the link status for each of those interfaces. With this ability, a user or an operating system could then choose the best available interface to use for communication. It would also be useful to have a means of predicting link failures before the link actually fails. Then the device would have ability to handoff transmission among the interfaces without disruption of the ongoing communication. In the event there is only a single interface with a useable link, a means of predicting that link's failure would allow the operating system or user to take some other appropriate action, such as sending an "about to be disconnected" message to the distant end, or adjusting position to obtain a better link quality.

This chapter discusses the challenges of sensing wireless links and the additional challenges involved in sensing wireless links in a technology-independent manner. Technology-independent link sensing provides the most flexibility for the latest communication devices with multiple or swappable network interfaces. This is followed by a discussion of predictable network link failures. Finally, the last section presents an overview of the Wireless Network Environment Sensor, which is a technology-independent sensor that detects, predicts and identifies faults in wireless links.

14.2   CHALLENGES IN SENSING MOBILE WIRELESS AD-HOC NETWORK LINKS

Link sensing in mobile wireless ad-hoc networks is more difficult than link sensing in wired (electrical or optical) backbone network links or even fixed wireless point-to-point links (such as a backbone microwave link). This section will discuss four characteristics of wireless ad-hoc networks that increase the difficulty of this problem: wireless links have limited bandwidth; ad-hoc network links often have low link utilization; the topology of an ad-hoc network is not known in advance; and the topology of a network of mobile hosts continuously changes.

14.2.1   Bandwidth

The bandwidth of a mobile wireless link is generally much less than a wired or fixed wireless network link. The popular wireless LAN standard, IEEE 802.11 has a 2Mbps maximum data rate and its high-speed extension has an 11Mbps maximum data rate. This is an order of magnitude slower than the popular wired LAN standard, IEEE 802.3 using Category 5 twisted pair cables (100BaseT), which runs at 100Mbps. Packet radios have an even lower data rate, as low as 9.6Kbps for some cellular standards. Any traffic injected onto a network from a network sensor robs bandwidth that would otherwise be consumed by the user's applications and so is considered by the user to be overhead. Considering that the mobile wireless network bandwidth is usually less than the user desires, any additional overhead from a network sensor would adversely impact network performance. The desire to minimize the impact on the underlying network performance favors passive monitoring of existing network traffic to determine link quality.

14.2.2 Link Utilization

However, passive monitoring of existing traffic adds complexity to a wireless link sensor for mobile ad-hoc networks. Unlike a backbone link that typically has a high utilization rate, some links in a mobile wireless network, especially stub links to hosts on the edge of the network, may have very low utilization rates. A wireless link sensor for mobile ad-hoc networks that uses passive monitoring must have a way of distinguishing a failed link that cannot pass traffic with a link that could be a high-quality link but happens to be unused.

14.2.3 Unknown Topology

The topology of an ad-hoc mobile wireless network is self-organizing. As hosts enter the network, the routing algorithm adjusts the network topology to incorporate these hosts. As hosts leave the network, the routing algorithm again must modify the network topology to route around those hosts that have left. In the most general case, there is no a priori knowledge of the initial network topology. To operate effectively in such a network, a sensor of mobile wireless ad-hoc network links cannot require or assume any information about the existence or location of network links.

14.2.4 Mobility

Wireless networks can be divided into two mutually exclusive types: fixed and mobile. In a fixed wireless network, the endpoints of the links do not move. An example of a fixed wireless network is a backbone link using point-to-point microwave transmission. Since the endpoints of the link are fixed and their location chosen to ensure line-of-sight between the endpoints, it is possible to use highly directional antennas and achieve a consistently high quality link providing an excellent grade of service. In contrast, in a mobile wireless network one or both ends of a wireless link can be displaced in space and network connectivity is desired while the network element (host) is moving. Since the hosts are mobile and their relative locations are unknown before establishing the network connection, omni-directional antennas are used. In addition, it is possible that due to obstructions in the network path, noise, destructive interference from multipath effects, or distance between hosts, wireless communication may be degraded or completely interrupted. Depending on the relative velocity of the hosts, network service may be interrupted quite suddenly: in less than a minute the link quality could go from excellent to non-existent. This is in contrast with wired backbone links, where some problems occur quite gradually and can be detected hours in advance of the actual fault [17].

Since the quality of a mobile wireless link is a function of both time and location, a link sensor cannot operate on the assumption that the link is normally in a high-quality state and faults are rare. This prevents a sensor from detecting faults based on a change in current link quality from the link quality in the recent past. In contrast, in a wired or fixed wireless network link faults are rare events and one can assume that the "normal" state of the link is a high quality link. Therefore, in a wired or fixed wireless link, any significant changes in the link state are faults.

Another implication of a network of mobile hosts is that the hosts are generally battery-powered and tend to have less computational power and smaller memory space than fixed hosts. Network sensing methodologies should be as small and computationally efficient as possible to minimize the impact on the device performance. Finally, in order to maximize battery life, a network sensing methodology should inject as little additional traffic into the network as possible, since transmitting data requires power.

14.3   TECHNOLOGY INDEPENDENCE

A host may have several types of wireless network interfaces that are used at different times over the course of a day, or even within a single communication. For instance, a PDA may have an 802.11 interface and a GPRS-enabled cell phone interface. A user may bring the PDA to a conference room for a meeting and use the 802.11 interface for network access. During the commute home, the user checks email via a GPRS-enabled cell phone interface. As data networks are upgraded or new networks are fielded, the list of available network interfaces increases.

If the number and type of network interfaces were fixed for a particular device, a network sensor made specifically for that type of interface, or even that brand and model of network interface, may be used. But if it is desired to monitor all devices currently in use and accommodate devices that may be used in the future, a technology-independent network sensor is required.

By definition, a technology-independent solution cannot draw its input from any information specific to the network device hardware, such as the received signal level or the noise power at the input of a RAKE receiver. Since such information can be very valuable in determining and predicting the network quality as seen by that specific device, the design decision not to use such information makes the wireless network-sensing problem significantly more difficult.

Another aspect of a technology-independent design is that it is independent of the platform on which it is operating. We define a platform as the combination of a mobile device and its operating system. For example, a platform might be a laptop or tablet computer running any one of several operating systems (such as a version of Windows, Linux, Mac OS, etc.), or a handheld device running an operating system specialized for a less computationally powerful device (such as Windows CE or Palm OS). A network sensor that operates on two or more of such platforms cannot acquire its input data in a way that is dependent on the characteristics of the operating system or the underlying hardware. For example, many operating systems keep track of statistics about the available network interfaces, such as the number of bytes transmitted, the number of bytes received, the packets discarded due to errors, etc. However, each operating system does so differently. A methodology that requires its input data to be in a location or format that is operating-system specific cannot be a technology-independent methodology. Finally, a network sensor that uses a computationally intensive algorithm may be impractical on a handheld device.

14.4 PREDICTABLE VS. NON-PREDICTABLE LINK FAILURES

In this section we examine the causes of wireless link failures and categorize them into predictable and non-predictable types. Link failures due to signal fading are found to be predictable. We analyze a model of a wireless link in terms of fast and slow fading and show that failures due to slow fading can be predicted in a technology-independent manner while failures due to fast fading are best addressed with a technology-specific solution within the wireless interface device itself.

14.4.1 Non-Predictable Failures

Mobile devices are usually small, lightweight, and battery-powered. In recent years they have become smaller and more rugged, but they are still quite fragile devices. However, due to their convenience and portability, they are often subjected to severe environmental stress such as shock from dropping or having objects dropped or placed on them. Also, the device may suddenly lose power from the batteries becoming discharged while the device is in use. In addition, the user can turn off or remove a network interface device, causing the wireless link to suddenly fail. We classify sudden failures of the mobile device itself or its wireless interface device(s) as unpredictable. For such faults, the goal of our methodology is to rapidly detect the fault once it has occurred.

14.4.2 Predictable Failures

There is another class of faults that are caused by a degradation of the wireless signal strength (without hardware malfunction). A wireless signal consists of a line-of-sight (directly transmitted) component and a non-line-of-sight (reflected) component. Both components of the wireless signal are affected by large-scale and small-scale path loss. Large-scale path loss occurs from free-space attenuation and shadowing (where a signal-impervious object blocks the wireless link). Small-scale path loss occurs from multipath interference, where many signals (some directly transmitted others reflected) are received, causing constructive and destructive interference of the signal at the receiver.

These different categories of path loss differ not only by the distance but also the timescale over which they affect the signal. Large-scale path loss operates over distances of meters and longer and has a timescale on the order of seconds when the hosts are moving at walking speed (approximately one to two metres per second). Small-scale path loss operates over much shorter distances and timescales, as discussed in [5] and [6].

We have found that predicting and compensating for small-scale path loss is well addressed at the hardware level, by such works as [2], [8], [5] and [6]. By adequately modeling the wireless channel, it is possible to not only predict signal attenuation, but also to modify the signal transmission to compensate for the channel variations. This is the purpose of the works discussed in section 5.1.

14.4.3   Failures Addressed by the Wireless Network Environment Sensor

We have conducted research on the problem of predicting and identifying faults due to large-scale path loss. We have created a methodology, called the Wireless Network Environment Sensor (WiNE Sensor) that detects, predicts, and identifies the cause of link failures in mobile wireless ad-hoc networks. We have shown through experiments on actual wireless networks using a mobile device carried by an experimenter moving at walking speed (approximately 1 m/s) that the faults discussed above are both predictable and identifiable. We found that wireless link failure due free-space attenuation can be predicted with approximately 42 seconds of warning time and failures due to shadowing predicted with approximately 31 seconds of warning time. We are also able to correctly distinguish link failures caused by free-space attenuation from those caused by shadowing. (See Section 6 and [15] for details.)

14.5   PREVIOUS WORK ON NETWORK SENSING

Our work, which is discussed in the next section, is the first work that addresses the problem of mobile wireless link sensing in a technology-independent manner. However, there is a large body of work that addresses sensing the wireless channel at the physical layer and using that information to adjust signal transmission parameters. In addition, the starting point for our development was work done by Thottan and Ji [17] and [18] to predict failures in wired backbone networks.

14.5.1   Wireless Channel Sensing

The problem of sensing the state of a wireless channel has been extensively studied for over 50 years [21], although the work of Cox in 1972 [4] is widely acknowledged to be the catalyst for study of wideband mobile radio channels. Starting in the mid 1980's, there has been an explosion in the growth of personal wireless devices, which has triggered additional research in the area of measuring and modeling wireless channels, especially in the 900MHz cellular band [4], [7], [11], [12], [20], and the 1.8GHz PCS band [9], [10], [14], [19]. Much of the work of wireless channel sensing has focused on characterizing the wireless channel. Characterization of the channel enables analytical and simulation model construction so that transmitter and receiver designs and source and channel coding techniques may be tested. However, none of the works cited above treat the problem of on-line, real-time link quality in an operational communications system.

Real-time channel estimation is required to implement on-line adaptation of the transmission system to the channel. Sensors designed specifically for estimating wireless channel quality usually require physical layer information obtained within the wireless transmitter or receiver, for example [2], [8], [5] and [6]. In [2], the authors describe a method of using the signal-to-interference plus noise ratio to measure channel quality. The authors use the discrete-time flat-fading model of the receiver output given by:

$$r_k = \alpha_k s_k + \gamma_k i_k + n_k \qquad (1)$$

where r_k is the received symbol at the kth time instant, α_k is the flat-fading channel coefficient, s_k is the transmitted symbol, γ_k is the interferer's fading coefficient, i_k is the interferer's transmitted symbol, and n_k is an additive white Gaussian noise process. The authors derive a metric that can be mapped to the per-symbol signal-to-interference plus noise ratio for two types of signaling: M-ary Phase Shift Keying and M-ary Differential Phase Shift Keying.

The authors in [2] assume that the measured or estimated channel coefficients are provided within the receiver. In [5] the authors describe a method of predicting such coefficients for a flat-fading channel. The authors of [5] denote the channel coefficients as c_k, using a model of a matched filter receiver given by (a model similar to that used by the authors of [2]).

$$y_k = c_k b_k + z_k \qquad (2)$$

where y_k is the matched filter and sampler output, c_k is the flat fading signal sampled at the data rate, b_k is the binary phase shift keying data sequence, and z_k is a complex additive white Gaussian noise process. The authors use an autoregressive model to predict the coefficients and show by simulation and experimentation that they can reliably predict such coefficients up to 10ms in advance. There are several adaptive transmission methods that can benefit from predicted channel state information. They demonstrate that an adaptive modulation method (M-ary Quadrature Amplitude Modulation) using predicted channel coefficients can achieve a significant bit-error-rate performance gain compared to an adaptive modulation method using traditional delayed channel state information.

These methods of physical layer channel sensing are complements to, rather than replacements for, our technology-independent wireless link sensing methodology. The method of acquiring the information to implement these physical-layer channel models (signal strength, signal-to-interference-plus-noise ratio, channel state, etc) is dependent upon the design of the network device and the information received pertains only to the channel used by that specific device. When incorporated into the hardware or firmware of the network interface device, they greatly increase the throughput of that specific device.

14.5.2   Technology-Independent Prediction of Wired Link Failures

Thottan and Ji have studied the problem of predicting faults in wired backbone networks [17] and [18]. They determined that it is possible to predict some types of network faults, such as a file server crash, hours in advance of the catastrophic failure. The methodology for network fault prediction that they developed happened to be independent of the physical layer interface used to transmit data.

Since faults are rare in wired links, one of their assumptions was that the wired link is normally in a high-quality, operational state. This allowed them to compare the current state of the link to a previous state and declare that any change in the state of the link corresponds to a fault. As discussed in section 2.4, in wireless links faults are not rare occurrences and there is no guarantee a mobile device will start out with a good quality link. Therefore, comparing the current state of the link to the recent past is not sufficient for determining link quality. Without additional information, a change of state of the link is merely that: a change. The change may be from a good quality link to a failing one, or vice-versa. To implement sensing for wireless links, our methodology models a high-quality link and compares the current state of the link to the model of the known good link. This model of a high-quality link we term a high-quality link baseline.

This fundamentally different approach of creating high-quality link baselines rather than using change detection had an added benefit. After developing our high-quality link baselines, we modified our baseline-generation algorithm to also model links that have failed due to signal strength attenuation or shadowing. Modifying our algorithm to create fault baselines gave us the ability not only to predict link faults, but to also identify why the fault was occurring. This is in marked contrast to the work of Thottan and Ji which could predict that some fault would occur, but could not determine the type of fault or the reason for the fault.

14.6   OUR METHODOLOGY

We developed a real-time, technology-independent sensor of faults in wireless network links.   Because wireless links have relatively low bandwidth and limited power budgets, our sensor derives its input from existing traffic, rather than adding new traffic that would consume power in transmission and further reduce the bandwidth available to the user. Because our sensor operates on existing traffic, yet links at the edges of ad-hoc networks may at times have low utilization rates, we implemented a mechanism to differentiate a failed link from an unused link.   Since an ad-hoc network has a time-varying topology, our sensor does not require specific knowledge of any other host.   Nor is there any kind of specialized network sensing management station or server.   Finally, since faults are not rare events in mobile wireless links, our methodology uses a system of baselines to which we compare the current link, rather than implementing a change detection scheme that can merely identify when the nature of the link has changed, but cannot determine whether the change is from good to poor quality or vice versa.

We implemented our methodology in software and conducted experiments on actual mobile wireless ad-hoc networks.   We conducted experiments using several laptop computers, from a Digital HiNote VP500 with a 166MHz Pentium processor and 32MB RAM to a Gateway Model 9300 Solo with a 500MHz Pentium III processor and 128MB RAM.   Some of the laptops used various versions of the Linux operating system while others used Microsoft Windows NT 4.0.   Experiments were conducted on three different wireless network interface devices: Lucent WaveLAN Turbo Bronze 802.11 LAN cards running at 2Mbps, Lucent WaveLAN Turbo Gold 802.11HR LAN cards running at 11Mbps, and Metricom Ricochet packet radio modems running at 56kbps.   On average, we predicted attenuation faults 42 seconds before the link actually failed and shadowing faults 31 seconds before link failure, when the mobile devices were moving at walking speed (~1 m/s).

14.6.1   Real-Time Prediction of Wireless Link Faults

Our methodology for real-time prediction of wireless link faults is called the Wireless Network Environment Sensor (WiNE Sensor).   A block diagram of the WiNE Sensor is shown in Figure 14.1.   The first algorithm to be executed is the Baseline Generation algorithm, which produces representations of a known-quality wireless link, which are then compared to a currently sensed link.   The procedure for creating the baselines is as follows: The Baseline Generation algorithm polls a host at a fixed interval using the Simple Network Management Protocol (SNMP) [16] and requests the current counter value for four SNMP MIB variables, interface inOctets, interface outOctets, ipInReceives and ipOutRequests. The numerical difference between successive counter values forms a time series of increments of the counter. This time series of increments is modeled by a piece-wise stationary AR process over a limited time interval. We use Akaike's Final Prediction Error scheme [1] to determine the optimum AR model order, which is an AR(1) model. We developed a variation on Akaike's FPE to determine the optimum window size, which is 16 [15]. The set of AR model parameters for each window forms the baseline for that MIB variable. There are two types of links used to create baselines: high-quality wireless links and wireless links recovering from link failure. Our methodology originally used baselines only to represent high-quality links, but we found that links recovering from link failure were not well represented by those baselines. The link was passing traffic well and should have been considered a good-quality link, but the WiNE Sensor was portraying the link as "failing or failed." Modeling the recovering link as a separate baseline produced the correct output.

The second algorithm, the MIB-Level Sensor Algorithm, collects the same four SNMP MIB variables on the currently sensed link. It also computes adjacent differences of the counter to form a time series. It computes AR model parameters using the same order and window size as that of the corresponding baseline. The algorithm then determines if the AR model parameters of the currently sensed link are statistically the same as the baseline parameters, using a generalized likelihood ratio 15]. The output of the MIB-level sensor algorithm is the generalized likelihood ratio: a numeric floating point value in the interval [0,1]. The generalized likelihood ratio for each MIB variable is sent to the fusion center.

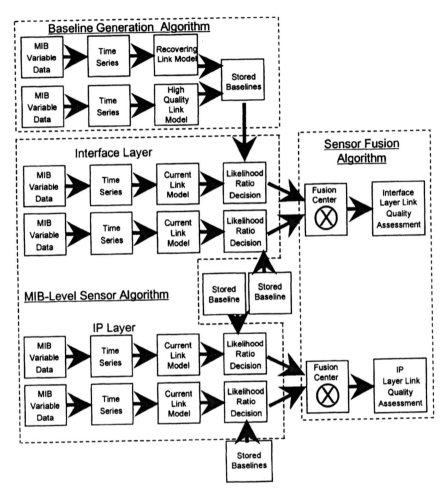

Figure –14.1. WiNE Sensor Block Diagram

The third algorithm is the Sensor Fusion algorithm, also discussed in depth in [15]. The Sensor Fusion algorithm combines multiple MIB-variable outputs into an overall estimate of the quality of the wireless link. By combining the outputs of several sensors, each of which has a partial picture of the link quality, we produce a more accurate representation of the link quality than any individual MIB-level sensor output. The output of the WiNE Sensor is a floating-point number in the range [0,1] that represents the link quality. The larger the number, the lower the link quality. We show that values below approximately 0.25 correspond to "good" quality links, values above approximately 0.8 are "failing" or "failed" links, and values between those thresholds are "fair" quality links.

14.6.2   Real-Time Identification of Wireless Link Faults

We extended our method of baselines to model failing links where the cause of failure was either due to attenuation or shadowing. We used this fault baseline generation algorithm in a second sensor, the Match WiNE Sensor, which identifies the cause of wireless link failures. This section presents a brief description of the algorithms contained in the Match WiNE Sensor. A detailed presentation can be found in [15]. The Match WiNE Sensor operates on the output of the WiNE Sensor, as shown in the block diagram in Figure 14.2. It consists of two algorithms: the first algorithm, the Fault Baseline Generator Algorithm, creates fault baselines from previously collected data where the cause of the fault is known. There is at least one fault baseline for each type of fault that the Match WiNE Sensor will identify.

The second algorithm, the Match WiNE Sensor Algorithm, identifies the cause of the failing link, based on current link information and its fault baselines. It receives as input the Link Quality Assessment from the WiNE Sensor. When the WiNE Sensor determines that a link is in the "Failing" state (the output of the WiNE Sensor is 0.81 or greater) but has not yet failed, the WiNE Match Sensor attempts to determine the cause of the link failure. The WiNE Match Sensor compares the current link to a library of known fault patterns (fault baselines) using a generalized likelihood ratio decision, in a similar manner to the way the WiNE Sensor determines the MIB variable output.

The output of the Match WiNE Sensor is the type of fault that the currently link most closely resembles, along with an index (from 0 to 1) of how close the current link matches that type of fault. A match of 0 is a perfect match; a match close to 1 is a very poor match.

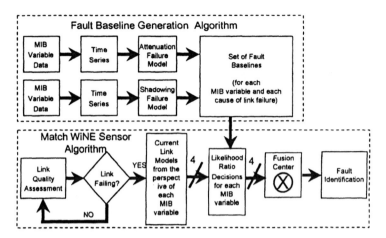

Figure –14.2. Match WiNE Sensor Block Diagram

14.6.3   Results from an Actual Wireless Network

This section presents results of an experiment conducted with the WiNE Sensor and WiNE Match Sensor on a Lucent WaveLAN Turbo Bronze ad-hoc wireless network, which is based on the IEEE 802.11 standard. In this experiment, a wireless link is formed by two laptops. One laptop is stationary; the other is mobile. Initially, the two laptops are next to each other. During the course of the experiment, the mobile laptop is carried at walking speed directly away from the stationary one, in an area where there is line-of-sight between the two laptops. The mobile laptop is carried away from the stationary one until the link fails (an attenuation fault). A short time later, the mobile laptop is carried back to the starting point, following the same path as the outgoing route.

In the results presented in Figure 14.3, the WiNE Sensor output is below 0.24, indicating a high-quality link, which is to be expected when the two hosts are close together in a low-noise environment. As the distance between the laptops increases, the level of the WiNE Sensor output also increases, indicating degradation of the link. Around 225 seconds into the experiment, the WiNE Sensor output rises above 0.81, indicating a failing link. However, the link doesn't actually fail until approximately 275 seconds into the experiment. In this example, one of many similar experiments, the WiNE Sensor provided 50 seconds of warning before link failure.

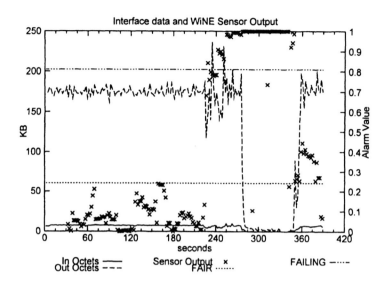

Figure −14.3. Interface Data and WiNE Sensor Link Quality Assessment for the WaveLAN
Bronze_66 Data Set (Attenuation Fault)

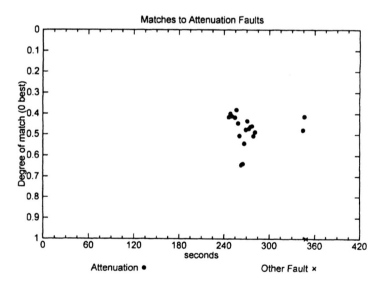

Figure −14.4. Match WiNE Sensor Outputs for the WaveLAN Bronze_66 Data Set
(Attenuation Fault)

While the link is failing (between 225 and 275 seconds), the outputs of
the WiNE Sensor are forwarded to the Match WiNE Sensor. Figure 14.4

shows that the Match WiNE Sensor determines that the link is failing due to an attenuation fault, which is correct.

14.7 CONCLUSION

This chapter discussed the challenges involved with sensing links in mobile wireless ad-hoc networks. The link sensing problem is more difficult than in a wired link due to the inherently limited bandwidth of wireless links, the possibility of low utilization of some links, the time-varying topology of ad-hoc networks, the relatively low computational power and limited battery power of mobile devices, and the nature of wireless links where faults are not rare occurrences.

For modern mobile devices that have several network interfaces and the possibility of hot-swapping network interfaces, the most flexible link sensor would be one that is technology-independent. However, technology-independent further complicates the link-sensing problem, since a technology-independent sensor cannot take advantage of specialized or proprietary information that is specific to a type of receiver design or to a particular operating system.

Despite these difficulties, we have developed an effective technology-independent sensor for wireless links in mobile ad-hoc networks. In experiments conducted on actual mobile wireless networks, our WiNE Sensor can rapidly detect all link failures, can predict link failures due to attenuation an average of 42 seconds in advance of the actual failure, and can predict link failures due to shadowing an average of 31 seconds in advance of the actual failure. Finally, we developed a methodology to identify the cause of those predictable faults. Our Match WiNE Sensor can distinguish between shadowing and attenuation as the cause of a wireless link failure.

REFERENCES

[1] Hirotugu Akaike, "Fitting Autoregressive Models for Prediction." *Ann. Inst. Stat. Math.*, vol. 21, no 2, 1969, 243-247.

[2] Krishna Balachandran, Srinivas R. Kadaba, and Sanjiv Nanda. "Channel Quality Estimation and Rate Adaptation for Cellular Mobile Radio." *IEEE Journal on Selected Areas in Communications*, Vol. 17, No. 7, July 1999, 1244-1256.

[3] J. Case, M Fedor, M. Schoffstall, and J. Davin. "A Simple Network Management Protocol (SNMP)." RFC 1157, May 1990.

[4] D. C. Cox. "Delay Doppler Characteristics of Multipath Propagation at 910MHz in a Suburban Mobile Radio Environment" *IEEE Transactions on Antennas and Propagation.* Vol. AP-20, No. 5, 1972, 625-635.

[5] Alexandria Duel-Hallen, Shengquan Hu, and Hans Hallen. "Long-Range Prediction of Fading Signals." *IEEE Signal Processing Magazine,* May 2000, 62-75.

[6] S. Hu, T. Eyceoz, A. Duel-Hallen, and H. Hallen. "Transmitter Antenna Diversity and Adaptive Signalling Using Long Range Prediction for Fast Fading DS/CDMA Mobile Radio Channels." *Proc. IEEE Wireless Commun. and Networking Conf.* WCNC '99, vol. 2, 1999, 824-828.

[7] S. Mockford and A. Turkmani. "Penetration Loss into Buildings at 900MHz." *Proc. IEE Colloquium on Propagation Factors and Interference Modeling for Mobile Radio Systems,* 1988, 1/1-1/7.

[8] Sanjiv Nanda, Krishna Balachandran, and Sarath Kumar. "Adaptation Techniques in Wireless Packet Data Services." *IEEE Communications Magazine.* January 2000, 54-64.

[9] C. Nche, A. M. D. Turkmani, A. A. Arowojolu. "Channel Sounder for PCN Networks." *IEE Colloquium on High Bit Rate UHF/SHF Channel Sounders-Technology and Measurement,* 1993, 5/1-5/6.

[10] C. Nche, D. G. Lewis, and A. M. D. Turkmani, "Wideband Characterization of Mobile Radio Channels at 1.8GHz." *Proc. 44th IEEE Vehicular Technology Conference.* Vol. 3, 1994, 1557-1561.

[11] W. G. Newhall, K. J. Saldanha, T. S. Rappaport. "Propagation Time Delay Spread Measurements at 915MHz in a Large Train Yard." *Proc. 46th IEEE Vehicular Technology Conference.* Vol. 2, 1996, 864-868.

[12] J. D. Parsons, D. A. Demery, A. M. D. Turkmani. "Sounding Techniques for Wideband Mobile Radio Channels: A Review." *IEE Proceedings- I, Communications, Speech, and Vision.* Vol. 138, No. 5, October 1991, 437-446.

[13] Bob O'Hara and Al Petrick. The IEEE 802.11 Handbook: A Designer's Companion. IEEE Press, 1999.

[14] Theodore S. Rappaport and Sandip Sandhu. "Radio-Wave Propagation for Emerging Wireless Personal-Communication Systems." *IEEE Antennas and Propagation Magazine.* Vol. 36, No. 5, October 1994, 14-24.

[15] Lisa A. Shay. *The Wireless Network Environment Sensor: A technology-independent sensor of faults in mobile wireless network links.* Ph.D. Dissertation, Rensselaer Polytechnic Institute, December 2002.

[16] William Stallings. *SNMP, SNMP V2 AND CMIP.* Reading, MA: Addison-Wesley, 1993.

[17] Marina Thottan and Chuanyi Ji, "Proactive Anomaly Detection Using Distributed Intelligent Agents." *IEEE Network*, Sep/Oct 1998, pp. 21-27.

[18] Marina Thottan and Chuanyi Ji. "Adaptive Thresholding for Proactive Network Problem Detection." *IEEE Int'l. Wksp. Sys. Mgmt.*, Newport, RI, 1998.

[19] A. M. D. Turkmani, A. F. Toledo. "Radio Transmission at 1800MHz into, and within, Multistory Buildings." *IEE Proceedings-1 Communications, Speech, and Vision*, Vol. 138, No. 6, December 1991, 577-584.

[20] A. M. D. Turkmani, D. A. Demery, J. D. Parsons. "Measurement and Modeling of Wideband Mobile Radio Channels at 900MHz." *IEE Proceedings-1 Communications, Speech, and Vision*, Vol. 138, No. 5, October 1991, 447-457.

[21] W. R. Young and L. Y. Lacy. "Echoes in Transmission at 450 Megacycles from Land-to-Car Radio Units." *Proc. IRE*, Vol. 38, 1950, 255-258.

INDEX

ABOUT THE EDITORS

Kia Makki is a Professor and Co-Director of the Telecommunications and Information Technology Institute at the Florida International University. He has numerous refereed publications in mobile and wireless networks and other related areas and his work has been extensively cited in books and papers. Dr. Makki has been Associate Editor, Editorial Board Member, and Guest Editor for special issues published in several international journals. He has been General/Program Chair of numerous conferences.

Niki Pissinou is a Professor and the Director of the Telecommunication & Information Technology Institute at the Florida International University. She has been the editor of eight journals including IEEE TKDE and guest editor of special issues published in seven journals.

Kami (Sam) Makki is a Faculty Member at the Queensland University of Technology in Brisbane, Australia. He is active in the fields of mobile and wireless communications, databases, and distributed systems and has a number of publications in these areas. He has also served as a technical program committee member for a number of IEEE and ACM sponsored technical conferences.

E.K. Park is a Professor and Head of the Software Architecture Discipline at the University of Missouri at Kansas City. His research publications have appeared in numerous journals and proceedings.

9 780792 372080